T0231095

Ecotoxicology of
METALS
in
INVERTEBRATES

Edited by

Reinhard Dallinger, Ph.D., University of Innsbruck, Austria
Philip S. Rainbow, Ph.D., Queen Mary and Westfield College,
University of London, United Kingdom

Proceedings of a session held at the First SETAC-Europe Conference
Sheffield, United Kingdom, April 7 -10, 1991

SETAC Special Publications Series

Series Editors

Dr. Thomas LaPoint
The Institute of Wildlife and Environmental Toxicology
Clemson University, U.S.A.
Dr. Peter W. Greig-Smith
Fisheries Laboratory
Ministry of Agriculture, Fisheries and Food, United Kingdom

Published in 1993 by
CRC Press
Taylor & Francis Group
6000 Broken Sound Parkway NW, Suite 300
Boca Raton, FL 33487-2742

International Standard Book Number-10: 0-87371-734-1 (Hardcover)
International Standard Book Number-13: 978-0-87371-734-2 (Hardcover)
Library of Congress Card Number 92-25660

Library of Congress Cataloging-in-Publication Data

Ecotoxicology of metals in invertebrates / edited by Reinhard Dallinger, Philip Rainbow.
    p. cm.
  Includes bibliographical references and index.
  ISBN 0-87371-734-1
    1. Invertebrates—Effect of metals on. 2. Metals—Toxicology.
    3. Metals—Environmental aspects. I. Dallinger, Reinhard. II. Rainbow, P. S.
QL364.E28 1992
592'.02'4—dc20   92-25560

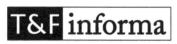

Taylor & Francis Group
is the Academic Division of T&F Informa plc.

Visit the Taylor & Francis Web site at
http://www.taylorandfrancis.com

and the CRC Press Web site at
http://www.crcpress.com

The SETAC Special Publications Series

The SETAC Special Publications Series was established by the Society of Environmental Toxicology and Chemistry to provide in-depth reviews and critical appraisals on scientific subjects relevant to understanding the impacts of chemicals and technology on the environment. The series consists of single- and multiple-authored/edited books on topics selected by the SETAC Board of Directors and the Council of SETAC-Europe for their importance, timeliness, and their contribution to multidisciplinary approaches to solving environmental problems. The diversity and breadth of subjects covered in the series will reflect the wide range of disciplines encompassed by environmental toxicology, environmental chemistry, and hazard/risk assessment. Despite this diversity, the goals of these volumes are similar; they are to present the reader with authoritative coverage of the literature, as well as paradigms, methodologies, controversies, research needs, and new developments specific to the featured topics. All books in the series are peer reviewed for SETAC by acknowledged experts.

The SETAC Special Publications will be useful to environmental scientists in research, research management, chemical manufacturing, regulation, and education, as well as students considering careers in these areas. The series will provide information for keeping abreast of recent developments in familiar areas and for rapid introduction to principles and approaches in new subject areas.

Thomas W. LaPoint
The Institute of Wildlife and Environmental Toxicology
Clemson University, U.S.A.

Peter W. Greig-Smith
Fisheries Laboratory
Ministry of Agriculture, Fisheries, and Food, U.K.

Series Editors, SETAC Special Publications

Foreword

The first scientific conference of SETAC-Europe was held from the 7th to the 10th of April 1991 at Sheffield, UK. European researchers and decision-makers from universities, institutes, industry, and governmental bodies met to discuss the latest developments in the various disciplines of environmental science, including ways in which scientific results might be applied for environmental improvement and sustainable development.

Two sessions, chaired by the editors of this book, were devoted to the subject of metals in invertebrate tissues. The considerable interest shown in these sessions suggested to the chairmen the excellent idea of assembling the papers presented on these two occasions to form a volume on ecotoxicology of metals in invertebrates. This they have done, at the same time adding some further material and organizing the papers in such a way as to present a balanced view of the major achievements and trends in this rapidly evolving field.

Initially, metal research in biology was mainly concerned with analyses. The availability of the highly sensitive analytical techniques of atomic absorption spectrophotometry, combined with micromethods for sample preparation, made it possible to analyze even very small invertebrates. Nowadays, however, the race for ppms has largely been replaced by physiological approaches emphasizing the kinetic and dynamic aspects of metal metabolism in animals.

Metal toxicity cannot be understood unless both the mechanisms involved in regulating physiological levels and the detoxification strategies employed in dealing with excessive amounts of essential elements are given consideration. This volume provides a variety of illustrations of this thesis, in studies from marine, freshwater, and terrestrial environments. The taxonomic diversity of the invertebrates provides us with excellent opportunities for investigating phylogenetic factors in metal regulation and for identifying groups particularly susceptible to metal intoxication.

The editors of this book are to be congratulated on the result of their efforts. With its emphasis on toxicology, the book builds on the foundation of existing physiological knowledge, at the same time putting this knowledge in a new perspective.

Nico M. van Straalen
SETAC-Europe

About the Editors

Reinhard Dallinger, Ph.D. is Associate Professor of the Department of Zoology (Animal Physiology) of the University of Innsbruck, Austria. He studied zoology and microbiology at the University of Innsbruck. He received his Ph.D. degree from the University of Innsbruck in 1978, his thesis focusing on copper metabolism in terrestrial isopods. Until 1980 he was employed by a private firm, working on projects concerning the composition of bark. In 1980 he joined the Department of Zoology as an Assistant Professor; he became Associate Professor in 1989.

Dr. Dallinger is a member of the following associations: Austrian Association of Toxicology, German Association of Zoologists, Society of Environmental Toxicology and Chemistry (SETAC-Europe), and the German Physiological Association. He has published more than 60 research papers and has been co-author of a book. He has received research grants from the Austrian "Fonds zur Förderung der wissenschaftlichen Forschung", the University of Innsbruck, public authorities, and private firms.

His present research concerns the effects of metals on terrestrial invertebrates and the mechanisms involved in metal storage and detoxification.

Philip S. Rainbow, Ph.D. is a Reader in Marine Biology in the School of Biological Sciences at Queen Mary and Westfield College, University of London. Dr. Rainbow received his B.A. degree from the University of Cambridge in 1972, and M.A. in 1976, after reading Natural Sciences (Zoology). He obtained his Ph.D. from the University of Wales in 1975, after research on barnacles in the Marine Science Laboratories, University College of North Wales (UCNW), Bangor. He has served as Lecturer and Reader at Queen Mary and Westfield College (previously Queen Mary College) since 1975.

Dr. Rainbow is a member of the Marine Biological Association of the United Kingdom, the Marine Biological Association of Hong Kong, and of the Estuarine and Coastal Sciences Association, a fellow of the Linnean Society of London, and a scientific fellow of the Zoological Society of London. He is also a university representative on the Executive Committee of the Field Studies Council. Dr. Rainbow has published more than 90 research papers and has been co-editor of three books and co-author of one. His research has been funded by NERC (Natural Environment Research Council, U.K.), the Royal Society, NATO Scientific Affairs Division, and private industry.

His current major research interests are in the significance of trace metal concentrations in marine and estuarine invertebrates, and in the mechanisms of uptake of dissolved metals by invertebrates.

Contributors

Dr. Thorvin Andersen
Section of Marine Zoology and
  Marine Chemistry
Department of Biology
University of Oslo
P.O. Box 1064
N-0316 Oslo, Norway

Dr. Andrew Bascombe
Middlesex University
(Centre for Urban Pollution
  Research)
Bounds Green Road
London N11 2NO, U.K.

Dr. Nina N. Belcheva
Institute of Marine Biology
Pacific Oceanological Institute
Russian Academy of Sciences
Vladivostok 690032, Russia

Dr. Burkhard Berger
Institut für Zoologie
(Abteilg. Zoophysiologie)
Universität Innsbruck
A-6020 Innsbruck, Austria

Eric A. J. Bleeker
Department of Aquatic Ecotoxicology
University of Amsterdam
Kruislaan 320
1098 SM Amsterdam
The Netherlands

Nicola Cardellicchio
Istituto Sperimentale
Talassografico "A. Cerruti", CNR
Via Roma 3
I-74100 Taranto, Italy

Dr. Emilio Carpenè
Dipartimento Biochimica (Sezione
  Biochimico Veterinaria)
Facoltà di Medicina Veterinaria
Università degli Studi di Bologna
Via Belmeloro 8/2
I-40126 Bologna, Italy

Dr. Victor P. Chelomin
Institute of Marine Biology
Pacific Oceanological Institute
Russian Academy of Sciences
Vladivostok 690032, Russia

Dr. Reinhard Dallinger
Institut für Zoologie
(Abteilg. Zoophysiologie)
Universität Innsbruck
A-6020 Innsbruck, Austria

Dr. Marina de Nicola
Dip. Genetica
Biologia Generale e Molecolare
Università di Napoli
Via Mezzocannone 8
I-80134 Napoli, Italy

Dr. Marianne H. Donker
Department of Ecology and
  Ecotoxicology
Vrije Universiteit
De Boelelaan 1087
NL-1081 HV, Amsterdam
The Netherlands

Dr. Graham Ellender
67 Wilfred Road
Ivanhoe East
Victoria 3079, Australia

Prof. Bryan Ellis
Middlesex University
(Centre for Urban Pollution
  Research)
Bounds Green Road
London N11 2NQ, U.K.

Eduard Felder
Institut für Zoologie
(Abteilg. Zoophysiologie)
Universität Innsbruck
A-6020 Innsbruck, Austria

Dr. Carmela Gambardella
Dip. Genetica
Biologia Generale e Molecolare
Università di Napoli
Via Mezzocannone 8
I-80134 Napoli, Italy

Mag. Susanne Gintenreiter
Institut für Zoologie
Universität Wien
Althanstrasse 14
A-1090 Wien, Austria

Alexandra Gruber
Institut für Zoologie
(Abteilg. Zoophysiologie)
Universität Innsbruck
A-6020 Innsbruck, Austria

Dr. Sandro M. Guarino
Dip. Genetica,
Biologia Generale e Molecolare
Università di Napoli
Via Mezzocannone 8
I-80134 Napoli, Italy

Dr. Stephen P. Hopkin
Department of Pure and Applied
  Zoology
University of Reading
Whiteknights P.O. Box 228
Reading RG6 2AJ, U.K.

Dr. Ketil Hylland
Section of Marine Zoology and
  Marine Chemistry
Department of Biology
University of Oslo
P.O. Box 1064
N-0316 Oslo, Norway

Dr. Ross Hyne
72 Wilson Street
Carlton North
Victoria 3054, Australia

Toralf Kaland
Section of Marine Zoology and
  Marine Chemistry
Department of Biology
University of Oslo
P.O. Box 1064
N-0316 Oslo, Norway

Dr. Paul L. Klerks
Department of Biology
University of Southwestern Louisiana
P.O. Box 42451
Lafayette, Louisiana 70504 USA

Michiel H. S. Kraak
Department of Aquatic Ecotoxicology
University of Amsterdam
Kruislaan 320
1098 SM Amsterdam
The Netherlands

Daphna Lavy
Department of Ecology and
  Ecotoxicology
Vrije Universiteit
De Boelelaan 1087
NL-1081 HV Amsterdam
The Netherlands

Dr. Jeffrey S. Levinton
Department of Ecology and
  Evolution
State University of New York at
  Stony Brook
Stony Brook, New York 11749, USA

Dr. Olga N. Lukyanova
Institute of Marine Biology
Pacific Oceanological Institute
Russian Academy of Sciences
Vladivostok 690032, Russia

Claudio Marra
Istituto Sperimentale
Talassografico "A Cerruti", CNR
Via Roma 3
I-74100 Taranto, Italy

Dr. A. John Morgan
School of Pure and Applied Biology
University of Wales
College of Cardiff
P.O. Box 915
Cardiff CF1 3TL, Wales, U.K.

Dr. John Morgan
Southern Science Ltd.
Kent Laboratory
Capstone Road, Chatham
Kent ME5 7QA, U.K.

Jürgen Moser
Institut für Zoologie
(Abteilg. Zoophysiologie)
Universität Innsbruck
A-6020 Innsbruck, Austria

Dr. Herbert Nopp
Institut für Zoologie
Universität Wien
Althanstrasse 14
A-1090 Wien, Austria

Dr. Johanna Ortel
Institut für Zoologie
Universität Wien
Althanstrasse 14
A-1090 Wien, Austria

Dr. David Pascoe
School of Pure and Applied Biology
University of Wales
College of Cardiff
P. O. Box 915
Cardiff CF1 3TL, Wales, U.K.

Gabriele Petri
Universität Oldenburg
Fachbereich Biologie (ICBM)
Postfach 2503
D-2900 Oldenburg, Germany

Dr. Philip S. Rainbow
School of Biological Sciences
Queen Mary and Westfield College
University of London
Mile End Road
London E1 4NS, U.K.

Dr. Michael Revitt
Middlesex University
(Centre for Urban Pollution
  Research)
Bounds Green Road
London N11 2NQ, U.K.

Dr. Greg D. Rippon
Alligator Rivers Region Research
  Institute
Office of the Supervising Scientist
P.O. Jabiru
Northern Territory 0886 Australia

Brian Shutes
Middlesex University
(Centre for Urban Pollution
  Research)
Bounds Green Road
London N11 2NQ, U.K.

Dr. Klaas R. Timmermans
Netherlands Institute for Sea
  Research
P. O. Box 59
NL-1790 AB Den Burg
Texel, The Netherlands

Merel Toussaint
Department of Environmental and
  Toxicological Chemistry
University of Amsterdam
Nw. Achtergracht
1018 VW Amsterdam
The Netherlands

Michael Turner
School of Pure and Applied Biology
University of Wales
College of Cardiff
P.O. Box 915
Cardiff CF1 3TL, Wales, U.K.

Dr. H. Erik van Capelleveen
Tauw Infra Consult b.v.
Postbus 479
7400 AL Deuente
The Netherlands

Dr. Nico M. van Straalen
Vrije Universiteit
Department of Ecology and
  Ecotoxicology
De Boelelaan 1087
NL-1081 HV, Amsterdam
The Netherlands

Carole Winters
School of Pure and Applied Biology
University of Wales
College of Cardiff
P.O. Box 915
Cardiff CF1 3TL, Wales, U.K.

Dr. Qin Xu
Department of Animal and
Plant Sciences
The University of Sheffield
P.O. Box 601
Sheffield, S10 2UQ, U.K.

Andrew Yarwood
JEOL (UK)
JEOL House
Watchmead
Welwyn Garden City
Herts AL7 1LT, U.K.

Dr. Gerd-Peter Zauke
Universität Oldenburg
Fachbereich Biologie (ICBM)
Postfach 2503
D-2900 Oldenburg, Germany

Contents

SECTION I

Marine Environments

General Considerations
1. The Significance of Trace Metal Concentrations in Marine Invertebrates

---

Marine Molluscs
2. Cadmium Bioaccumulation in the Scallop *Mizuhopecten yessoensis* from an Unpolluted Environment

3. Accumulation and Subcellular Distribution of Metals in the Marine Gastropod *Nassarius reticulatus*

4. Metallothionein in Marine Molluscs

Marine Crustacea
5. Metal Concentrations in Antarctic Crustacea: the Problem of Background Levels

6. Effects of Cadmium on Survival, Bioaccumulation, Histopathology, and PGM Polymorphism in the Marine Isopod *Idotea baltica*

CHAPTER 1

The Significance of Trace Metal Concentrations in Marine Invertebrates

Philip S. Rainbow

TABLE OF CONTENTS

0-87371-734-1/93/$0.00 + $.50
© 1993 by Lewis Publishers

3

I. INTRODUCTION

Trace metals occur in nature, for example dissolved in seawater, in very low concentrations (Table 1), yet are capable of exerting biological effects at these concentrations or at concentrations within only a few orders of magnitude of those listed. Trace metals may also be referred to as heavy metals by some authors, and strictly need to be defined according to objective chemical criteria.[1] Furthermore, the term trace metal may or may not be restricted to those heavy metals which have been identified as playing an essential role in the metabolism of some organism. Pragmatically in this acount, the term trace metal is used synonymously with that of heavy metal, whether essential or nonessential, and will refer to any of the metals so listed in Table 1. All such metals, essential or not, are toxic above a threshold bioavailability.

Trace metals are taken up and accumulated by marine invertebrates to tissue and body concentrations usually much higher on a wet weight basis than concentrations in the surrounding seawater.[3-5] Metals may be taken up by marine invertebrates from solution, from food, pinocytotically as in the lamellibranch bivalve gill or ascidian pharynx, or by esoteric routes including the blood sinuses of the foot of particular gastropod molluscs or via the nephridiopores of narcotized polychaetes.[4] In general, solution and food represent the major routes of metal uptake by marine invertebrates, the order of priority varying with many factors including species, food type, relative concentrations (or strictly bioavailabilities) of metal in food and water, physicochemical parameters of the aquatic medium, etc. It is rarely necessary to invoke the need for active transport mechanisms for the uptake of trace metals across cell membranes, facilitated diffusion being sufficient together with the protein binding of incoming metals, which provides a sink with no significant back diffusion of metals.[6] Thus, metals continue to enter marine invertebrates against an apparent concentration gradient, contributing to high accumulated body concentrations.

II. ACCUMULATION STRATEGIES

It is a striking feature of the literature that body concentrations of accumulated trace metals, essential or nonessential, vary greatly between invertebrates, often between closely related species (Table 2).[7] Thus mussels and caridean decapods typically have low body concentrations of zinc in comparison to their respective relatives, the oysters and barnacles (Table 2).

A key to appreciating the significance of different metal concentrations in marine invertebrates is an understanding of the metal accumulation strategies adopted.[15-17] Metal accumulation strategies fall along a gradient from the strong accumulation of all metal taken up, to the regulation of the body metal concentration to an approximately constant level by balancing excretion to metal uptake; intermediate strategies include degrees of net accumulation when metal uptake

**Table 1. Dissolved Concentration
Ranges of Metals in Seawater at
35 ppt Salinity**

Major metals		
Sodium	468	mmol kg^{-1}
Magnesium	53.2	mmol kg^{-1}
Calcium	10.3	mmol kg^{-1}
Potassium	10.2	mmol kg^{-1}
Strontium	90	μmol kg^{-1}
Minor metals		
Lithium	25	μmol kg^{-1}
Rubidium	1.4	μmol kg^{-1}
Barium	32–150	nmol kg^{-1}
Molybdenum	110	nmol kg^{-1}
Trace metals		
Vanadium	20–35	nmol kg^{-1}
Arsenic	15–25	nmol kg^{-1}
Nickel	2–12	nmol kg^{-1}
Zinc	0.05–9	nmol kg^{-1}
Copper	0.5–6	nmol kg^{-1}
Chromium	2–5	nmol kg^{-1}
Manganese	0.2–3	nmol kg^{-1}
Iron	0.1–2.5	nmol kg^{-1}
Selenium	0.5–2.3	nmol kg^{-1}
Antimony	1.2	nmol kg^{-1}
Cadmium	0.001–1.1	nmol kg^{-1}
Tungsten	500	pmol kg^{-1}
Lead	5–175	pmol kg^{-1}
Cobalt	10–100	pmol kg^{-1}
Silver	0.5–35	pmol kg^{-1}
Gold	25	pmol kg^{-1}
Tin	1–12	pmol kg^{-1}
Mercury	2–10	pmol kg^{-1}

From Bruland, K.W., *Chemical Oceanography,* Vol. 8, 1983, 57.

exceeds excretion, weak net accumulation also being known as partial regulation.[15-17]

Zinc accumulation in barnacles and the caridean decapod *Palaemon elegans* exemplify the two extremes of the gradient of metal accumulation strategies. Barnacles like *Elminius modestus* are strong net accumulators of zinc with high zinc uptake rates and no significant zinc excretion.[8,15] The accumulated zinc is stored in a detoxified form — in granules bound to pyrophosphate[18] — in the body tissue without access to the exterior.[8] The barnacle body content of zinc therefore increases over time with zinc exposure. *P. elegans* as a regulator of

**Table 2. Ranges of Typical Accumulated Concentrations of
Zinc (μg Zn g^{-1} Dry Weight) in Crustaceans and
Bivalve Molluscs**

a) CRUSTACEANS

	Body concentration range	Ref.
Barnacles		
Semibalanus balanoides	1,220–113,250	8
Elminius modestus	4,900–11,700	8
Balanus amphitrite	2,726–11,990	9
Caridean Decapods		
Palaemon elegans	76–138	10
Pandalus montagui	119–214	10
Crangon crangon	56–174	10

b) BIVALVE MOLLUSCS

	Soft tissue concentration range	
Oysters		
Ostrea edulis	660–3,280	11
Crassostrea virginica	322–12,675	11
Saccostrea cucullata	430–8,629	12
Mussels		
Mytilus edulis	14–500	11
Perna viridis	77–164	13
Septifer virgatus	74–116	14

From Rainbow, P.S., *Proc. Bioaccumulation Workshop,* in press.

body zinc maintains an approximately constant body concentration of zinc over a wide range of zinc bioavailabilities.[15,19] The zinc uptake rate of *P. elegans* increases with dissolved zinc concentration[20] and with particular physicochemical changes[21,22] but the body zinc concentration remains unchanged, thereby indicating that zinc excretion has been increased to match uptake.[20-22] Zinc turnover can be significant during regulation, for more than 30% of the total body zinc in *P. elegans* is replaced in 10 days at 10°C on exposure to 100 μg Zn l^{-1}, with no increase in body zinc concentration.[20] A third crustacean, the amphipod *Echinogammarus pirloti,* is a weak accumulator of zinc; it has a low zinc uptake rate comparable to that of *P. elegans* but does not excrete any of the zinc taken up.[15]

Thus the different concentrations of zinc in barnacles and caridean decapods reported in Table 2 can be explained by the different zinc accumulation strategies of the crustaceans. Barnacles have high zinc uptake rates, no significant zinc excretion, and detoxified storage of accumulated zinc. *P. elegans* has a low zinc uptake rate and matches zinc excretion to uptake, thereby regulating the zinc body concentration. It is relevant that if the external zinc concentration to which

P. elegans is exposed is increased, there is eventually a point where zinc uptake exceeds maximum zinc excretion and net zinc accumulation begins (the point of regulation breakdown).[15,19,20] After regulation breakdown, mortality of *P. elegans* follows at an accumulated body zinc concentration as low as 200 μg Zn g⁻¹ (approximately double the regulated body concentration),[19] suggesting that much of the extra accumulated zinc remains metabolically available to play a toxic role.

In the parallel case of the bivalve molluscs (Table 2), oysters accumulate high concentrations of zinc in detoxified granules,[23] while mussels excrete much accumulated zinc in granules from the kidney.[24] Thus, oysters are strong accumulators of zinc (e.g., Phillips and Yim[14]) and mussels are weak net accumulators or partial regulators of zinc.[10,14]

Rainbow and White[15] also investigated the accumulation strategies of the decapod, amphipod, and barnacle for another essential metal, copper, and for the nonessential metal, cadmium. The decapod appeared to regulate the body concentration of copper, but verification of regulation awaits the separate identification of incoming and already accumulated metal,[17] a difficult procedure in the absence of a suitable copper radiotracer. The amphipod and barnacle were net accumulators of copper and all three crustaceans were net accumulators of the nonessential metal cadmium.[15] Indeed, regulation as a metal accumulation strategy does seem to be restricted to essential metals.[16]

Although the concept of the regulation of total body metal content is useful pragmatically, it is difficult to envisage a specific feedback regulatory mechanism operating at the whole-animal level.[5] It must be remembered, therefore, that body metal contents are summations of the contents of the constituent tissues or organs;[5] regulation is clearly possible at the latter level with a particular organ acting as a detoxified storage site for excess body metal (e.g., hepatopancreas or equivalent), or as a site for its excretion (e.g., gill or kidney). The mussel kidney both temporarily stores and excretes zinc,[24] and the ventral caecal cells of stegocephalid amphipod crustaceans similarly concentrate and then excrete iron derived from the diet as crystals of ferritin.[25]

III. CELLULAR COMPARTMENTALIZATION AND TISSUE-SPECIFIC PARTITIONING

In the case of essential metals, each tissue potentially contains two components of metal — metal in a (required) metabolically available form, and that detoxified. Nonessential metals must be detoxified to avoid the possibility of toxic action. Mechanisms of cellular detoxification of trace metals[26,27] include the binding of some (e.g., Cu, Cd, Hg, Ag, and Zn) to metallothioneins[28,29] and many to metabolically inert granules,[30-34] in either case removing the metals from metabolic access to sensitive vital cellular components.

Why then do different tissues contain different trace metal concentrations? Clearly some organs act as dump sites, temporary or permanent, for detoxified metals, essential and nonessential, and such organs will have extremely high metal concentrations. Other tissues in the same animal may well have lower and finely controlled concentrations of the same metal.[5] Any metabolically active tissue requires a background concentration of an essential metal to meet essential requirements — for example, in enzymes. The blood of marine invertebrates often contains respiratory pigments such as copper-bearing hemocyanin or iron-bearing hemoglobin, chlorocruorin, or hemerythrin. It is possible, therefore, to make theoretical estimates of the essential trace metal requirements of organisms, as attempted by Pequegnat et al.,[35] White and Rainbow,[36,37] and Depledge.[38] Such estimates are considered in detail below, but few studies have as yet succeeded in identifying the metabolic roles of particular metals in particular tissues. A notable exception is the work of Chan,[39] which throws considerable light on the significance of different zinc concentrations in crustacean muscles.

Pioneering work by Bryan[40,41] showed that structurally different crustacean muscles have different zinc concentrations — fast contracting muscles have low zinc concentrations but slow contracting muscles have high zinc concentrations. Chan[39] was able to show that the distribution of soluble zinc in crustacean muscles is remarkably simple (Table 3). In effect, all soluble zinc is distributed between only three components:

1. Bound to hemocyanin present as a blood contaminant and therefore able to be discounted.
2. Bound to the enzyme arginine kinase, and
3. Bound to metallothionein saturated with zinc.

Thus all muscle cytosolic zinc is effectively either bound to the enzyme arginine kinase or is present in the detoxified storage protein metallothionein. Fast contracting muscles (e.g., leg muscles, Table 3) have low zinc concentrations with little zinc bound to arginine kinase while slow contracting muscles (e.g., claw muscles, Table 3) have more zinc in total and bound to arginine kinase. Zinc inhibits arginine kinase which catalyzes the formation of arginine phosphate, an energy storage compound available for the rapid rephosphorylation of ADP during muscle contraction. Zinc may, therefore, regulate the activity of arginine kinase, and metallothionein possibly controls the intracellular availability of zinc.[39] Thus we have a handle on the functional significance of different zinc concentrations in different crustacean muscles, but we have a long way to go before we can draw similar conclusions for other metals and other tissues.

Trace metals may also play structural roles in some tissues, as in the case of zinc and manganese in the jaws of nereid polychaetes[42,43] and copper in the jaws of glycerid polychaetes.[44]

Table 3. Mean Concentrations (±1 SD) of Zn Expressed as μg g⁻¹ Dw of Original Muscle Tissue Extracted (Homogenized in 2 vol 0.05 *M* Tris, 0.02 *M* CaCl₂, 0.1 m*M* Mercaptoethanol, pH 7.8, Centrifuged 11,500 g) Bound to Hemocyanin (Blood Contaminant), the Enzyme Arginine Kinase, and Metallothionein in Different Muscles of the Crabs *Carcinus maenas* and *Liocarcinus depurator*

	Claw muscles		Cephalothoracic leg muscles	
Carcinus maenas				
Hemocyanin	21.5 ± 3.5	(25.2)	10.9 ± 1.1	(21.4)
Arginine kinase	14.0 ± 2.9	(16.5)	4.2 ± 2.3	(8.2)
Metallothionein	38.5 ± 8.3	(45.3)	18.9 ± 3.8	(37.1)
		87.0		65.7
Total extract of muscle	85.0 ± 5.9		51.0 ± 6.7	
Original muscle	275 ± 36.4		200 ± 45.6	
Liocarcinus depurator				
Hemocyanin	10.0 ± 1.5	(12.6)	13.7 ± 3.2	(34.7)
Arginine kinase	18.6 ± 2.4	(23.3)	7.5 ± 2.8	(18.9)
Metallothionein	31.9 ± 6.1	(40.1)	18.9 ± 4.7	(47.8)
		76.0		101.4
Total extract of muscle	79.5 ± 10.2		39.5 ± 8.8	
Original muscle	189 ± 30.5		99.2 ± 31.3	

Note: Values in parentheses represent percentage of total zinc concentration in each muscle extract — n = 3 in each case.

From Chan, H.M., Ph.D. thesis, University of London, 1990.

IV. METABOLIC REQUIREMENTS

As indicated above, it is possible to make theoretical estimates of the metabolic requirements of invertebrates for essential metals in an attempt to identify these components when interpreting total essential metal concentrations in invertebrate tissues or bodies. White and Rainbow[36] built on the original work of Pequegnat et al.[35] and calculated theoretical estimates for the metabolic requirements of zinc and copper in molluscs and crustaceans. White and Rainbow[37] extended these calculations to iron and manganese. White and Rainbow[36,37] estimated the weights of metal-containing enzymes in tissue, and thence the weight of enzyme-associated metal in typical invertebrate tissue. As stressed by the authors themselves, it is necessary to make many assumptions in these calculations, particularly on the relative abundance of different enzymes.

White and Rainbow[36] also considered the amount of copper bound to the respiratory pigment hemocyanin in mollusc and crustacean bodies in their calculations of total metabolic requirements. Zinc is bound to hemacyanin and, given the possibility that it might be essential to hemocyanin functioning,[45] they also calculated a required body load of zinc to this effect. Chan,[39] however, has shown that stoichiometric ratios of zinc to hemocyanin vary widely intraspecifically and interspecifically in decapod crustaceans, casting doubt on any essential role for zinc here.

Depledge[38] has criticized (with justification) the assumptions of White and Rainbow[36] in a reevaluation of the metabolic requirements of copper and zinc in decapods, considering the different metal requirements of different tissues. In particular, Depledge[38] stressed that the estimates of White and Rainbow[36] should apply only to metabolizing tissue, and that they had not allowed for the weight contribution of the presumably metabolically inactive exoskeleton. This criticism is accepted and the recalculations shown in Table 4 are made in this light. These recalculations also discount any role for zinc in the functioning of hemocyanin (Table 4).

Table 4 considers three different decapod crustaceans — the carideans *Palaemon elegans* and *Pandalus montagui,* and the crab *Carcinus maenas.* Knowledge of the percentage distributions of dry weight between exoskeleton and total soft tissue, and of blood volumes, allows the calculation of the distribution of body weight into exoskeleton, blood, and other soft tissues (assumed metabolically active). In the first instance, estimates of enzyme-associated metal contents are applied only to the soft tissue component that excludes blood, and of hemocyanin copper only to the blood itself, but these calculations are then extended to the total soft tissue which includes the blood.

As detailed in Table 4 the theoretical enzymatic Zn contributions to whole-body zinc concentrations vary from 8.9 to 20.0 μg Zn g^{-1}, while measured body zinc concentrations in crustaceans from the Firth of Clyde, Scotland varied from 57.5 to 96.6 μg Zn g^{-1}. In terms of soft tissue concentrations, estimates of enzymatic Zn are 31.9 to 33.3 μg Zn g^{-1}, in comparison with observed values of 57.5 to 159 μg Zn g^{-1}. Thus, theoretical estimates of the contributions of enzyme-associated zinc are low in comparison with observed values, but are of the same order.

In the case of copper, total theoretical copper concentrations (enzyme-associated plus hemocyanin-associated) vary from 31.3 to 38.1 μg Cu g^{-1}, in comparison with measured values of 30.2 to 110 μg Cu g^{-1}. In terms of soft tissues, theoretical values fall in the range 63.5 to 112 μg Cu g^{-1}, measured values being 58.6 to 140 μg Cu g^{-1}. Agreement of theoretical and observed copper concentrations is remarkable given the assumptions made in the theoretical calculations.[36] The observation that, for both zinc and copper, measured values usually exceed the theoretical estimates may reflect the presence of some storage capacity for the essential metals above immediate requirements.

In conclusion, such theoretical estimates should not be taken too literally, but they do provide an approximate guideline (to less than an order of magnitude?)

Table 4. Recalculation of Theoretical Estimates of Zinc and Copper Requirements of 3 Decapod Crustaceans, and Comparisons with Measured Concentrations

	Palaemon elegans	Ref.	Pandalus montagui	Ref.	Carcinus maenas	Ref.
% Distribution dry weight						
Exoskeleton	52.2	46	40.0	47	72.0	39
Combined soft tissues (inc. blood)	47.8		60.0		28.0	
Combined soft tissue wt (inc. blood) of 1 g dry wt crustacean	0.478 g		0.600 g		0.280 g	39
Blood volume (% fresh wt)	15	38	15	38	20	39
Fresh wt:dry wt ratio	3.5:1	38	3.5:1	38	3.5:1	38
Blood vol. of 1 g dw crustacean (ml)	0.52		0.52		0.70	
Protein content of blood (mg ml^{-1})	40	48	40	48	30	39
If dry wt of blood results from protein content						
Blood dry wt of 1 g dw crustacean	0.01 g		0.021 g		0.021 g	
Dry wt of soft tissue (exc. blood) of 1 g dw crustacean	0.457 g		0.579 g		0.259 g	
Zinc						
Theoretical enzymatic Zn content of metabolizing soft tissue (excludes blood) = 34.5 µg Zn g^{-1} dw				36		
Enzymatic Zn content of soft tissue (exc. blood) of 1 g dw crustacean (µg Zn)	15.8	46	20.0		8.9	
Theoretical enzymatic Zn contribution to total body Zn concn (µg Zn g^{-1} dw)	15.8		20.0		8.9	
Theoretical enzymatic Zn contribution to Zn concn of combined soft tissue (inc. blood) (µg Zn g^{-1} dw)	33.1		33.3		31.9	
Measured Zn concns (crustaceans from Firth of Clyde, Scotland)						
Total ± 1 SD (µg Zn g^{-1} dw)	80.6 ± 4.7	46	57.5 ± 7.8	47	96.6 ± 8.6	49
Combined soft tissue (inc. blood) (µg Zn g^{-1} dw)	87.0	46	57.5	47	159 ± 17.8	49
Copper						
Theoretical enzymatic Cu content of metabolizing soft tissue (excludes blood) = 26.3 µg Cu g^{-1} dw				36		
(i) Enzymatic Cu content of soft tissue (exc. blood) of 1 g dw crustacean (µg Cu)	12.1		15.2		6.8	

Table 4 (continued). Recalculation of Theoretical Estimates of Zinc and Copper Requirements of 3 Decapod Crustaceans, and Comparisons with Measured Concentrations

	Palaemon elegans	Ref.	Pandalus montagui	Ref.	Carcinus maenas	Ref.
Theoretical enzymatic Cu contribution to total body Cu conc (µg Cu g⁻¹ dw)	12.1		15.2		6.8	
(ii) Hemocyanin Cu content of blood						
Blood Cu concn (all in hemocyanin) (µg Cu ml⁻¹)	44	38	44	38	35	39
Total blood Cu content of 1 g dw crustacean (µg Cu)	22.9		22.9		24.5	
Contribution of hemocyanin Cu to total body Cu concn (µg Cu g⁻¹ dw)	22.9		22.9		24.5	
(i) + (ii) Total theoretical Cu concn in body (µg Cu g⁻¹ dw)	35.0		38.1		31.3	
Blood contribution of Cu to combined soft tissue (inc. blood) (µg Cu g⁻¹ dw combined soft tissue)	47.9		38.2		87.5	
Enzymatic contribution of Cu to combined soft tissue (inc. blood) (µg Cu g⁻¹ dw combined soft tissue)	25.3		25.3		24.3	
Total theoretical Cu concn in combined soft tissue (inc. blood) (µg Cu g⁻¹ dw combined soft tissue)	73.2		63.5		112	
Measured Cu concns (crustaceans from Firth of Clyde, Scotland)						
Total ± 1 SD (µg Cu g⁻¹ dw)	110 ± 12.7	46	57.4 ± 18.9	47	30.2 ± 5.9	49
Combined soft tissue (inc. blood) (µg Cu g⁻¹ dw)	140	46	73.5	47	58.6 ± 27.6	49

when interpreting body concentrations of essential metals in marine invertebrates. It is relevant, therefore, that the regulated body concentrations of zinc and copper in caridean decapods like *P. elegans* are of the same order as the theoretical calculations in Table 4. Thus, much of the body metal of such regulators probably remains in metabolically available form to fulfil essential roles. Remember that a relatively small increase in accumulated zinc in *P. elegans* was lethal,[19] indicating the metabolic availability of zinc in this decapod (cf. detoxified storage in barnacles).

Thus, regulated body concentrations may themselves provide good estimates of required metabolic concentrations, and indeed show light on intraspecific and interspecific differences — the caridean decapods *P. montagui* (70.2 μg Zn g^{-1}), *P. elegans* (77.1 μg Zn g^{-1}), and *Palaemonetes varians* (96.4 μg Zn g^{-1}) regulated body zinc concentrations to different levels at 10°C.[22,47] Similarly, *P. elegans* regulates its body zinc concentration to different levels at different temperatures — 75.8 μg g^{-1} at 5°C, 77.1 μg g^{-1} at 10°C, 87.1 μg g^{-1} at 15°C, and 90.6 μg g^{-1} at 20°C — probably in reflection of different metabolic needs.[50]

V. BACKGROUND CONCENTRATIONS

When trying to assess the significance of a trace metal concentration in a marine invertebrate, it is attractive to consider the concept of a background concentration — that is, the normal or typical metal concentration found in an invertebrate collected from a habitat remote from anthropogenic (or other atypical) input of metals. Identification of a background concentration might then allow an investigator to recognize not only instances of raised metal bioavailability (e.g., in the use of biomonitors to determine sites of metal pollution) but also cases of possible metal deficiency.

The papers of Zauke and Petri[51] and of Hopkin,[52] presented later in this volume, both address this concept. Hopkin, for example, combines the concepts of calculated theoretical requirements[36-38] and background concentrations (intraspecific comparison of measured body and organ concentrations) to suggest the possibility of copper deficiency in individuals of the woodlouse *Oniscus asellus* from one site out of many in southwestern England. The concept of background concentrations is implicit in the analysis of the results of biomonitoring programs[53] — biomonitors with "high" accumulated metal contents taken to be indicative of regions with locally raised bioavailabilities of metal. "High" accumulated body burdens of metal are recognized by comparison with other data collected simultaneously. It might be logical, therefore, to expect that many of the measured metal concentrations of the biomonitor in the survey would fall into a relatively narrow band indicative of the "background concentration" — the biomonitors from metal hotspots having metal concentrations raised to different degrees.

In practice, biomonitoring data do not always show such a distribution. For example, the data set may show a graduation from low to high accumulated concentrations with no obvious narrow band equivalent to the "background concentration". Thus, data in Tables 5 and 6 for barnacles[9] and amphipods,[54] respectively, show gradients of accumulated concentrations, with no clear "background concentration" apparent. It is relevant that, in both cases, allowance has already been made for size effects by use of Analysis of Covariance.[9,54]

It is clear, therefore, that even after allowance for size, there are often too many biological variables that need to be accounted for before such data might be expected to allow identification of a background concentration. It is almost inevitable that local differences in habitat will affect parameters like the feeding rates, respiration rates, and therefore scope for growth of a biomonitor. Thus, factors other than comparative metal bioavailabilities will have affected final accumulated metal concentrations. The lack of a clearly identifiable background concentration of the biomonitor does not usually prevent an investigator from drawing conclusions about raised metal bioavailabilities in biomonitors at the top of lists such as illustrated in Tables 5 and 6, but does create doubt when considering the significance of metal concentrations further down the list. It becomes impossible, therefore, to identify sites of only slightly raised metal bioavailability, when the raised accumulated metal concentrations of biomonitors from such sites are lost in the noise of variable metal concentrations caused by other biological factors affecting samples from nonpolluted sites.

The data in Tables 5 and 6 have been collected over different geographical scales (albeit over relatively tight time scales). The Hong Kong sites (Table 5) are less than 50 km apart, but do vary from wave-exposed to sheltered, from oceanic to affected by estuarine influence. The Scottish sites (Table 6) are spaced up to a maximum of 150 km apart. In either case it is not possible to assume uniformity of parameters affecting feeding, growth, etc. between sites — an almost inevitable conclusion for any biomonitoring survey.

Two further examples illustrate this point. On a vast geographical scale, concentrations of both zinc and copper in the pelagic euphausiid (*Meganyctiphanes norvegica* (Table 7) are significantly higher in specimens collected from the northeastern Atlantic, remote from anthropogenic influence, than in those from the Firth of Clyde close to shore.[54] It is almost certain, therefore, that factors other than metal pollution have caused these differences; such factors could include growth rates, nature of food, season, etc.

On a much smaller geographical scale, samples of the littoral amphipod *Platorchestia platensis* from Hong Kong exemplify the possible role of diet in the establishment of a typical background concentration in an invertebrate (Table 8). Concentrations of zinc in samples of *P. platensis* from four sites fall into three categories — concentrations in amphipods from Wu Kai Sha are raised above those in amphipods from Hoi Ha Wan and Cape D'Aguilar, with even lower concentrations in specimens from Mai Po (Table 8).[56] In fact, data from other biomonitors have confirmed that Cape D'Aguilar, facing the prevailing

Table 5. Concentrations of Copper (µg g^{-1} Dry Wt) in the Bodies of Three Species of Barnacles, Each of Standardized Dry Weight, as Estimated from Best Fit Double Log Regressions Between Accumulated Body Metal Concentration (y) and Body Dry Weight (x), With 95% Confidence Limits (CL)

Barnacle (standardized body dry wt) Site	Capitulum mitella (0.05 g)			Tetraclita squamosa (0.02 g)			Balanus amphitrite (0.004 g)		
	Concn	CL	ANCOVA	Concn	CL	ANCOVA	Concn	CL	ANCOVA
Chai Wan Kok							3472	3950 / 3052	A
Kwun Tong							2574	3414 / 1940	B
North Point				203	243 / 170	A	1010	1387 / 869	C
Hung Hom	545	1290 / 375	A	94.9	129 / 69.9	B			
Kowloon Pier				80.7	92.0 / 70.7	B			
Stonecutters	537	1170 / 247	A						
Queens Pier	154	435 / 54.4	B	80.1	91.3 / 70.3	B			
Causeway Bay	132	232 / 74.6	B	69.2	98.2 / 48.7	B			
Hang Hau							486	639 / 373	D
Reef Island	36.5	53.4 / 25.0	C	30.9	41.3 / 23.2	C			
Rennies Mill							304	368 / 251	E

Table 5 (continued). Concentrations of Copper (μg g^{-1} Dry Wt) in the Bodies of Three Species of Barnacles, Each of Standardized Dry Weight, as Estimated from Best Fit Double Log Regressions, Between Accumulated Body Metal Concentration (y) and Body Dry Weight (x), With 95% Confidence Limits (CL)

Barnacle (standardized body dry wt)	Capitulum mitella (0.05 g)			Tetraclita squamosa (0.02 g)			Balanus amphitrite (0.004 g)		
	Concn	CL	ANCOVA	Concn	CL	ANCOVA	Concn	CL	ANCOVA
Tung Chung				14.9	24.3 / 9.1	D	295	13670 / 6.4	E
Wu Kai Sha							213	244 / 187	F
Tai Po Kau							142	235 / 86.2	G
Sha Tin							116	136 / 98.5	H
Lai Chi Chong							59.3	195 / 18.0	H
Cape D'Aguilar	29.2	35.6 / 24.1	D						

Note: Samples were all collected between 8 and 16 April, 1986 from sites in Hong Kong. Sites are ranked in approximate order of decreasing copper contamination. Different letters in the ANCOVA column for one species indicate significant differences ($p < 0.05$) in metal concentrations between samples by Analysis of Covariance of best fit regressions.

From Phillips, D.J.H. and Rainbow, P.S., *Mar. Ecol. Prog. Ser.*, 49, 83, 1988.

Table 6. *Orchestia gammarellus:* **Estimates of Copper Concentration (μg g^{-1} dry wt) with 95% Confidence Limits (CL) for Western Scotland Amphipods of 10 mg Standardized Dry Weight as Derived From Best-Fit Double Log Regressions of Copper Concentration Against Dry Weight**

Site	Copper concentration	95% CL	ANCOVA
Whithorn	132	118, 148	A
Loch Long	129	118, 142	A
Auchencairn	95.2	87.7, 104	B,C
Holy Loch	93.2	81.3, 107	B,C
Carsethorn	91.8	67.6, 125	B,C
Creetown	87.2	70.1, 109	B,C
Helensburgh	86.7	72.5, 104	B,C
Brodick	85.8	78.2, 94.0	B,C
Erskine	84.0	76.2, 92.5	B
Millport	77.5	69.4, 86.4	C,D
Powfoot	76.2	64.0, 90.6	C,D
Loch Ryan	75.6	67.3, 84.9	C,D
Strachur	71.9	63.7, 81.2	D
Ardwell	71.5	46.4, 110	D
Girvan	66.0	54.4, 79.9	D

Note: Samples sharing any common letter in the ANCOVA column are not significantly different by Analysis of Covariance. All samples were collected between 2/26/88 and 3/23/88.

From Rainbow, P.S., Moore, P.G., and Watson, D., *Estuarine Coastal Shelf Sci.*, 28, 567, 1989.

currents and winds of the South China Sea, receives clean oceanic water and is one of the most metal-free sites in Hong Kong[57] (see also Table 5). Hoi Ha Wan is similarly remote from anthropogenic influence. Mai Po, on the other hand, lies at the inner end of Deep Bay which appears to be affected by metals in the discharge of the nearby Pearl River.[56,57] One explanation of the data is that the amphipods from Cape D'Aguilar and Hoi Ha Wan are showing zinc concentrations typical of *P. platensis* feeding on the cast-up macrophytic alga *Sargassum hemiphyllum,* itself with clean "background" zinc concentrations. *P. platensis* at Mai Po, however, are feeding on fallen leaves of the mangrove tree *Kandelia candel.* It is possible that the differential bioavailabilities of zinc in the two vegetation types — a marine seaweed and leaves from essentially a terrestrial tree — have caused different accumulated zinc concentrations in the amphipod. Ironically then, the best estimate of a background zinc concentration in *P. platensis* is that provided by the amphipods from Cape D'Aguilar and Hoi Ha Wan feeding in a more typical way.

Where does this leave the concept of a background concentration? It is certainly possible to recognize high accumulated metal concentrations in marine

Table 7. *Meganyctiphanes norvegica:* Metal Concentrations (μg g^{-1}) of Euphausiids Collected From the N.E. Atlantic (July 1985) or the Firth of Clyde, Scotland (October 1984)

	NE Atlantic	Firth of Clyde
Zinc		
Mean	102	43.0
SD	54.3	7.7
Range	40.1–281	26.9–62.5
n	29	30
Copper		
Concn (0.05 g)	57.5	35.8
95% CL	39.6–83.3	31.2–41.3
Range	8.8–67.2	30.8–72.6
n	29	30

Note: Figures are quoted either as the mean with standard deviation in the absence of a size effect on metal concentration (zinc), or as the metal concentration of a 0.05 g euphausiid estimated from the best fit regression line of log concentration vs. log dry weight with 95% confidence limits, when there is a significant size effect in at least one population (copper).

From Rainbow, P.S., Moore, P.G., and Watson, D., *Estuarine Coastal Shelf Sci.,* 28, 567, 1989.

Table 8. *Platorchestia platensis:* Zinc Concentrations (μg g^{-1} Dry Wt) in Standardized 5 Mg Dry Wt Amphipods From 4 Sites in Hong Kong (April 1989) Estimated From Best Fit Double Log Regressions of Metal Concentration Against Dry Weight

Site	Zn concn.	95% Confidence limits	ANCOVA
Wu Kai Sha	354	583, 354	A
Hoi Ha Wan	199	242, 163	B
Cape d'Aguilar	193	284, 131	B
Mai Po	109	141, 83.7	C

Note: Samples sharing any letter in the ANCOVA column are not significantly different.

From Rainbow, P.S., in *The Marine Flora and Fauna of Hong Kong and S. China,* Proc. 4th Mar. Biol. Workshop, in press.

invertebrates and thereby recognize sites of metal pollution, given all the necessary caveats in the design of biomonitoring programs.[53] Identification of a background concentration in a particular species is much more difficult, for we encounter noise in our data created by extraneous parameters affecting other aspects of the biology of the invertebrate. Such variability will tend to prevent the establishment of a narrow band of low accumulated metal concentrations identifiable as background.

It is also apparent that although members of a particular taxon of invertebrates may share the same accumulation strategy for one metal, it is certainly not valid to make interspecific comparisons of absolute body metal concentrations. Thus, although all barnacles appear to share a strong net accumulation strategy for metals like copper and zinc (see earlier), it is not possible to define a high or low accumulated metal concentration for barnacles in general. Table 5 illustrates this point well. A body copper concentration of 203 μg g^{-1} is high in the case of *Tetraclita squamosa* but would be low for *Balanus amphitrite*. It is probable that such differences come about by differential uptake rates and growth rates. Any assessment of the significance of a metal concentration can only be made intraspecifically.

Similarly, deep sea caridean decapods like *Systellaspis debilis*[37,58] have mean cadmium concentrations (about 12 μg Cd g^{-1} dry weight), an order of magnitude greater than those of British coastal decapods like *P. elegans* or *Crangon crangon*.[10] Such concentrations in the British coastal species would almost certainly be indicative of cadmium enrichment,[19] but are perfectly normal in the deep sea species. Cadmium body concentrations in the coastal penaeid decapod *Metapenaeopsis palmensis* from Hong Kong are so low as to be below detection limits ($<$0.06 to $<$0.032 μg Cd g^{-1} dry weight)[10] and probably result from the interaction between cadmium uptake rate and the rapid growth rate of tropical penaeids.

In a similar exploration of the concept of background concentrations, Zauke and Petri[51] ask whether the metal concentrations in mesopelagic and benthic crustaceans from the Antarctic Ocean are suitable as background levels, as an aid to the interpretation of data from areas influenced by man. In concluding that the metal concentrations of Antarctic crustaceans are not suitable to this end, Zauke and Petri[51] dissect in more detail some of the points raised here, stressing that biological variables can influence accumulated metal concentrations significantly.

VI. CONCLUSIONS

To conclude, progress is being made on the interpretation of the significance of metal concentrations in marine invertebrates. We are now often in a position to recognize a high or low accumulated concentration in a particular species, and to avoid the pitfall of generalizing between even closely related species. An understanding of accumulation strategies[17] is an essential prerequisite to this end.

At the tissue level there is an appreciation of the division of metal into metabolically available and detoxified components, with an expanding literature base on the latter. Some light is being shown, therefore, on why metal concentrations vary between tissue types or even between the same tissue in different species. Still, little is known on the functional significance of the difference between body metal concentrations of even closely related species, including between regulators, the study of which should clarify differential metabolic requirements given the absence of significant accumulated stores of detoxified metal.

It is difficult to define a background concentration of a metal, even intraspecifically, given the role of biological variables, and impossible interspecifically.

REFERENCES

1. **Nieboer, E. and Richardson, D. H. S.,** The replacement of the nondescript term heavy metals by a bologically and chemically significant classification of metal ions, *Environ. Pollut.,* B 1, 3, 1980.
2. **Bruland, K. W.,** Trace elements in seawater, in *Chemical Oceanography,* Vol. 8, Riley, J. P. and Chester, R., Eds., Academic Press, London, 1983, 157.
3. **Rainbow, P. S.,** The significance of trace metal concentrations in decapods, *Symp. Zool. Soc. London,* 59, 291, 1988.
4. **Rainbow, P. S.,** Heavy metal levels in marine invertebrates, in *Heavy Metals in the Marine Environment,* Furness, R. W. and Rainbow, P. S., Eds., CRC Press, Boca Raton, FL, 1990, chap. 5.
5. **Depledge, M. H. and Rainbow, P. S.,** Models of regulation and accumulation of trace metals in marine invertebrates, *Comp. Biochem. Physiol.,* 97C, 1, 1990.
6. **Rainbow, P. S. and Dallinger, R.,** Metal uptake, regulation, and excretion in freshwater invertebrates, this volume, chap. 7.
7. **Rainbow, P. S.,** The significance of accumulated heavy metal concentrations in marine organisms, in *Proc. Bioaccumulation Workshop,* February, 1991, Sydney, in press.
8. **Rainbow, P. S.,** Heavy metals in barnacles, in *Biology of Barnacles,* Southward, A. J., Ed. A. A. Balkema, Rotterdam, 1987, 405.
9. **Phillips, D. J. H. and Rainbow, P. S.,** Barnacles and mussels as biomonitors of trace elements: a comparative study, *Mar. Ecol. Prog. Ser.,* 49, 83, 1988.
10. **Rainbow, P. S.,** Trace metal concentrations in a Hong Kong penaeid prawn, *Metapenaeopsis palmensis* (Haswell), in *The Marine Flora and Fauna of Hong Kong and Southern China, Hong Kong, 1986,* Proc. 2nd Int. Mar. Biol. Workshop, Morton, B. Ed., Hong Kong University Press, Hong Kong, 1990, 1221.
11. **Eisler, R.,** *Trace Metal Concentrations in Marine Organisms,* Pergamon Press, Oxford, 1981.
12. **Phillips, D. J. H.,** The Rock Oyster *Saccostrea glomerata* as an indicator of trace metals in Hong Kong, *Mar. Biol.,* 53, 353, 1979.

13. **Phillips, D. J. H.**, Organochlorines and trace metals in green-lipped mussels *Perna viridis* from Hong Kong waters: a test of indicator ability, *Mar. Ecol. Prog. Ser.*, 21, 251, 1985.
14. **Phillips, D. J. H. and Yim, W. W.-S.**, A comparative evaluation of oysters, mussels and sediments as indicators of trace metals in Hong Kong waters, *Mar. Ecol. Prog. Ser.*, 6, 285, 1981.
15. **Rainbow, P. S. and White, S. L.**, Comparative strategies of heavy metal accumulation by crustaceans; zinc, copper and cadmium in a decapod, an amphipod and barnacle, *Hydrobiologia*, 174, 245, 1989.
16. **Phillips, D. J. H. and Rainbow, P. S.**, Strategies of trace metal sequestration in aquatic organisms, *Mar. Environ. Res.*, 28, 207, 1989.
17. **Rainbow, P. S., Phillips, D. J. H., and Depledge, M. H.**, The significance of trace metal concentrations in marine invertebrates: a need for laboratory investigation of accumulation strategies, *Mar. Pollut. Bull.*, 21, 321, 1990.
18. **Pullen, J. H. S. and Rainbow, P. S.**, The composition of pyrophosphate heavy metal detoxification granules in barnacles, *J. Exp. Mar. Biol. Ecol.*, 150, 249, 1991.
19. **White, S. L. and Rainbow, P. S.**, Regulation and accumulation of copper, zinc and cadmium by the shrimp, *Palaemon elegans, Mar. Ecol. Prog. Ser.*, 8, 95, 1982.
20. **White, S. L. and Rainbow, P. S.**, Regulation of zinc concentration by *Palaemon elegans* (Crustacea:Decapoda): zinc flux and effects of temperature, zinc concentration and moulting, *Mar. Ecol. Prog. Ser.*, 16, 135, 1984.
21. **Nugegoda, D. and Rainbow, P. S.**, Effect of a chelating agent (EDTA) on zinc uptake and regulation by *Palaemon elegans* (Crustacea:Decapoda), *J. Mar. Biol. Assoc. U.K.*, 68, 25, 1988.
22. **Nugegoda, D. and Rainbow, P. S.**, Effects of salinity changes on zinc uptake and regulation by the decapod crustaceans *Palaemon elegans* and *Palaemonetes varians*, *Mar. Ecol. Prog. Ser.*, 51, 57, 1989.
23. **George, S. G., Pirie, B. J. S., Cheyne, A. R., Coombs, T. L., and Grant, P. T.**, Detoxification of metals by marine bivalves; ultrastructural study of the compartmentalisation of copper and zinc in the oyster, *Ostrea edulis, Mar. Biol.*, 45, 147, 1978.
24. **George, S. G. and Pirie, B. J. S.**, Metabolism of zinc in the mussel *Mytilus edulis* (L.): a combined ultrastructural and biochemical study, *J. Mar. Biol. Assoc. U.K.*, 60, 575, 1980.
25. **Moore, P. G. and Rainbow, P. S.**, Ferritin crystals in the gut caeca of *Stegocephaloides christianiensis* Boeck and other Stegocephalidae (Amphipoda:Gammaridea): a functional interpretation, *Philos. Trans. R. Soc. London, Ser. B*, 306, 219, 1984.
26. **Viarengo, A.**, Heavy metals in marine invertebrates: mechanisms of regulation and toxicity at the cellular level, *CRC Crit. Rev. Aquat. Sci.*, 1, 295, 1989.
27. **George, S. G.**, Biochemical and cytological assessments of metal toxicity in marine animals, in *Heavy Metals in the Marine Environment*, Furness, R. W. and Rainbow, P. S., Eds., CRC Press, Boca Raton, FL, 1990, 123.
28. **Engel, D. W. and Brouwer, M.**, Metallothionein and metallothionein-like proteins: physiological importance, *Adv. Comp. Environ. Physiol.*, 5, 53, 1989.
29. **Carpenè, E.**, Metallothionein in marine molluscs, this volume, chap. 4.

30. **Mason, A. Z. and Nott, J. A.**, The role of intracellular biomineralised granules in the regulation and detoxification of metals in gastropods with special referecne to the marine prosobranch *Littorina littorea, Aquat. Toxicol.*, 1, 239, 1989.

31. **Brown, B. E.**, The form and function of metal-containing 'granules' in invertebrate tissues, *Biol. Rev.*, 57, 621, 1982.

32. **Simkiss, K., Taylor, M., and Mason, A. Z.**, Metal detoxification and bioaccumulation in molluscs, *Mar. Biol. Lett.*, 3, 187, 1982.

33. **Taylor, M. G. and Simkiss, K.**, Inorganic deposits in invertebrate tissues, *Environ. Chem.*, 3, 102, 1984.

34. **Taylor, M. G. and Simkiss, K.**, Structural and analytical studies on metal ion-containing granules, in *Chemical Perspectives in Biomineralisation*, Mann, S., Webb, J., and Williams, R. J. P., Eds., VCH Publishers, Weinheim, Germany, 1989, 428.

35. **Pequegnat, J. E., Fowler, S. W., and Small, L. F.**, Estimates of the zinc requirements of marine organisms, *J. Fish. Res. Board. Can.*, 26, 145, 1969.

36. **White, S. L. and Rainbow, P. S.**, On the metabolic requirements for copper and zinc in molluscs and crustaceans, *Mar. Environ. Res.*, 16, 215, 1985.

37. **White, S. L. and Rainbow, P. S.**, Heavy metal concentrations in the mesopelagic decapod crustacean *Systellaspis debilis, Mar. Ecol. Prog. Ser.*, 37, 147, 1987.

38. **Depledge, M. H.**, Re-evaluation of metabolic requirements for copper and zinc in decapod crustaceans, *Mar. Environ. Res.*, 27, 115, 1989.

39. **Chan, H. M.**, Aspects of the Biology of Zinc in Crabs with Particular Emphasis on the Shore-Crab, *Carcinus maenas*, (L.), Ph.D. thesis, University of London, 1990.

40. **Bryan, G. W.**, Zinc regulation in the lobster *Homarus vulgaris*. I: tissue zinc and copper concentrations, *J. Mar. Biol. Assoc. U.K.*, 44, 549, 1964.

41. **Bryan, G. W.**, Zinc concentrations of fast and slow contracting muscles in the lobster, *Nature (London)*, 213, 1043, 1967.

42. **Bryan, G. W. and Gibbs, P. E.**, Zinc — a major inorganic component of nereid polychaete jaws, *J. Mar. Biol. Assoc. U.K.*, 59, 969, 1979.

43. **Bryan, G. W. and Gibbs, P. E.**, Metals in nereid polychaetes: the contribution of metals in the jaws to the total body burden, *J. Mar. Biol. Assoc. U.K.*, 60, 641, 1980.

44. **Gibbs, P. E. and Bryan, G. W.**, Copper — the major component of glycerid polychaete jaws, *J. Mar. Biol. Assoc. U.K.*, 60, 205, 1980.

45. **Martin, J.-L. M., van Wormhoudt, A., and Ceccaldi, H. J.**, Zinc-hemocyanin binding in the hemolymph of *Carcinus maenas, Comp. Biochem. Physiol.*, 58A, 193, 1977.

46. **White, S. L. and Rainbow, P. S.**, A preliminary study of Cu-, Cd- and Zn-binding components in the hepatopancreas of *Palaemon elegans* (Crustacea:Decapoda), *Comp. Biochem. Physiol.*, 83C, 111, 1986.

47. **Nugegoda, D. and Rainbow, P. S.**, Zinc uptake and regulation by the sublittoral prawn *Pandalus montagui* (Crustacea:Decapoda), *Estuarine Coastal Shelf Sci.*, 26, 619, 1988.

48. **Depledge, M. H. and Bjerregaard, P.**, Haemolymph protein composition and copper levels in decapod crustaceans, *Helgol. Wiss. Meeresunters.*, 43, 207, 1989.

49. **Rainbow, P. S.**, Accumulation of Zn, Cu and Cd by crabs and barnacles, *Estuarine Coastal Shelf Sci.*, 21, 669, 1985.

50. **Nugegoda, D. and Rainbow, P. S.,** The effect of temperature on zinc regulation by the decapod crustacean *Palaemon elegans* Rathke, *Ophelia,* 27, 17, 1987.
51. **Zauke, G.-P. and Petri, G.,** Metal concentrations in Antarctic Crustacea: the problem of background levels, this volume, chap. 5.
52. **Hopkin, S. P.,** Deficiency and excess of copper in terrestrial isopods, this volume, chap. 18.
53. **Phillips, D. J. H. and Rainbow, P. S.,** *The Biomonitoring of Trace Aquatic Contaminants,* Elsevier Applied Science, London, in press.
54. **Rainbow, P. S., Moore, P. G., and Watson, D.,** Talitrid amphipods as biomonitors for copper and zinc, *Estuarine Coastal Shelf Sci.,* 28, 567, 1989.
55. **Rainbow, P. S.,** Copper, cadmium and zinc concentrations in oceanic amphipod and euphausiid crustaceans, as a source of heavy metals to pelagic seabirds, *Mar. Biol.,* 103, 513, 1989.
56. **Rainbow, P. S.,** The talitrid amphipod *Platorchestia platensis* as a potential biomonitor of copper and zinc in Hong Kong: laboratory and field studies, in *The Marine Flora and Fauna of Hong Kong and S. China,* Proc. 4th Mar. Biol. Workshop, Morton, B., Ed., Hong Kong University Press, in press.
57. **Rainbow, P. S.,** Biomonitoring of heavy metal pollution and its application in Hong Kong waters, in *Proc. Int. Conf. Mar. Biol. Hong Kong and S. China Sea, 1990,* Morton, B., Ed., Hong Kong University Press, in press.
58. **Ridout, P. S., Rainbow, P. S., Roe, H. S. J., and Jones, H. R.,** Concentrations of V, Cr, Mn, Fe, Ni, Co, Cu, Zn, As, Cd in mesopelagic crustaceans from the north east Atlantic Ocean, *Mar. Biol.,* 100, 465, 1989.

CHAPTER 2

Cadmium Bioaccumulation in the Scallop *Mizuhopecten yessoensis* From an Unpolluted Environment

Olga N. Lukyanova, Nina N. Belcheva, and Victor P. Chelomin

TABLE OF CONTENTS

0-87371-734-1/93/$0.00 + $.50
© 1993 by Lewis Publishers

I. INTRODUCTION

The ability of bivalves to accumulate dissolved heavy metals and to store them for a long period in their tissues allows the use of these animals in bio-monitoring programs investigating marine pollution. It should be borne in mind that certain species are known to accumulate selected metals under unpolluted conditions. For example, the Japanese scallop *Mizuhopecten yessoensis,* as well as many other members of the Pectinidae, accumulates cadmium in the soft tissues to a higher degree than most other bivalve molluscs.[1-3] In bivalves, tissue cadmium concentrations either decrease with growth, remain unchanged, or increase with age.[4-7] Most previous investigators have determined trace metal concentrations in whole individuals, although the principal metal accumulating organs — excretory and digestive organs — are but a small part of the whole body mass. Determination of the amount of metal in isolated organs allows one to establish how their metal concentrations depend on age and size. Dependence of cadmium and other metal concentrations in isolated organs of *M. yessoensis* on size has, so far, not been investigated. We therefore determined the amounts of cadmium, zinc, and copper in organs of scallops of six size groups (shell size: 2 to 17 cm; age: 1 to 8 years), collected from apparently clean sites. Additionally, we studied metal distributions among membrane fractions and cytoplasmic proteins, and made a chromatographic analysis of these proteins in scallops of different ages.

II. MATERIALS AND METHODS

The scallops *M. yessoensis* were collected in July 1984 from the southern part of Peter the Great Bay, Sea of Japan, where the environmental concentration of cadmium[8] was about $0.1 \ \mu g \ l^{-1}$. The molluscs were divided into six groups according to shell size and age:[9]

Group I: 2.0 to 2.5 cm, 1 year old, n = 50;
Group II: 5.5 to 7.5 cm, 2 years old, n = 5;
Group III: 7.7 to 9.0 cm, 3 to 4 years old, n = 21;
Group IV: 10.0 to 12.0 cm, 4 to 5 years old, n = 15;
Group V: 13.0 to 14.0 cm, 6 years old, n = 28;
Group VI: 16.0 to 17.0 cm, 8 years old, n = 9.

For analysis, the following organs and tissues were taken: mantle, hepato-pancreas, kidney, gill tissues, foot muscle, and gonads. Individual organs from scallops of each age group were dissected out and cut into pieces. Average samples for each organ were prepared by pooling organ pieces from several animals. In groups I and II all the animals contributed to an average sample; in other groups eight individuals in each group were used. Three replicates (2 to 3 g wet weight) of an average sample of each organ were taken to determine cadmium, zinc, and copper concentrations.

For separation of subcellular fractions, 8 g of wet tissue were simultaneously taken (3 replicates) from each average sample of hepatopancreas, kidney, and gill tissues. The tissues were homogenized in 0.05 M Tris-HCl buffer, pH 7.5, with 0.25 M sucrose, 0.01 M MgCl$_2$, and 0.5 M NaCl for membrane stabilization. Nuclei, mitochondria, microsomes, and cytosol were separated using the method of Fleischer and Kervina.[10] The protein concentrations of the subcellular fractions were determined using a modified Lowry method.[11] Freshly dissected hepatopancreas tissues (approximately 5 g wet weight) pooled from 5 3-year-old scallops and separately from 5 8-year-old scallops were homogenized and cytoplasmic proteins were isolated as previously described.[12] The distribution of cadmium among protein fractions in the hepatopancreas cytosol of *M. yessoensis* was examined by gel chromatography. Hepatopancreas tissue from 3- and 8-year-old scallops (approximately 5 g wet weight in each case) were homogenized in 3 volumes of 0.025 M Tris-HCl buffer, pH 7.5, containing 10^{-3} M PMSF (phenylmethyl sulphonyl fluoride) in an ice bath and centrifuged at 20,000 g for 30 min at 4°C. The supernatants were heated for 10 min at 60°C and the denatured proteins removed by spinning for 10 min in a refrigerated centrifuge (4°C). Approximately 0.8 ml of the clear supernatant was applied to a 73 × 1.6 cm Sephadex G-75 column and eluted with the same buffer at 4°C at a rate of 0.4 ml min^{-1}. Absorbance of all eluted fractions was monitored at 254 nm using a LKB Type 2151 UV/VIS monitor. Fractions (3 to 6 ml) were collected for analysis of cadmium concentration. The tissue metallothionein level was determined by the competitive Hg-binding assay[13] with the modifications of Lobel and Payne.[14] In order to determine metal contents, samples of wet tissues, subcellular fractions, and proteins were dried at 85°C. The dried samples were wet acid ashed (HNO$_3$ + HClO$_4$, 2:1 v/v) at 180°C. The dry residues were dissolved in 0.1 N HCl and analyzed by flame atomic absorption spectrophotometry (Shimadzu model AA-610S, Japan) with deuterium background correction.

III. RESULTS AND DISCUSSION

The amounts of cadmium in the organs of the scallop *M. yessoensis* in all size groups differed considerably, ranging from 2 to 6 μg g^{-1} dry weight in muscle, mantle, and gills, to 25 to 400 μg g^{-1} in hepatopancreas, and from 100 to 600 μg g^{-1} dry weight in kidney (Figure 1). In the mantle and gills no age-dependent increases in cadmium concentration were observed, but both the hepatopancreas and kidney exhibited a significant age-related increase in metal concentration. The apparent decrease in cadmium concentration of the muscle with size is not significant.

Age/size-related variations in cadmium concentrations have been documented earlier for some marine invertebrates. For example, in the Antarctic scallop *Adamassium colbecki* collected in unpolluted areas of the Ross Sea, the

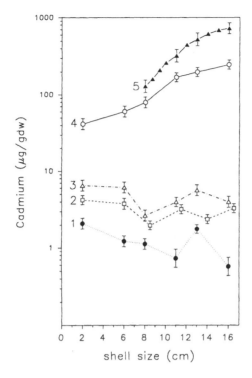

FIGURE 1. *Mizuhopecten yessoensis.* Cd concentrations in 1: muscle; 2: mantle; 3: gills; 4: hepatopancreas; 5: kidney. Concentration; $\mu g\ g^{-1}$ dry weight; cm: shell size (indication of age).

cadmium concentration was 142 $\mu g\ g^{-1}$ in the digestive gland and 11.6 $\mu g\ g^{-1}$ in the kidney.[15] With an increase in scallop body weight from 0.7 to 4.0 g, the concentrations of cadmium, zinc, and manganese in whole soft tissues decreased.[15] In *Placopecten magellanicus* from various sites along the U.S. Atlantic coast the cadmium concentration in the digestive gland increased as the animal grew.[16] In *P. magellanicus* collected from U.S. North Atlantic waters the cadmium concentration in the whole visceral mass was 2.7 to 27 $\mu g\ g^{-1}$,[17] in the digestive gland 94 $\mu g\ g^{-1}$, and in the kidney 62.6 $\mu g\ g^{-1}$.[16] In polluted areas the cadmium concentration in the digestive gland increased greatly. In *Mytilus edulis* from various sites in the Bay of Saint Lawrence, organ cadmium concentrations decreased with increasing shell size (0.8 to 6.5 cm).[18] This tendency is distinctly pronounced in immature individuals. Adults do not concentrate cadmium with age.[18] In the gastropod mollusc *Haliotis tuberculata,* cadmium concentrations remained unchanged in foot and whole soft tissue, but increased in viscera with increase in weight.[19] Among vertebrates, consistent increases in cadmium concentration with age have been observed in the kidney and liver of both the horse and man: 1.1 to 168 $\mu g\ g^{-1}$ wet weight in horse kidney and 11 to 60 $\mu g\ g^{-1}$ wet weight in human kidney.[20] The above data indicate that organs

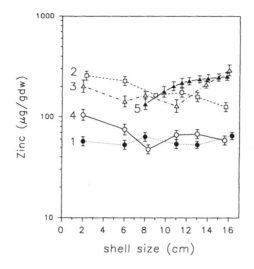

FIGURE 2. *Mizuhopecten yessoensis;* Zn concentrations; see Figure 1 for explanation of symbols.

of the scallop *M. yessoensis* contain considerably higher cadmium concentrations than other species.

Unlike the case of cadmium, the concentration of zinc in the hepatopancreas fell, but not significantly with increase in size (Figure 2). Concentrations of zinc rose significantly in the kidney and gills (Figure 2). Borchardt et al.[21] showed that the concentration of zinc in the soft tissues of bivalves from clean regions either decreased with bivalve growth or were not dependent on shell size. Bernhard and Andreae[6] found no growth-related changes in the zinc concentrations of organs of the oysters *Ostrea edulis* and *Crassostrea gigas,* which typically have high soft tissue levels of zinc. In *M. yessoensis* the variation in mantle zinc concentration is also not significant (Figure 2).

Maximum copper concentrations were detected in the hepatopancreas and kidney of *M. yessoensis,* the highest being 1370 μg g^{-1} in the hepatopancreas of 1-year-old scallops. As for many other species, the highest copper concentrations in the scallop organs were found in the youngest individuals. Copper concentrations decreased in 2-year-old scallops and then changed very little in all groups and all organs. The fall in copper concentration of all organs between 1- and 2-year-old scallops is dramatic (Figure 3), and not seen in the cases of cadmium and zinc. It is possible that the high copper concentrations of the youngest scallops reflect an atypical metabolic requirement for the essential metal copper, but supporting evidence is lacking.

Analysis of the intracellular distributions of cadmium in the hepatopancreas and gills of different age groups showed that the greatest portions of metal were accumulated in the cytosol (Table 1). This is true for all groups, and it is worth mentioning that cadmium in the cytosol accounted for 71.1% of total hepato-

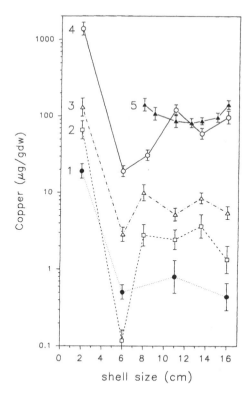

FIGURE 3. *Mizuhopecten yessoensis;* Cu concentrations; see Figure 1 for explanation of symbols.

pancreas cadmium in 1-year-old scallops and 98.8% in 8-year-olds. When the cadmium concentration is expressed per milligram of protein, this difference is even more pronounced: 0.12 μg mg^{-1} for 1-year-olds and 1.4 μg mg^{-1} for 8-year-old molluscs. Similar ratios were found for the gills, although the cadmium concentration here was almost an order of magnitude lower than in the hepatopancreas. In the kidney of 6-year-old molluscs, cadmium was also concentrated mainly in the cytoplasmic fraction.

Table 2 shows the relative distribution of zinc in the hepatopancreas and gills between membrane-bound and cytosolic fractions. This ratio is higher in the gills for age group III, but higher in the hepatopancreas for age group VI.

The copper content in the gills was almost evenly distributed between cytosolic and membrane fractions, and this proportion showed no systematic change with age. In the hepatopancreas, the greater part of the copper was located in the cytosol, the proportion remaining at a similar level in all the age groups (Table 2).

In spite of the apparent age-related changes in the distribution of cadmium between membrane structures and cytosol, the bulk of the cadmium remains

Table 1. **Subcellular Distributions of Cadmium in the Hepatopancreas, Gills, and Kidney of the Scallop *Mizuhopecten yessoensis***

Group of animals	% of total tissue metal				μg mg^{-1} Protein			
	N	M	MS	CT	N	M	MS	CT
Hepatopancreas								
I	9.60	14.9	3.8	71.1	0.10	0.10	0.22	0.12
II	0.48	1.0	0.5	98.0	0.03	0.04	0.02	0.52
III	0.46	1.0	0.7	97.9	0.04	0.05	0.09	0.63
IV	0.17	0.6	0.5	98.7	0.02	0.05	0.01	0.56
V	0.20	0.6	0.4	98.8	0.06	0.10	0.10	1.40
Gills								
II	5.6	10.6	6.1	77.9	0.01	0.01	0.02	0.02
III	13.8	9.5	4.8	71.8	0.02	0.01	0.02	0.02
V	3.7	n.d.	n.d.	96.2	0.01	—	—	0.20
Kidney								
V	3.8	10.5	1.5	84.0	0.25	0.48	0.15	1.15

Note: N: nucleus
M: mitochondria
MS: microsomes
CT: cytosol
n.d.: not detected
— signifies no data

Average values of three determinations are given; error ±3%.

Table 2. ***Mizuhopecten yessoensis:* Relative Distributions of Cd, Zn, and Cu Between Whole Membrane and Cytosolic Fractions (Membrane-Bound Metal/Cytosolic Metal) in the Hepatopancreas and Gills of Scallops of Age Groups I to VI**

Age group	Hepatopancreas			Gills	
	I	III	VI	III	VI
Cadmium	0.395	0.020	0.012	0.286	0.040
Zinc	0.153	0.109	0.081	0.138	0.043
Copper	0.053	0.077	0.048	1.110	0.920

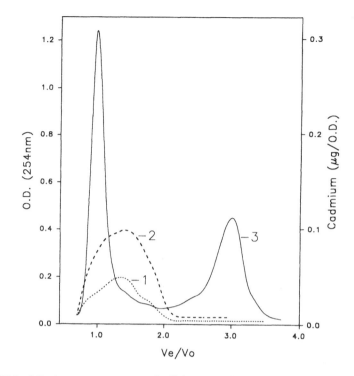

FIGURE 4. *Mizuhopecten yessoensis;* Cd content (µg/ml) in cytoplasmic proteins from hepatopancreas of 1: 3-year-old scallops, 2: 8-year-old scallops; 3: optical density of eluted proteins of both 3- and 8-year-old scallops. V_e — effluent volume; V_o — void volume.

bound to cytosolic elements. Separation of cytosolic proteins by gel chromatography on a Sephadex G-75 column (Figure 4) showed that almost the entire cytoplasmic cadmium is found in the region of high-molecular-weight proteins (HMWPs). Calibration of the column by standard proteins enabled us to suggest that cadmium is mostly bound to proteins with molecular weights from 40,000 to 60,000 Da. Similar results have been obtained for *Pecten maximus*[22] where the greater part of the cadmium was bound to a 55,000 Da protein. Binding of cadmium to proteins corresponding in molecular weight to metallothioneins was not found. It also remains possible that the apparent absence of metallothioneins is an experimental artifact, for separations were carried out in the absence of reducing agents. However, using the Hg-saturating method for metallothionein determination we detected a small amount of these proteins which slightly increased with age (from 1.2 nmol/mg in 3-year-olds to 1.6 nmol/mg of protein in 8-year-olds). These differences can be accounted for by the high sensitivity of the method for metallothionein determination by competitive Hg binding. The present results do not rule out the possibility that *M. yessoensis* possesses proteins with a molecular mass corresponding to metallothioneins, which, if present, do not seem to play a significant role in age-dependent cadmium accumulation in scallops inhabiting areas with background cadmium levels.

Comparison of the distribution of cadmium among cytoplasmic proteins in the hepatopancreas of 3- and 8-year-olds (Figure 4) showed that cadmium accumulation with age is associated with a considerable increase in the metal level in HMWPs. Binding of cadmium to HMWPs (45,000 Da) was also observed in *Placopecten magellanicus* subjected to this metal.[23] Metallothionein-like proteins were identifiable only by the use of additional methods of concentration and separation by HPLC and they were absent in the control animals.[23]

The participation of HMWPs in cadmium binding is a characteristic of not only members of *Pectinidae,* but also of other bivalves such as *Mercenaria mercenaria,*[24] *Macoma baltica,*[25] and *Crassostrea virginica.*[26] The physiological role of these proteins in cadmium metabolism is still not clear. The present results and the literature data lead us to suggest that in molluscs, particularly in the Pectinidae, these proteins have the same biological significance as metallothioneins do in most animals. Age-dependent cadmium accumulation in the hepatopancreas and kidney of *M. yessoensis* is evidence for the absence of any effective mechanisms for regulating cadmium concentrations at the whole body level. Such age-dependent accumulation is present when the net excretion rate of cadmium is lower than the net uptake rate. Involvement of different molecular weight proteins in cadmium binding has previously been shown in aquarium experiments with *M. yessoensis* exposed to elevated cadmium concentrations.[27]

IV. CONCLUSIONS

Thus in *Mizuhopecten yessoensis* there are different mechanisms for the net accumulation of the essential trace metals zinc and copper, and the nonessential metal cadmium. The patterns of metal accumulation in the organs reflect the presence of at least two systems of net accumulation of metal at this level. At one extreme, the metal concentration of an organ remains constant over development (or even falls if the rate of organ growth exceeds the rate of metal accumulation). This constancy of organ concentration can be considered as the regulation of its metal level. At the other extreme, tissues such as the kidney and hepatopancreas of *M. yessoensis* accumulate high concentrations of cadmium with age, even if environmental cadmium levels are relatively low. This organ accumulation of a toxic metal must be associated with its detoxification. The very high age-dependent cadmium concentrations as detoxified stores of the metal in the hepatopancreas or kidney of *M. yessoensis* are striking in comparison with other molluscs and appear to be a characteristic feature of the metal biology of pectinid scallops.

REFERENCES

1. **Boyden, C. R.**, Effect of size upon metal content of shellfish, *J. Mar. Biol. Assoc. U.K.*, 57, 675, 1977.
2. **Pesch, G. G. and Stewart, N. E.**, Cadmium toxicity to three species of estuarine invertebrates, *Mar. Environ. Res.*, 3, 145, 1980.
3. **Khristoforova, N. K.**, Peculiarities of trace metal composition of soft tissues of some commercial molluscs from Peter the Great Bay, in *Proc. 4th Symp., Sea Farming*, Vladivostok, 1983, 198, (in Russian).
4. **Bryan, G. W.**, The occurrence and seasonal variation of trace metals in the scallops *Pecten maximus* (L.) and *Chlamys opercularis* (L.), *J. Mar. Biol. Assoc. U.K.*, 53, 145, 1973.
5. **Boyden, C. R.**, Trace element content and body size in molluscs, *Nature, London*, 251, 311, 1974.
6. **Bernhard, M. and Andreae, M. O.**, Transport of trace metals in marine food chains, in *Changing Metal Cycles and Human Health*, Nriagu, J. O., Ed., Springer-Verlag, Berlin, 1984, 143.
7. **Szefer, P. and Szefer, K.**, Occurrence of ten metals in *Mytilus edulis* (L.) and *Cardium glaucum* (L.) from the Gdansk Bay, *Mar. Pollut. Bull.*, 16, 446, 1985.
8. **Patin, S. A., Morozov, N. R., Romanteeva, A. S., Melnikova, R. M., and Borisenko, G. S.**, Trace metals in the ecosystem of Sea of Japan, *Geochimia*, 3, 423, 1979, (in Russian).
9. **Silina, A. V.**, Determination of age and growth rate of Yeso scallop *Patinopecten yessoensis*, by the sculpture of its shell surface, *Sov. J. Mar. Biol.*, 4, 827, 1978.
10. **Fleischer, S. and Kervina, M.**, Subcellular fractionation of rat liver, *Methods Enzymol.*, 31, 6, 1974.
11. **Marcwell, M. A., Haas, S. M., Brieber, L. L., and Tolbert, N. E.**, A modification of the Lowry procedure to simplify protein determination in membrane and lipoprotein samples, *Anal. Biochem.*, 87, 206, 1978.
12. **Evtushenko, Z. S., Lukyanova, O. N., and Khristoforova, N. K.**, Biochemical changes in selected body tissues of the scallop *Patinopectin yessoensis* under long-term exposure to low cadmium concentration, *Mar. Ecol. Prog. Ser.*, 20, 165, 1984.
13. **Piotrowski, J. K., Balanowska, W., and Sapota, A.**, Evaluation of metallothionein content in animal tissues, *Acta Biochim. Pol.*, 20, 207, 1973.
14. **Lobel, P. B. and Payne, J. F.**, The mercury-203 method for evaluating metallothioneins: interference by copper, mercury, oxygen, silver and selenium, *Comp. Biochem. Physiol.*, 86, 37, 1987.
15. **Mauri, M., Orlando, E., Nigro, M., and Regoli, F.**, Heavy metals in the Antarctic scallops *Adamussium colbecki*, *Mar. Ecol. Prog. Ser.*, 67, 27, 1990.
16. **Uthi, J. F. and Chou, C. L.**, Cadmium in sea scallop (*Placopecten magellanicus*) tissues from clean and contaminated areas, *Can. J. Fish. Aquat. Sci.*, 44, 91, 1987.
17. **Greig, R. A., Wenzloff, D. R., Mackenzie, C. L., Merril, A. S., and Zdanowicz, V. S.**, Trace metals in sea scallops, *Placopecten magellanicus*, from eastern United States, *Bull. Environ. Contam. Toxicol.*, 19, 326, 1978.
18. **Cossa, D., Bourget, E., and Piuze, J.**, Sexual maturation as a source of variation in the relationship between cadmium concentration and body weight of *Mytilus edulis* (L.), *Mar. Pollut. Bull.*, 10, 174, 1979.

19. **Bryan, G. W., Potts, G. W., and Forster, G. R.**, Heavy metals in the gastropod mollusc *Haliotis tuberculata*, *J. Mar. Biol. Assoc. U.K.*, 57, 379, 1977.
20. **Elinder, C. G. and Nordberg M.**, Critical concentration of cadmium estimated by studies on horse kidney metallothionein, in *Biological Roles of Metallothionein*, Foulkes, E. C., Ed., Elsevier, New York, 1982, 37.
21. **Borchardt, T., Buerherdt, S., Hablizeb, H., Karte, L., and Zeitner, R.**, Trace metal concentrations in mussels: comparison between estuarine, coastal and off-shore regions in the south- eastern North Sea from 1983 to 1986, *Mar. Ecol. Prog. Ser.*, 42, 17, 1988.
22. **Stone, H. C., Wilson, S. B., and Overnell, J.**, Cadmium-binding proteins in the scallop, *Pecten maximus, Environ. Health Perspect.*, 65, 189, 1986.
23. **Fowler, B. A. and Gould, E.**, Ultrastructural and biochemical studies of intra-cellular metal-binding patterns in kidney tubule cells of the scallop *Placopecten magellanicus* following prolonged exposure to cadmium or copper, *Mar. Biol.*, 97, 207, 1988.
24. **Carmichael, N. G., Squibb, K. S., Engel, D. W., and Fowler, B. A.**, Metals in the molluscan kidney: uptake and subcellular distribution of [109]Cd, [54]Mn and [65]Zn by the clam, *Mercenaria mercenaria, Comp. Biochem. Physiol.*, 65A, 203, 1980.
25. **Langston, W. J. and Zhou, M.**, Cadmium accumulation, distribution and elim-ination in the bivalve *Macoma baltica:* neither metallothionein nor metallothionein-like proteins are involved, *Mar. Environ. Res.*, 21, 225, 1987.
26. **Roesijadi, G. and Klerks, P. L.**, Kinetic analysis of cadmium binding to metal-lothionein and other intracellular ligands in oyster gills, *J. Exp. Zool.*, 251, 1, 1989.
27. **Evtushenko, Z. S., Belcheva, N. N., and Lukyanova, O. N.**, Cadmium accu-mulation in organs of the scallop *Mizuhopecten yessoensis*. 2. Subcellular distri-bution of metals and metal-binding proteins, *Comp. Biochem. Physiol.*, 83C, 377, 1986.

CHAPTER 3

Accumulation and Subcellular Distribution of Metals in the Marine Gastropod *Nassarius reticulatus* L.

Toralf Kaland, Thorvin Andersen, and Ketil Hylland*

TABLE OF CONTENTS

* Author to whom all correspondence should be addressed.

0-87371-734-1/93/$0.00 + $.50
© 1993 by Lewis Publishers

I. INTRODUCTION

Aquatic organisms are totally enveloped in their medium; in most cases, they use it as a source of respiratory oxygen, drink it, respond osmotically to it, and use it as a source of trace minerals. Aquatic biota may take up or excrete some substances actively from the complex mixture of dissolved organic and inorganic components present in any natural water mass, while others simply diffuse in or out along concentration gradients. The entry of many trace metals into marine organisms from solution is generally thought to be by passive facilitated diffusion. It is debatable whether it is an exacting task for the organism to take up and retain sufficient quantities of essential trace metals such as Zn, Cu, and Fe while preferably limiting the uptake and retention of those not needed, e.g., Hg, Cd, and Pb.[1]

Essential metals may also be toxic when present internally at concentrations beyond the limits of cellular regulatory processes. To avoid toxicosis by metals, any of three main strategies may be chosen by an organism: the uptake may be limited, excretion increased, or the metal immobilized within the tissues. Excess intracellular metals, be they essential or nonessential, are handled in different ways by different marine invertebrates. They may be included in granules,[2,3] immobilized in tertiary lysosomes,[4] or bound to cytosolic components such as metallothioneins (MT), other metal-binding proteins (MBP),[5] high molecular weight proteins,[6] amino acids,[7] and nonpeptides.[8] For various unknown reasons, some species of marine organisms accumulate one or a few metals, e.g., polychaetes of the genus *Melinna* store Cu in their branchiae,[9,10] the high levels of Zn or Cd in scallops, and As in the polychaete *Tharyx marioni*.[11]

Marine molluscs have a large capacity for accumulation of metals such as Cu, Zn, and Cd, and have therefore commonly been suggested as biomonitors of metal contamination.[12] Furthermore, the aforementioned MBPs in marine invertebrates, and specifically molluscs, have in the last decade come to attention as potential biological indicators of environmental metal exposure.[13,14]

Before applying such methods in field situations, more information is needed on endo- and exogenous factors influencing such indices, and also laboratory-type data on accumulation rates and induction kinetics.[15] Much information is presently available on such factors in bivalves, especially *Mytilus edulis*. In the present study we therefore wanted to evaluate the accumulation and subcellular distribution of Cd, Cu, and Zn in the hepatopancreas, gills, intestine, and foot of the prosobranch gastropod *Nassarius reticulatus*. As an initial approach, we exposed the gastropod to dissolved Cu, Zn, and Cd at concentrations equivalent to about 25% of 96-h LC_{50} values.

II. MATERIALS AND METHODS

A. Chemicals

Nitric acid, acetic acid, and mercaptoethanol came from Merck, Darmstadt, Germany. Nitric acid was double distilled from an analytical grade before use. Trizma base, blue dextran, AMP, albumin, glutathione, lactoglobulin, myoglobin, and metallothionein from horse kidney were all bought from Sigma Chemical Co., St. Louis, MO, U.S. Superose 12 preparative grade column material was obtained from Pharmacia Fine Chemicals, Uppsala, Sweden. Copper, Zn, and Cd standard solutions for atomic absorption spectrophotometry were bought from Teknolab A/S, Drøbak, Norway.

B. Experimental Animals

N. reticulatus, the netted dog whelk, were collected in November 1990 on a sandy beach in the outer Oslo Fjord at Drøbak, Norway. The gastropods were maintained in seawater of 34 ‰ salinity at 10°C and at a constant photoperiod of 12L:12D for 1 week before the start of the experiments. Groups of 24 gastropods, of shell length between 18 and 24 mm, were separately exposed to 50 μg/l Cu, 1500 μg/l Zn, and 900 μg/l Cd in 3 aquaria, and a fourth group kept in clean seawater. All gastropods were exposed for 20 days and the water changed every second day. The gastropods were not fed during the exposures.

C. Preparation of Subcellular Fractions

The posterior part of the hepatopancreas (digestive gland), anterior intestine, gills, and the soft part of the foot were pooled from eight gastropods in each exposure group. Three such replicates were taken from each group for analysis. Tissues were weighed and homogenized in ice-cold 100 mM Tris-acetate buffer, pH 8.1, with 5 mM mercaptoethanol and 0.1 mM (phenylmethylsulfonyl fluoride) PMSF. Samples were homogenized using 15 up-and-down strokes with a glass-teflon homogenizer (Potter-Elvehjem) rotating at 750 rpm. The homogenates were centrifuged at 10,000 g for 30 min. This first pellet, hereafter referred to as the "mitochondrial" fraction, is enriched with mitochondria, nuclei, cell membranes, Golgi vesicles, and peroxisomes. The first supernatant was recentrifuged at 100,000 g for 60 min yielding the microsomal fraction, containing mainly endoplasmatic reticulum and ribosomes. Both pellets were transferred to PTFE containers for digestion before metal analysis. The final supernatant, cytosol, was stored at −80°C until use. Some nuclei will unavoidably break in the homogenization step, and dissolved nuclear components will therefore be found in the cytosolic fractions.[16] Care was taken to avoid thawing-refreezing cycles and cytosols used in the experiments were only frozen once.

After the posterior part of hepatopancreas, anterior intestine, gills, and foot of the gastropod were removed, the remaining tissues were pooled for metal analyses. The remaining tissues were not fractionated as described above.

D. Digestion of Samples and Determination of Metal Concentrations

The three subcellular fractions (mitochondrial, microsomal, and cytosolic) and the remaining tissues were wet-ashed and analyzed for Cu, Zn, and Cd by flame or electrothermal atomic absorption spectrophotometry (AAS) using a Varian SpectrAA 10. The mitochondrial and microsomal fractions, as well as the remaining tissue from the dissection, were digested with nitric acid in PTFE containers, heated in a microwave oven (150 W, 2 h), and then appropriately diluted with double-distilled water. The cytosolic fractions were mixed 1:1 (v/v) with nitric acid and digested at 70°C for 2 h on a heating block. All samples were stored in polyethylene bottles until metal analysis.

E. Gel Permeation Chromatography

Aliquots (100 μl) of hepatopancreas cytosol were applied to a column packed with Superose 12 preparative grade (1 × 30 cm) and eluted with 100 mM Tris buffer, pH 8.1, at a flow rate of 0.2 ml min^{-1}. Before application, the cytosols were filtered through a 0.22 μm Millex GV filter (Millipore). The optical absorbance of the eluents was continuously monitored at 254 nm and 280 nm. Eluate fractions (1.0 ml) were collected and analyzed immediately for Cu, Zn, and Cd. The Superose 12 column was calibrated with blue dextran (200 kDa), bovine albumin (66 kDa), lactoglobulin (36.5 kDa), myoglobin (17.5 kDa), metallothionein from horse kidney (6.5 kDa), and glutathione (620 Da).

F. Statistical Methods

Differences between the control and each of the three metal-exposed groups with regard to the concentrations of Cu, Zn, and Cd in each subcellular fraction were tested using one-way analysis of variance (ANOVA) on log$_e$-transformed values, with a subsequent comparison between the control and exposed groups using Dunnett's test.[17] The level of significance was set to 0.05 for the rejection of H_0: no difference between groups.

III. RESULTS AND DISCUSSION

A. Accumulation and Subcellular Distribution of Cu

All tissues accumulated some Cu following exposure to 50 μg/l of the metal in the water, although the increases in the foot and 'remainder' fractions were minor (Figure 1A). The gill appeared the main target organ with a fivefold accumulation over control values, which could possibly also explain the high toxicity of Cu to *Nassarius reticulatus* (96 h LC_{50} <250 ppb).[27]

In unexposed tissues, most of the Cu present was found in mitochondrial and microsomal pools. However, increased Cu content as a result of exposure to the metal was largely sequestered in the cytosolic compartment of all four tissues examined. There were significantly increased levels of Cu in the mitochondrial fraction of all four tissues studied (Tables 1 to 4).

In hepatopancreas of both Cu-exposed and unexposed *N. reticulatus*, Cu was found to be associated with numerous cytosolic components. The major component, of apparent M_r similar to mammalian MT (10 kDa), also sequestered excess Cu following exposure (Figures 2 and 3). The remaining components were both of higher and lower apparent molecular size. Different cytosolic Cu-binding components have been described previously, both in marine bivalves[18-20] and gastropods.[21] In the latter study of *Buccinum tenuissimum*, most of the cytosolic Cu was found to associate with the environmentally induced Cd-binding protein, of similar apparent size to that observed in the present study.

B. Accumulation and Subcellular Distribution of Zn

The extra accumulation of Zn in the four tissues examined was much less than that of Cu and Cd (Figure 1B), even though Zn concentration (and free ion concentration) in the seawater was by far the highest of the three. It is therefore possible that the accumulated concentration of Zn is regulated in *N. reticulatus*, even at extreme exposures (1.5 mg/l for 20 days). Zinc levels were elevated in the intestine and foot of Zn-exposed gastropods, but this increase was differently distributed in the cells of different tissues. In intestine, the increase was associated with the mitochondrial and cytosolic fractions (Table 3), whereas all three fractions of foot contained significantly elevated levels of Zn (Table 4). The cytosol appeared to be the major Zn compartment in all four tissues, both in the Zn-exposed and control gastropods.

Nearly all cytosolic Zn in the hepatopancreas of Zn-exposed and unexposed *N. reticulatus* eluted with very low molecular weight components (M_r <1000 Da), althouth there were minor amounts bound to high molecular weight components and associated with the Cu- / Cd-binding 10 kDa protein (Figures 3 and

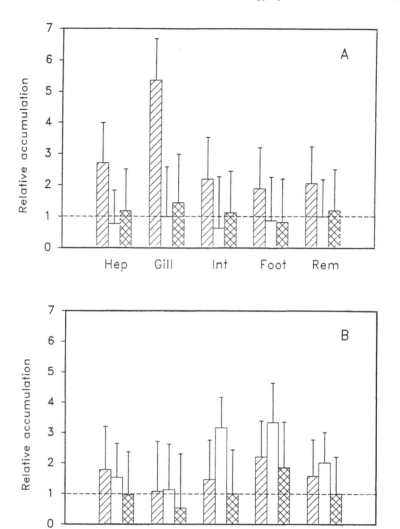

FIGURE 1. Accumulation of Cu (diagonally hatched), Zn (blank), Cd (cross-hatched) in different organs relative to control levels. The given values are summed concentrations from the subcellular fractions of each tissue. Hep — hepatopancreas, Int — intestine, Rem — remainder. A. Gastropods exposed to 50 μg/l Cu, B. Gastropods exposed to 1.5 mg/l Zn, C. Gastropods exposed to 0.9 mg/l Cd. Vertical lines denote one standard deviation.

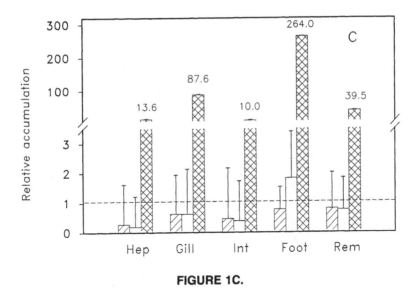

FIGURE 1C.

4). Similar cytosolic distribution patterns have been observed for zinc in bivalves, e.g., *Ostrea edulis*,[7] where it was suggested that most of the Zn present was associated with taurine and homarine. Most of the cytosolic Zn in the digestive glands of other marine gastropods investigated, *Littorina littorea*[22] and *Murex trunculus*,[23] was, however, found to be associated with components of higher molecular weight.

C. Accumulation and Subcellular Distribution of Cd

The only nonessential metal of the three, Cd, was accumulated to levels up to two orders of magnitude higher in the tissues of Cd-exposed compared to control gastropods (Figure 1C). Furthermore, the Cd contents of the three subcellular fractions in all four tissues were also significantly increased compared to controls (Tables 1 to 4). As for Zn, Cd was mainly sequestered into the cytosol in all four tissues.

In the hepatopancreas cytosol of unexposed gastropods, Cd was sequestered into two pools: a minor one of high molecular weight, and a major one corresponding to the relative apparent molecular size of mammalian MT, i.e., 10 kDa (Figure 5). Similar cytosolic Cd distributions have been reported in other marine gastropods e.g, *Littorina littorea*[22] and *Murex trunculus*.[23] Interestingly, in Cd-exposed gastropods a third, very low molecular weight Cd pool was evident. It is conceivable that this presumably more labile Cd pool may give rise to toxic manifestations of the metal, this latter more metabolically available metal having 'spilt over' from the detoxified Cd bound to metallothionein-like proteins.[24] A similar effect was observed in the bivalve *Macoma balthica*, in which increased levels of Cu, Ag, and Zn were found associated with very low molecular weight components when there were high metal concentrations in the metallothionein-like pool.[25]

Table 1. Concentration of Cu, Zn, and Cd in Subcellular Fractions From Hepatopancreas of *Nassarius reticulatus* Exposed to Cu, Zn, and Cd in Seawater

Fraction	Control (mean ± sd)	Cu-exposed (mean ± sd)	Zn-exposed (mean ± sd)	Cd-exposed (mean ± sd)
Cu				
Mitochondrial	32.5 ± 11.0	52.2 ± 6.2*	66.1 ± 15.4*	8.4 ± 0.8*
Microsomal	10.3 ± 2.9	10.7 ± 3.5	13.3 ± 5.7	2.2 ± 0.3*
Cytosolic	5.1 ± 1.7	64.7 ± 9.3*	6.1 ± 1.9	2.5 ± 2.0
Zn				
Mitochondrial	109.0 ± 31.5	67.4 ± 15.5	188.2 ± 37.0*	18.0 ± 2.4*
Microsomal	17.2 ± 1.8	14.4 ± 6.1	26.9 ± 1.5	2.8 ± 1.0*
Cytosolic	747.3 ± 48.8	591.9 ± 34.9	1122.5 ± 162.5*	146.9 ± 30.2*
Cd				
Mitochondrial	0.22 ± 0.05	0.19 ± 0.04	0.21 ± 0.06	4.60 ± 1.38*
Microsomal	0.09 ± 0.03	0.05 ± 0.02	0.06 ± 0.01	1.10 ± 0.52*
Cytosolic	2.75 ± 0.32	2.41 ± 0.60	2.75 ± 0.95	36.35 ± 8.52*

Note: Each value is the mean of three pooled samples of eight individuals. All values in μg/g wet weight of original hepatopancreas. Values significantly different from control are indicated by * ($p < 0.05$, Dunnett's test).

Table 2. Concentration of Cu, Zn, and Cd in Subcellular Fractions From Gills of Nassarius reticulatus Exposed to Cu, Zn, and Cd in Seawater.

Fraction	Control (mean ± sd)	Cu-exposed (mean ± sd)	Zn-exposed (mean ± sd)	Cd-exposed (mean ± sd)
Cu				
Mitochondrial	4.5 ± 0.8	25.0 ± 1.0*	4.2 ± 2.8	2.1 ± 1.0
Microsomal	3.9 ± 1.9	7.2 ± 1.3	5.1 ± 1.7	1.7 ± 0.9
Cytosolic	0.2 ± 0.1	13.2 ± 4.6*	0.3 ± 0.2	1.5 ± 0.5*
Zn				
Mitochondrial	17.2 ± 4.0	11.0 ± 5.6	12.2 ± 2.7	6.4 ± 2.2
Microsomal	2.0 ± 1.4	0.8 ± 0.5	0.7 ± 0.2	1.0 ± 0.7
Cytosolic	20.3 ± 10.5	27.3 ± 14.9	30.3 ± 4.8	16.5 ± 6.8
Cd				
Mitochondrial	0.03 ± 0.01	0.17 ± 0.06*	0.06 ± 0.02	4.81 ± 1.08*
Microsomal	0.01 ± 0.01	0.01 ± 0.01	0.02 ± 0.02	0.25 ± 0.12*
Cytosolic	0.28 ± 0.11	0.28 ± 0.15	0.09 ± 0.05*	22.85 ± 7.59*

Note: Each value is the mean of three pooled samples of eight individuals. All values in μg/g wet weight of original gill. Values significantly different from control are indicated by * ($p < 0.05$, Dunnett's test).

Table 3. Concentration of Cu, Zn, and Cd in Subcellular Fractions From Intestine of *Nassarius reticulatus* Exposed to Cu, Zn, and Cd in Seawater

Fraction	Control (mean ± sd)	Cu-exposed (mean ± sd)	Zn-exposed (mean ± sd)	Cd-exposed (mean ± sd)
Cu				
Mitochondrial	14.6 ± 4.3	38.9 ± 10.0*	25.3 ± 3.1	8.2 ± 5.0
Microsomal	8.1 ± 1.0	8.2 ± 2.0	6.9 ± 1.3	1.2 ± 0.5*
Cytosolic	1.5 ± 0.4	5.9 ± 0.7*	2.8 ± 0.8*	2.6 ± 0.7
Zn				
Mitochondrial	27.5 ± 5.2	21.9 ± 4.5	60.6 ± 13.1*	10.5 ± 1.2*
Microsomal	9.2 ± 2.7	2.0 ± 1.3*	10.1 ± 2.4	0.9 ± 0.6*
Cytosolic	189.3 ± 28.8	127.2 ± 58.4	640.0 ± 83.9*	78.1 ± 18.5*
Cd				
Mitochondrial	0.50 ± 0.18	0.93 ± 0.36	0.57 ± 0.08	5.13 ± 2.18*
Microsomal	0.08 ± 0.05	0.10 ± 0.03	0.11 ± 0.06	0.32 ± 0.06*
Cytosolic	4.66 ± 0.76	4.91 ± 1.03	4.65 ± 1.55	48.84 ± 17.44*

Note: Each value is the mean of three pooled samples of eight individuals. All values in $\mu g/g$ wet weight of original intestine. Values significantly different from control are indicated by * ($p < 0.05$, Dunnett's test).

Table 4. Concentration of Cu, Zn, and Cd In Subcellular Fractions From Foot of *Nassarius reticulatus* Exposed to Cu, Zn, and Cd in Seawater

Fraction	Control (mean ± sd)	Cu-exposed (mean ± sd)	Zn-exposed (mean ± sd)	Cd-exposed (mean ± sd)
Cu				
Mitochondrial	2.9 ± 0.2	14.5 ± 5.3*	8.8 ± 0.1*	5.0 ± 0.9*
Microsomal	17.2 ± 2.4	16.5 ± 2.7	34.4 ± 5.3*	9.2 ± 3.4*
Cytosolic	0.7 ± 0.03	8.7 ± 2.7*	2.3 ± 0.6*	2.0 ± 0.5*
Zn				
Mitochondrial	16.7 ± 5.4	15.5 ± 4.6	55.6 ± 10.2*	41.0 ± 18.1*
Microsomal	3.5 ± 0.8	7.0 ± 3.0	10.8 ± 0.8*	4.7 ± 2.4
Cytosolic	30.5 ± 7.6	21.5 ± 7.5	99.9 ± 12.7*	49.6 ± 16.6
Cd				
Mitochondrial	0.07 ± 0.01	0.10 ± 0.04	0.17 ± 0.03*	21.6 ± 2.3*
Microsomal	0.01 ± 0.002	0.02 ± 0.01	0.02 ± 0.004	2.1 ± 0.4*
Cytosolic	0.23 ± 0.02	0.15 ± 0.05	0.42 ± 0.22	59.4 ± 7.0*

Note: Each value is the mean of three pooled samples of eight individuals. All values in µg/g wet weight of original foot tissue. Values significantly different from control are indicated by * ($p < 0.05$, Dunnett's test).

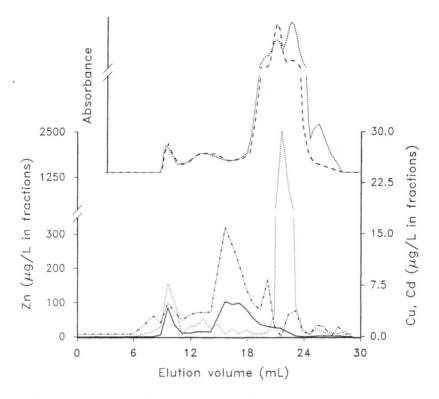

FIGURE 2. Cytosolic distribution of metal-binding components in hepatopancreas from *N. reticulatus* kept in seawater (control). Upper panel: 280 nm (long dashes), 254 nm (short dashes); Lower panel: Cu (dot-dot-dash), Zn (dotted line), Cd (solid line). Mammalian MT would elute at 17 ml.

D. Interactions Between Metals

The most obvious interaction between the three elements was the tissue-dependent effect of Cd on the contents of Cu and Zn in subcellular fractions. In the hepatopancreas, Cu and Zn concentrations in all three subcellular pools of Cd-exposed animals were lowered compared to controls, all significantly except the change in cytosolic Cu (Table 1). The effects of Cd on the other metals in intestine and foot appeared more specific: the Cu contents of mitochondrial and microsomal pools decreased, whereas the cytosolic Cu content increased. The latter effect may have been caused by an induction by Cd of the Cu- / Cd-binding 10 kDa protein described above. In the intestine, the Zn concentrations in all three pools decreased significantly following Cd exposure, whereas mitochondrial Zn levels in the foot increased. One of the reasons why *N. reticulatus* appears untroubled by the high Cd concentration used in this study, 0.9 mg/l, may be the lack of change in gill Cu and Zn contents following that exposure. There appears to be a large binding capacity for Cd, probably inducible

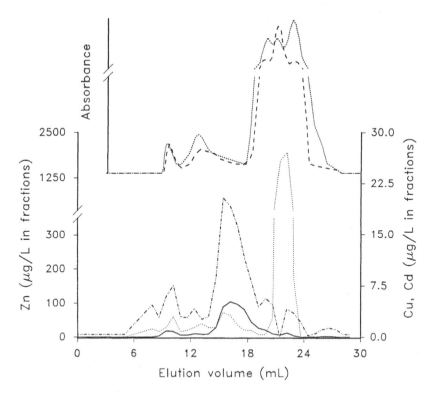

FIGURE 3. Cytosolic distribution of metal-binding components in hepatopancreas from *N. reticulatus* kept in 50 μg/l Cu. Details as in Figure 2.

and predominantly cytosolic, in the gills of *N. reticulatus,* that might protect against acute toxicity of the metal.

Interactions following exposure to Cu and Zn were not as dramatic as those described above for Cd. Zinc exposure signficantly increased the Cu contents of some tissues and fractions, i.e., all fractions in the foot and the mitochondrial fraction in the hepatopancreas. The only apparent effect of Cu exposure on Zn and Cd, as yet unexplained, was the concomitant decrease in intestinal microsomal Zn concentration.

IV. CONCLUDING REMARKS

Nassarius reticulatus accumulated all three metals following exposure through the water. The gastropod may be regulating Zn accumulation, as only minor increases were observed even at very high ambient Zn exposures. To the contrary, there appeared to be little control over Cd accumulation. Similarly, Cu accumulated in all tissues, but at the comparatively low water concentrations used

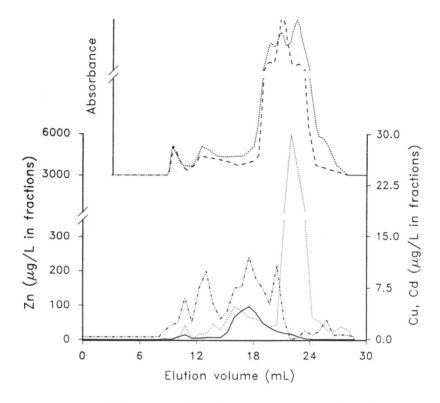

FIGURE 4. Cytosolic distribution of metal-binding components in hepatopancreas from *N. reticulatus* kept in 1.5 mg/l Zn. Details as in Figure 2.

any regulatory capacity of the gastropod might not have been saturated. Some authors have stressed the importance of intracellular granules for detoxification and as a means of rendering metals inaccessible for higher trophic levels.[26] From the low Zn and Cd content in the mitochondrial fraction, it appears unlikely that substantial amounts of Cd or Zn would be associated with such granules in *N. reticulatus*.

V. ABSTRACT

The subcellular and cytosolic distribution of Cu, Zn, and Cd in different organs of a marine gastropod, *Nassarius reticulatus,* the netted dog whelk, has been studied. This littoral, scavenging gastropod is found on both hard and soft substrata along the coasts of northern Europe. Gastropods were collected in the outer Oslo Fjord, at an uncontaminated site. Groups of 100 snails each were exposed to 50 µg/l Cu, 1500 µg/l Zn, and 900 µg/l Cd in the water, and the same number kept in clean seawater. The digestive gland, intestine, gills, and

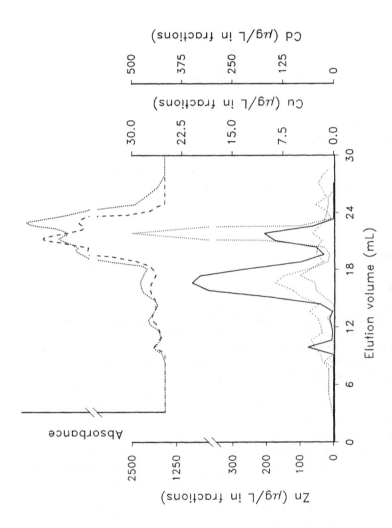

FIGURE 5. Cytosolic distribution of metal-binding components in hepatopancreas from *N. reticulatus* kept in 0.9 mg/l Cd. Details as in Figure 2.

soft part of the foot were pooled in three groups, each of eight snails, and homogenized. The homogenates were centrifuged at 10,000 g, the supernatant again at 100,000 g, and all fractions analyzed for Cd, Cu, and Zn by atomic absorption spectrophotometry. The cytosols were further fractionated on a Superose 12 column (1 × 30 cm), and the metal content of the collected fractions determined. Copper, Zn, and Cd accumulated in all tissues in netted dog whelk exposed to those metals, but distribution patterns varied according to metal and tissue. Most of the cytosolic Cd and Cu were recovered in a protein of approximate M_r 10 kDa, whereas the main Zn pool was associated with components of M_r <1000 Da. The possible significance of the observed organ specificity for metal accumulation and detoxification was discussed.

ACKNOWLEDGMENTS

Part of this study was supported by a grant to Ketil Hylland from the Norwegian Fisheries Research Council, which is gratefully acknowledged.

REFERENCES

1. **Depledge, M. H. and Rainbow, P. S.,** Models of regulation and accumulation of trace metals in marine invertebrates, *Comp. Biochem. Physiol.,* 97C, 1, 1990.
2. **Nott, J. A. and Langston, W. J.,** Cadmium and the phosphate granules in *Littorina littorea, J. Mar. Biol. Assoc., U.K.,* 69, 219, 1989.
3. **George, S. G.,** Biochemical and cytological assessments of metal toxicity in marine animals, in *Heavy Metals in the Marine Environment,* Furness, R. W. and Rainbow, P. S., Eds., CRC Press, Boca Raton, FL, 1990, 126.
4. **George, S. G.,** Heavy metal detoxication in the mussel *Mytilus edulis* — composition of Cd-containing kidney granules (tertiary lysosomes), *Comp. Biochem. Physiol.,* 76C, 53, 1983.
5. **Engel, D. W. and Brouwer, M.,** Metallothionein and metallothionein-like proteins: physiological importance, *Comp. Environ. Physiol.,* 5, 53, 1989.
6. **Nolan, C. V. and Duke, E. J.,** Cadmium accumulation and toxicity in *Mytilus edulis:* involvement of metallothioneins and heavy-molecular weight protein, *Aquat. Toxicol.,* 4, 153, 1983.
7. **Coombs, T. L.,** The nature of zinc and copper complexes in the oyster *Ostrea edulis, Mar. Biol.,* 28, 1, 1974.
8. **Wong, V. W. and Rainbow, P. S.,** A low molecular Zn-binding ligand in the shore crab *Carcinus meanas, Comp. Biochem. Physiol.,* 87C, 203, 1987.
9. **Gibbs, P. E., Bryan, G. W., and Ryan, V. P.,** Copper accumulation by the polychaete *Melinna palmata:* an antipredation mechanism?, *J. Mar. Biol. Assoc. U.K.,* 61, 707, 1981.

10. **Eriksen, K. D. H.**, Metal-Binding in Cytosolic Extracts of Polychaetes: Qualitative and Quantitative Studies of Six Taxa Collected at Locations of Different Metal Levels, Cand. Sci. thesis, University of Oslo, 1988.
11. **Gibbs, P. E., Langston, W. J., Burt, G. R., and Pascoe, P. L.,** *Tharyx marioni* (Polychaeta): a remarkable accumulator of arsenic, *J. Mar. Biol. Assoc. U.K.,* 63, 313, 1983.
12. **Phillips, D. J. H.,** Use of macroalgae and invertebrates as monitors of metal levels in estuaries and coastal waters, in *Heavy Metals in the Marine Environment,* Furness, R. W. and Rainbow, P. S., Eds., CRC Press, Boca Raton, FL, 1990, chap. 6.
13. **Lobel, P. B. and Payne, J. F.,** An evaluation of mercury-203 for assessing the induction of metallothionein-like proteins in the mussels exposed to cadmium, *Bull. Environ. Contam. Toxicol.,* 33, 144, 1984.
14. **Pavicic, J., Skreblin, M., Raspor, B., and Branica, M.,** Metal pollution assessment of the marine environment by determination of metal-binding proteins in *Mytilus* sp., *Mar. Chem.,* 22, 235, 1987.
15. **Rainbow, P. S., Phillips, D. J. H., and Depledge, M. H.,** The significance of trace metal concentrations in marine invertebrates. A need for laboratory investigation of accumulation strategies, *Mar. Pollut. Bull.,* 21, 321, 1990.
16. **Deutscher, M. P., Ed.,** Guide to protein purification, *Methods in Enzymology,* 182, 203, 1990.
17. **Dunnett, C. W.,** A multiple comparison procedure for comparing several treatments with a control, *J. Am. Stat. Assoc.,* 50, 1096, 1955.
18. **Fayi, L. and George, S. G.,** Purification of very low molecular weight Cu-complexes from the european oyster, in *Marine Pollution and Physiology: Recent Advances,* Vernberg, F. J , Thurberg, F. P., Calabrese, W. B., and Vernberg, W. B., Eds., University of South Carolina Press, Columbia, 1986, 145.
19. **Harrison, F. L. and Lam, J. R.,** Partitioning of copper among copper-binding proteins in the mussel *Mytilus edulis* exposed to soluble copper, *Mar. Environ. Res.,* 16, 151, 1985.
20. **Viarengo, A., Pertica, M., Mancinelli, G., Zanicchi, G., and Orunesu, M.,** Rapid induction of copper-binding proteins in the gills of metal exposed mussels, *Comp. Biochem. Physiol.,* 67C, 215, 1980.
21. **Dohi, Y., Ohba, K., and Yoneyama, Y.,** Purification and molecular properties of two cadmium-binding glycoproteins from the hepatopancreas of a whelk, *Buccinum tenuissimum, Biochim. Biophys. Acta,* 745, 50, 1983.
22. **Langston, W. J. and Zhou, M.,** Evaluation of the significance of the metalbinding proteins in the gastropod *Littorina littorea, Mar. Biol.,* 92, 505, 1986.
23. **Dallinger, R., Carpenè, E., Dallavia, G. J., and Cortesi, P.,** Effects of cadmium on *Murex trunculus* from the Adriatic Sea. Accumulation of metal and binding to a metallothionein-like protein, *Arch. Environ. Contam. Toxicol.,* 18, 554, 1989.
24. **Brown, D. A.,** Toxicology of Trace Metals: Metallothionein Production and Carcinogenesis, Ph.D. thesis, University of British Columbia, Vancouver, 1978.
25. **Johansson, C., Cain, D. J., and Luoma, S. N.,** Variability in the fractionation of Cu, Ag, and Zn among cytosolic proteins in the bivalve *Macoma balthica, Mar. Ecol. Prog. Ser.,* 28, 87, 1986.
26. **Nott, J. A. and Nicolaidou, A.,** Transfer of metal detoxification along marine food chains, *J. Mar. Biol. Assoc. U.K.,* 70, 905, 1990.
27. **Kaland, T.,** Unpublished data.

CHAPTER 4

Metallothionein in Marine Molluscs

Emilio Carpenè

TABLE OF CONTENTS

0-87371-734-1/93/$0.00 + $.50
© 1993 by Lewis Publishers

I. INTRODUCTION

Metallothioneins (MTs) are low molecular weight proteins (6 to 7 kDa), with a high content of certain trace metals (seven equivalents of Zn and/or Cd per mole MT). Their characteristic amino acid composition (high cysteine content, about $1/_3$ of the total, but no aromatic amino acids or histidine), and their sequences and spectroscopic properties make these proteins unique. The first protein to contain all these features was first detected in equine kidney cortex[1] and subsequently purified and characterized by Kägi and Vallee.[2] If a polypeptide contains all these features then it can be designated as a classic metallothionein (MTs of class I), even if the term is often used for other metal organic complexes differing in some respects from classical MTs (MTs of class I and II).[3]

Since the original discovery of MT, several metalloproteins with biochemical properties in common with MT have been identified in molluscs. These proteins have been reported either as MTs or MT-like proteins, or less specifically as metal-binding proteins. In the present review, the term MTs will be used in its broadest sense, thus possibly including in this class proteins that on further characterization will turn out not to be true MTs. In spite of the fact that Mollusca is one of the most widely represented phyla as far as the number of species is concerned (most of them are marine), in comparison to Chordata, few studies have been carried out on molluscan MTs. This is still more surprising considering that much research has been carried out on the biology of trace metals in molluscs. Furthermore, the wide range of adaptations to different habitats shown by molluscs make them interesting models for biochemical studies on metal metabolism. Recent work on MT amino acid sequences in molluscs[4-6] has allowed us to be sure that true MTs are synthesized in these invertebrates. In view of such recent findings this author believes that a review of the presence of MTs in marine Mollusca is of importance, not only to stimulate interest in more work in this field but to provide further information on the function of MTs.

II. METALLOTHIONEIN INDUCTION IN MOLLUSCS

One of the characteristics of MT biosynthesis in general is the rapid induction of this protein by a wide range of chemically differing agents.

Induction of MTs and MT-like proteins in molluscs can be obtained in the laboratory by exposing the animals to heavy metals like Cd,[7-9] Cu,[10] and Hg[11] in solution.

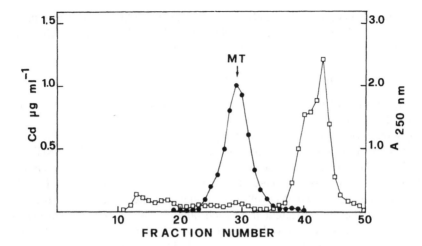

FIGURE 1. Sephadex G-75 gel filtration chromatography of a cytosolic sample obtained from kidney of *Scapharca inaequivalvis*. The molluscs were taken from the Adriatic Sea and were not exposed to Cd in the laboratory; the environmental Cd concentration is not known. Column 0.9 × 95 cm, eluent 0.02 *M* Tris-HCl, pH 8.6. Fractions were monitored for Cd (-●-●-●-), and absorbance at 250 nm (-□-□-□-).

Molluscs taken from the field can also contain MTs. Howard and Nickless[12] found that in the marine gastropod *Patella vulgata* major proportions of the water-soluble Cd and Cu were associated with a protein with properties similar to mammalian MT.

In the gastropod *Murex trunculus* (hepatopancreas and kidney),[13] and in the kidney of the bivalve *Scapharca inaequivalvis* from the northern Adriatic Sea, Cd is naturally sequestered by low molecular weight ligands (Figure 1). In contrast, studies on *Murex trunculus* from the Mediterranean Sea have shown that high molecular weight proteins have a greater importance for the sequestering of cadmium.[14]

However, most work has been done on the characterization of molluscan MTs experimentally induced in the laboratory by different heavy metals. In this phylum the first MTs to be well characterized were those from the digestive gland of the bivalve *Mytilus edulis*,[9] and from the soft parts of the same species[15] previously exposed to Cd. The purified proteins showed many of the characteristics of mammalian MTs, meeting in part the criteria for classification as MTs. Subsequently, a Cu-thionein[10] and an Hg-thionein[11] were also isolated from *Mytilus galloprovincialis* and *Mytilus edulis*, respectively.

We have studied Cd-induced proteins in several species of Adriatic molluscs (*Mytilus galloprovincialis, Murex trunculus,* and *Venus gallina*),[13,16,17] including recent studies on the hemoglobin-containing bivalve *Scapharca inaequivalvis*,[18] which together with *Murex trunculus* will be discussed in detail later. Surprising data have also come from a study of the relationship between anaerobic metabolism and MTs in *Mytilus galloprovincialis,* where MTs could be chromatographically resolved only in the presence of mercaptoethanol.[19]

A. Turnover and Lipofuscin

In the gills of the oyster *Crassostrea virginica* two metallothioneins were induced by exposure to Cd. The two proteins were identical except for a hydrophobic blocked N-terminal (acetyl group) on one (CvNAcMT). The rate of synthesis of blocked MT was greater than that of unblocked MT (CvMT) over the first 4 days of exposure, declining by day 7 to the same rate. Apparent rates of turnover (4.3 to 6.9 d) were similar to those of mammalian CdMTs. The turnover rates of the two MTs following removal of the oysters to Cd-free water were not statistically different from each other.[20] This does not mean that the metal itself is quickly removed from the tissues. Probably as it is liberated from MT, it is again chelated to newly synthesized MT, making Cd turnover different from MT turnover.

In *Mytilus edulis* no Cd-containing proteins similar to MT were found in lipofuscin granules (tertiary lysosomes) from the kidney. However, both primary and secondary lysosomes contained a Cd-binding protein with an elution volume characteristic of MT. Probably primary lysosomes take up MT which, in turn, is degraded in secondary lysosomes, and the Cd is finally bound to other ligands in tertiary lysosomes. A large proportion of the Cd was found associated with ligands of even lower molecular weight than MT.[21]

The level of Cu-thioneins in *Mytilus galloprovincialis* rapidly decreased during a detoxification period of 10 days; the metal was accumulated in the lysosomal system and finally excreted by exocytosis.[22] Cu, however, differs from Cd in that in tertiary lysosomes Cu is bound to a protein[23] which resembles a Cu-MT isolated previously.[10] The amino acid composition, however, is poor in glycine and rich in aspartic and glutamic acid; this difference could render the protein more susceptible to catabolic processes.[22]

B. Tissue Distribution

In molluscs high levels of metallothioneins (deduced from the chromatographic patterns) are produced in nonparenchymatous tissues, such as in the mantle,[24] foot and muscle of mussels,[16] and the gills of oyster.[25] In environmentally low level metal exposures, kidney MT has been demonstrated by gel filtration and flame atomic absorption spectrophotometry in *Scapharca inaequivalvis* kidney (Figure 1)[18] and *Murex trunculus* kidney and hepatopancreas,[13] whereas in the other tissues it was undetectable. MT is probably accumulated at subacute levels in the kidney as has been demonstrated in mammals; when Cd concentration is artificially raised in the water, other tissues are stimulated to produce MT as evidenced by metal profiles of gel chromatography.[18] The characteristic Cd sequestration in *Mytilus edulis* can be related to the histological and physiological properties of the molluscan tissues which are in direct contact with the surrounding environment.[16]

C. *Scapharca* and *Murex* Metallothioneins

The *Scapharca inaequivalvis* and *Murex trunculus* MTs will be discussed in more detail because they have been the objects of most of our recent work. *S. inaequivalvis* is an interesting mollusc containing erythrocytes whose hemoglobins have been well characterized.[26] Moreover, it is suspected that its wide distribution could be connected with a well-developed anaerobic metabolism which makes the bivalve successful in anoxic waters that are associated, for example, with red tides. The northern Adriatic area is also influenced by the river Po, which drains a highly industrialized and intensively farmed valley, bringing to the sea several pollutants including heavy metals. In this environment a protective system based on MT biosynthesis and anaerobic metabolism might be essential for survival of the species.[27]

Recently, we subjected 150 specimens of *S. inaequivalvis* (50 to 60 mm) to nominal concentrations of 0 (control) and 0.5 μg Cd ml^{-1} for up to 28 days. Some preexposed specimens were also maintained, after the exposure period, in aquaria without Cd for up to 56 days. The bivalves were divided into four different groups: three groups were exposed to Cd and the last one was held in the aquaria as a control. Water temperature and salinity varied between 17 and 19°C and 33 and 35‰, respectively, during this period. From each group, six animals were removed weekly and their tissues (gills, mantle, kidney, and hepatopancreas) were dissected. Concentrations of Cd in the mineralized tissues were dissolved in 1 *M* HCl and were measured by flame atomic absorption spectrophotometry. Part of the gill and hepatopancreas tissues was reserved for MT isolation following the procedure previously reported.[9] The total amount of Cd contained in the fractions corresponding to the MT peak was computed and considered to be a useful measurement for the MT content. Cd was also determined in the pellets obtained after the centrifugation step.

When exposed to Cd this mollusc accumulates high levels of it in the mantle, gills, hepatopancreas, kidney; an example from the mantle is shown in Figure 2. After the exposure was suspended, the Cd concentration decreased very slowly (Figure 2), resembling the behavior shown in mammals. The metal was distributed between the supernatant and the precipitate obtained after centrifugation at 100,000 g. In the supernatant of the gills most of the Cd is bound to a low molecular weight protein that probably is MT (Figure 3).

When the total Cd present in the fractions eluting at the MT peak was plotted against time, the gills showed there was an approximately linear accumulation of Cd bound to MT during the period of exposure. Afterwards, there was a decrease when the molluscs were returned to sea water without Cd (Figure 4); similar results were obtained for the hepatopancreas. Changes of Cd concentrations in the precipitate obtained from a homogenate of the gills after centrifugation at 100,000 g were also present, as shown in Figure 5. In common with other

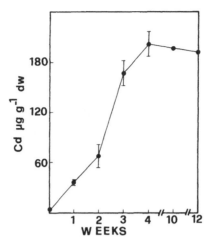

FIGURE 2. Cd accumulation pattern in the mantle of *Scapharca inaequivalvis* exposed to 0.5 μg Cd/ml sea water. Values are expressed as μg Cd/g dry weight; error bars are standard deviations. After four weeks molluscs were kept in natural sea water without Cd. Water temperature was 18 ± 1°C.

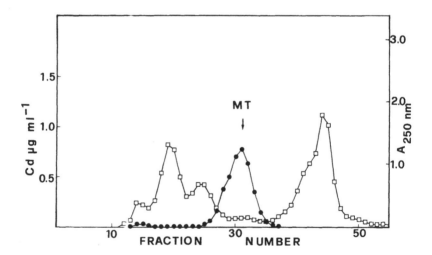

FIGURE 3. Sephadex G-75 gel filtration chromatography of a cytosolic sample from the gills of *Scapharca inaequivalvis*. The cytosol was prepared after the molluscs were exposed to 0.5 μg Cd/ml sea water for 3 weeks. Fractions were monitored for Cd (-●-●-●-), and absorbance at 250 nm (-□-□-□).

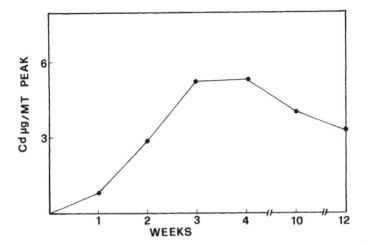

FIGURE 4. Variations of the total Cd contained in the MT peak from the gills of *Scapharca inaequivalvis* exposed to 0.5 µg Cd/ml sea water. The peak was isolated by Sephadex G-75 gel filtration chromatography; the last two values refer to animals returned to natural sea water without Cd.

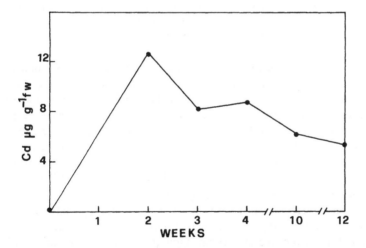

FIGURE 5. Variations of Cd in the pellets of gills from *Scapharca inaequivalvis* exposed to 0.5 µg Cd/ml sea water. The pellets were obtained by centrifugation of the crude homogenate at 100,000 g for 30 min. Data are expressed as µg/g fresh weight. The last two values refer to animals returned to natural sea water without Cd.

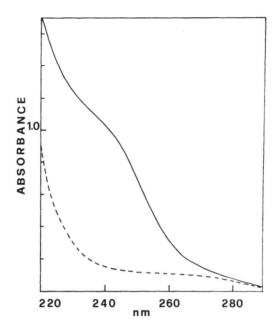

FIGURE 6. UV spectrum of partially purified Cd-MT from hepatopancreas of *Scapharca inaequivalvis* exposed to 0.5 μg Cd/ml sea water. The isoform was isolated by DEAE-cellulose chromatography. Untreated MT (————), MT treated with HCl (- - - - -).

bivalves, *Scapharca inaequivalvis* showed a strong capacity to synthesize Cd-binding proteins. After this first investigation, some animals were exposed to Cd to obtain enough tissue for further steps in the purification and characterization of the metal-binding protein(s). Fractions of the hepatopancreas corresponding to the elution volume of MT on Sephadex G75 were applied to a DEAE column and eluted with a linear gradient of Tris-HCl. The results obtained by DEAE fractionation showed the presence of at least two isoforms. The two isoforms were named MT1 and MT2, according to their elution order. MT1 was clearly dominant with respect to MT2 (data not shown).

Fractions corresponding to MT1 were concentrated by diaflo and scanned in the UV absorbance region, where they showed spectra characteristic of MT; there was low absorbance at 280 nm, high absorbance at 250 nm, and loss of the shoulder at 250 nm on acidification (Figure 6).

The purity of kidney MT was tested by HPLC and only one isoform was separated. The sample was previously concentrated by absorption on DEAE-cellulose and elution in a single step.

These preliminary results show that *S. inaequivalvis* is poorer in MT isoforms than is *Mytilus edulis*.[9,28] All the results obtained indicate that the isolated protein could be a MT of class I. However only the amino acid composition or sequence

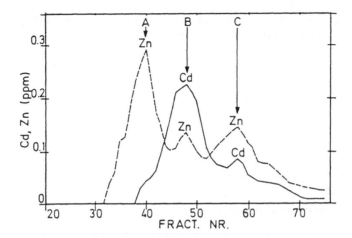

FIGURE 7. Ion exchange chromatography (DEAE-cellulose) on pooled low molecular weight Cd fractions of Sephadex G-75 from kidney and hepatopancreas cytosol of *Murex trunculus* exposed to Cd. Symbols: ———— Cd; - - - - - Zn. Arrows indicate the cadmium- and zinc-containing peaks A, B, and C.

can provide an unambiguous characterization. When Cd is chelated by MT it is well blocked and it is not free to interact with other ligands of higher molecular weight which play essential roles in cell homeostasis.

Similar studies were carried out on *Murex trunculus,* a carnivorous gastropod, relatively common in the Adriatic Sea, mostly toward the Yugoslav rocky coast which apparently may be less affected by pollution. Surprisingly, as in the case of the kidney of *S. inaequivalvis, M. trunculus* also contains MT in the hepatopancreas and kidney even when taken from waters considered unpolluted;[13] Bouquegneau et al. previously found high amounts of cadmium in the soft parts of *M. trunculus* from unpolluted areas of the Mediterranean Sea.[14] The apparent molecular weight of *M. trunculus* Cd-MT was approximately 11 kDa, and upon ion exchange chromatography the Cd ligands split into three distinct peaks with variable metal contents (Figure 7).

The chromatographic patterns, amino acid compositions, and UV spectra confirmed that in Cd-exposed animals the isolated molecules belong to the large group of metallothionein-like proteins. However, the capacity of the detoxification function could be limited by a "spillover" effect and the progressive increase of Cd in the granular fraction of the tissues of cadmium-exposed animals.[13]

We investigated the subcellular distribution of Cd in two other Adriatic molluscs: the bivalve *Venus gallina* and the cephalopod *Sepia officinalis.* The first was artificially exposed to Cd and produced a low molecular weight binding protein.[17] The second had a naturally high level of Cd in the large hepatopancreas

(5 to 10 μg Cd/g fresh weight), but so far we have failed to isolate a specific ligand.[29] As discussed in detail below for the case of *Mytilus galloprovincialis*, this could be an artifact introduced during the isolation procedure, perhaps due to the lack of a thiol protective agent in the buffer. Overall, all results point to the production of thionein-like proteins by molluscs as being a rather common process. However, qualitative and quantitative differences between species can exist; some populations of the tellinid clam, *Macoma balthica,* fail to produce MT-like proteins and Cd is mainly bound to high molecular weight proteins.[30]

III. MOLECULAR PROPERTIES

A. Purification and Artifacts

The procedure to isolate metallothioneins from molluscs is generally similar to the one used for mammals, i.e., tissue homogenization followed by centrifugation at 100,000 g and fractionation of the supernatant by gel filtration on Sephadex G-75. However, some minor modifications have been adopted by several researchers: deoxygenation of the buffer with an inert gas, introduction of sucrose in the homogenization step, 0.02% NaN_3 as a preservative, solvent and heat treatment, and protease inhibitors. Unfortunately, little attention was paid to improving the technique. In molluscs, seasonal variations of physiological and biochemical parameters could interfere differentially with the isolation of MT.

Stone et al.[31] tested the effects of mercaptoethanol on the isolation of a Cd-binding component of 55 kDa in the scallop *Pecten maximus*. On treatment with 50 mM mercaptoethanol this component was halved in apparent molecular weight as estimated by Sephadex G-100 gel filtration chromatography. The researchers ask whether there is any sequence homology between the 20 kDa fragment produced by mercaptoethanol and the 20 kDa and 10 kDa proteins isolated from *Mytilus edulis*. It is now known that at least a homology exists between the *M. edulis* MTs themselves, and also that the 20 kDa isoform is a dimeric structure kept together by a shared Cd. The components of the 20 kDa class were shown to possess linked peptides consisting of 71 amino acids, which were distinct from the 72 amino acid peptides of the 10 kDa class.[28] Perhaps the bridging function of Cd could be extended to more complex forms. At the moment it is not clear if the *M. edulis* metal complexes are arranged in metal-thiolate clusters containing three and four divalent metal atoms, respectively, in clusters A and B. This is generally the stoichiometry so far reported for the MTs of class I. An exception is constituted by crab MT, which has three divalent metal atoms in each cluster. These stoichiometries are not conserved when the complexed metal is a monovalent metal like Cu.[32]

Related purification problems are the difficulties encountered by Roesijadi and Drum,[33] who discussed the use of mercaptoethanol in isolating a Hg-binding component in *M. edulis*. More recently, Roesijadi et al.[34] described some dif-

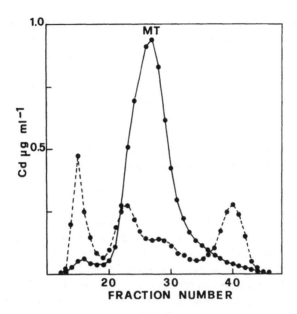

FIGURE 8. Sephadex G-75 gel filtration chromatography of two cytosolic samples from the gills of *Mytilus galloprovincialis.* The cytosols were prepared after the molluscs were exposed to 0.5 μg Cd/ml sea water for 7 days. Homogenization and elution were carried out with two different buffers: a) Tris-HCl 20 m*M* pH 8.6 without mercaptoethanol; b) Tris-HCl 20 m*M* pH 8.6 containing 5 m*M* mercaptoethanol. Fractions were monitored for Cd in both procedures: a) without mercaptoethanol (- ● - ● - ●); b) in presence of mercaptoethanol (-●-●-●-●-).

ficulties in the purification of MT from molluscs which can explain the conflicting findings of different authors. From our own experience, using the same procedure previously described for *M. edulis,*[9,24] we were unable to repeat our previous results in the related species *Mytilus galloprovincialis,* and a good metallothionein yield[19] was obtained only with the use of mercaptoethanol (following the work of Minkel et al.[35] for rat liver MT). In our most recent experiments, without mercaptoethanol, part of the Cd was eluted with the high molecular weight fraction, but the introduction of mercaptoethanol in the buffer shifted the Cd to a fraction corresponding to MT (Figure 8). In this case, the "spillover" hypothesis is not valid, and some data reported in the literature as "spillover" effects could be considered artifacts connected with the oxidation of MT to polymeric forms.

Discrepancies between different data on MT in molluscs could be linked not only to different procedures, but also to changes in physiological and environmental parameters. Roesijadi and Fellingham[36] found a different resistance to Hg in mussels collected at different times of the year. Those collected in the late spring, when mussels would be under increasing natural physiological stress, were less resistant in comparison to those collected in winter. In this field, research is still open to many questions. What is the influence of sex and temperature on MT synthesis? What role is played by season on the variations of concentrations of physiological metal ligands which in molluscs probably vary more than in mammals? Do natural antioxidants like glutathione prevent oxidation of MTs? In *M. galloprovincialis* a reduction of antioxidant defense systems occurs during winter.[37] Moreover, mollusc tissues are very rich in taurine, an amino acid that is also suspected to function as an antioxidant,[38] and coincidentally its concentration is higher in *S. inaequivalvis* than in *M. edulis*.[39]

B. Isoforms

Characteristically, MTs display genetic polymorphism which results often, but not always, in isoforms of different charge. These isoforms are easily separated by ion exchange chromatography. Considering the noticeable number of reports on Cd-binding protein isoforms isolated upon DEAE chromatography, a certain degree of genetic polymorphism could also exist in molluscs.[9] Recently, Mackay[28] found that in the common mussel *M. edulis,* the two low molecular weight Cd binding proteins of 10 and 20 kDa were resolved into 4 and 3 isoforms, respectively, by anion-exchange chromatography. MT heterogeneity is often more complex when investigated by HPLC, but in *S. inaequivalvis* we were unable to obtain such patterns and only one isoform was separated.[18] Studies of the genetics and regulation of MT expression in molluscs have been recently initiated by Unger et al.;[40] in *Crassostrea virginica* two cDNA sequences encoding MT were identified.

C. Amino Acid Composition and Sequence

The term ''isometallothioneins'' should apply only to isoforms of MTs arising from genetically determined differences in primary structure, and not to forms that differ only in metal composition or are a modification of the same primary sequence.[3] Thus the amino acid composition may confirm the presence of isoforms, but this is not necessarily conclusive. Similarly, the absence of isoforms after DEAE chromatography will not exclude the presence of isoforms which are linked to noncharged amino acids. Only by sequencing purified metallothioneins or isolating their genes can the isoforms be identified with certainty.

Several amino acid compositions of molluscan MTs have been published, and the differences between these and mammalian MTs (mostly in the cysteine content) have raised the question whether molluscs could have specific MTs. However the discrepancies are probably due to purification procedures (as described above). The recent work of Roesijadi et al.[34] on oyster MT sequences and Mackay[28] on mussel MT sequences confirm the presence of Class I MTs in these animals. These results are also important from an evolutionary point of view, because the oyster MTs show a greater similarity to vertebrate MTs than to other invertebrate MTs. All cysteines in the first 27 residues of the oyster metallothioneins align with those in the mammalian forms.[34] Moreover, 21 cysteinyl residues were arranged in 9 Cys-X-Cys motifs, 5 as Cys-Lys-Lys and a single as Cys-X-X-Cys.[40] The two chromatographically isolated MTs were virtually identical in amino acid composition, (29% cysteine, 16% glycine, and 13% lysine). However, in the form named CdBP2, the NH_2-terminal was blocked by an undetermined moiety which may have been an acyl group. The NH_2-terminal block in CdBp2 is probably the only feature that distinguishes the two proteins from each other. The function of this modification is currently being explored.[34]

D. Electrophoresis

Some difficulties have been encountered in electrophoretic analysis of MT from mussels,[9] but, more recently, Roesijadi et al.[34] obtained reliable results with the Ornstein-Davis technique in the presence of SDS. It is difficult to say if these discrepancies are due to species differences or to different degrees of purification.

IV. BIOLOGICAL RELEVANCE AND ECOTOXICOLOGICAL SIGNIFICANCE

It is generally assumed that MTs play a central role in metal homeostasis. They can function as storage proteins for essential metals such as Zn and Cu and prevent deficiencies, or as chelating agents to bind toxic metals such as Cd and Hg and the above essential elements when they are present in excess. This hypothesis is supported by many studies, but since the discovery of equine MT in 1957[1] the functional significance of MTs has continued to be a topic of discussion. For example, many nonmetallic agents can also induce MT synthesis; this protein may have other cellular functions.[41] In a living organism many chemical structures and reactions are regulated by metals of the transition series. Consequently, the concentration of "free" trace elements must fluctuate around an optimum which is critical for the success of the individual animal. Each cellular compartment also has its optimum concentration, and it is reasonable to suspect that metal concentrations are in a state of continuous flux, often deviating from the optimum. In this complex situation, MTs could act as a buffer system taking up metals when they are in excess and releasing them when they decrease in available concentration.

The work of Udom and Brady,[42] showing MT to be an excellent metal donor for metalloenzymes, is rather attractive in this context. At the moment, Cd-MT is considered only as a nontoxic form of the metal, but some investigators also suspect a biological function for Cd. The absence of MT synthesis in response to heavy metal contamination does not necessarily imply lower tolerance to metals. In British populations of *Macoma balthica* the lack of MT was associated with reduced accumulation rates/permeability to Cd which could "compensate" for the absence of a detoxifying protein.[30]

Relatively little is known about metal metabolism; how the different elements are transported across the cell membranes; how many metal binding ligands exist and what the role of granules is in metal ion storage. Detoxification by MT may be just one of several systems available for selection during evolution to control the activities of free metal ions intracellularly. When molluscs are exposed to sublethal doses of heavy metals there is often a linear accumulation; the fluxes may not be controlled and when the metals reach the cytoplasm they are trapped by MT, whereas intracellular concentrations of light cations as Na^+ or Ca^{++} are tightly controlled. It is not necessary to propose energy-requiring pumps for Zn/Cd uptake, even if the presence of active transport systems cannot *a priori* be excluded. Metal detoxification by MT is an inducible process which might be relatively economical for the cell because it may be turned on and off when necessary. When molluscs live in a contaminated environment specific mechanisms can also evolve: oysters from a copper contaminated area when exposed to Cd accumulated less Cd than those from an uncontaminated area.[43] Moreover, in the oysters from the polluted area, Cd was bound only to very low molecular weight ligands. The authors of this research were unable to prove if these ligands were of genetic or epigenetic origin.[43] In oysters, homarine has been proposed as a possible low molecular weight Zn-binding ligand.[44] In the crustacean *Carcinus maenas,* however, a Zn-binding complex of nonproteinaceous origin containing carboxylic groups has been found;[45] incidentally in the same species difficulties were encountered in isolating MT in the absence of reducing conditions.[46]

After the recent discovery that the *Mytilus edulis* MTs are of the class I type,[28] and considering both the worldwide distribution of the mussel and how easily it can be reared, this species can be considered as a useful experimental model for studies in biochemistry and ecotoxicology. Moreover, the concentration of *M. edulis* MTs might be useful as an indicator of metal exposure.[47] The ability of molluscs to concentrate large amounts of heavy metals without any apparent ill effects could make these animals potentially very dangerous for their predators, which may lack the ability to synthesize MTs and therefore be unable to cope with a (for them) lethal heavy metal load. Environmental adaptation is not a new phenomenon which has arisen in response to man-made effects; rather, existing mechanisms for regulating metal ion levels have been made use of by these invertebrates. It cannot be excluded that at some future date these mechanisms may enable an increase in apparently undesirable trace elements in the biosphere to be exploited in some positive way. It is not inconceivable that a new metal complex of biological relevance might be selected to carry out a biological function.

ACKNOWLEDGMENTS

The author is grateful to the following: Dr. T.L. Coombs and D. A. Rowlerson, for helpful comments on an earlier draft; and to Dr. R. Serra for the final preparation of the manuscript.

REFERENCES

1. **Margoshes, M. and Vallee, B. L.,** A cadmium protein from equine kidney cortex, *J. Am. Chem. Soc.,* 79, 4813, 1957.
2. **Kägi, J. H. R. and Vallee, B. L.,** Metallothionein: a cadmium and zinc-containing protein from equine renal cortex, *J. Biol. Chem.,* 236, 2435, 1961.
3. **Fowler, B. A., Hildebrand, C. E., Kojima, Y., and Webb, M.,** Nomenclature of metallothionein, in *Metallothionein II,* Kägi, J. H. R. and Kojima, Y., Eds., Birkhauser, Boston, 1987.
4. **Mackay, E. A., Dunbar, B., Davidson, I., and Fothergill, J. E.,** Polymorphism of cadmium-induced mussel metallothionein, in *Proc. 22nd Annu. Meeting of the Swiss Society for Experimental Biology, Experientia,* 46 pp. A36, Abstr. Nr. 310, 1990.
5. **Roesijadi, G., Vestling, M. M., Murphy, C. M., Klerks, P. L., and Fenselau, C. C.,** Structure and behavior of acetylated and non-acetylated form of cadmium-induced metallothioneins of a mollusc, *Biochim. Biophys. Acta,* 1074, 230, 1991.
6. **Unger, M. E., Chen, T. T., Murphy, C. M., Vestling, M. M., Fenselau, C. C., and Roesijadi, G.,** Primary structure of molluscan metallothionein deduced from PCR-amplified cDNA and mass spectrometry of purified proteins, *Biochim. Biophys. Acta,* 1074, 371, 1991.
7. **Casterline, J. L., Jr. and Yip, G.,** The distribution and binding of cadmium in oyster, soybean, and rat liver and kidney, *Arch. Environ. Contam. Toxicol.,* 3, 319, 1975.
8. **Noël-Lambot, F.,** Distribution of cadmium, zinc, and copper in the mussel *Mytilus edulis.* Existence of cadmium-binding proteins similar to metallothioneins, *Experientia,* 32, 324, 1976.
9. **George, S. G., Carpenè, E., Coombs, T. L., Overnell, J., and Youngson, A.,** Characterization of cadmium-binding proteins from mussels, *Mytilus edulis* (L.), exposed to cadmium, *Biochim. Biophys. Acta,* 580, 225, 1979.
10. **Viarengo, A., Pertica, M., Mancinelli, G., Zanicchi, G., and Orunesu, M.,** Rapid induction of copper-binding proteins in the gills of metal exposed mussels, *Comp. Biochem. Physiol.,* 67C, 215, 1980.
11. **Roesijadi, G. and Hall, R. E.,** Characterization of mercury-binding proteins from the gills of marine mussels exposed to mercury, *Comp. Biochem. Physiol.,* 70C, 59, 1981.
12. **Howard, A. G. and Nickless, G.,** Protein binding of cadmium, zinc, and copper in environmentally insulted limpets *Patella vulgata, J. Chromatogr.,* 104, 457, 1975.

13. **Dallinger, R., Carpenè, E., Dalla Via, G. J., and Cortesi, P.,** Effects of cadmium on *Murex trunculus* from the Adriatic Sea. I. Accumulation of metal and binding to a metallothionein-like protein, *Arch. Environ. Contam. Toxicol.,* 18, 554, 1989.

14. **Bouquegneau, J. M., Martoja, M., and Truchet, M.,** Localisation biochimique du cadmium chez *Murex trunculus* L. (Prosobranche Neo-gasteropode) en milieu naturel non pollue et apres intoxication experimentale, *C.R. Acad. Sci. Paris,* 296, 1121, 1983.

15. **Frazier, J. M., George, S. G., Overnell, J., Coombs, T. L., and Kägi, J. H. R.,** Characterization of two molecular weight classes of cadmium binding proteins from the mussel, *Mytilus edulis* (L), *Comp. Biochem. Physiol.,* 80C, 257, 1985.

16. **Carpenè, E., Cattani, O., Hakim, G., and Serrazanetti, P.,** Metallothionein from foot and posterior adductor muscle of *Mytilus galloprovincialis, Comp. Biochem. Physiol.,* 74C, 331, 1983.

17. **Cortesi, P., Carpenè, E., De Zwaan, A., Cattani, O., and Isani, G.,** The effects of cadmium accumulation on anaerobic metabolism in bivalve molluscs, in *Proc. 19th FEBS Meet.,* Rome, 1989.

18. **Carpenè, E., Cattani, O., Isani, G., and Cortesi, P.,** Unpublished data.

19. **Carpenè, E. and Serra, R.,** Unpublished data.

20. **Roesijadi, G., Vestling, M. M., Murphy, C. M., Klerks, P. L., and Fenselau, C. C.,** Characterization of two molluscan metallothioneins (MT): structure and rates of synthesis and turnover, in *Proc. 6th Int. Symp. Responses Mar. Organisms Pollut.,* Woods Hole Oceanographic Institution, Woods Hole, Massachusetts, 1991.

21. **George, S. G.,** Heavy metal detoxication in the mussel *Mytilus edulis* — composition of Cd-containing kidney granules (tertiary lysosomes), *Comp. Biochem. Physiol.,* 76C, 53, 1983.

22. **Viarengo, A., Zanicchi, G., Moore, M. N., and Orunesu, M.,** Accumulation and detoxication of copper by the mussel *Mytilus galloprovincialis* Lam.: a study of the subcellular distribution in the digestive glands cells, *Aquat. Toxicol.,* 1, 147, 1981.

23. **Viarengo, A., Pertica, M., Canesi, L., Mazzucotelli, A., Orunesu, M., and Bouquegneau, J. M.,** Purification and biochemical characterization of a lysosomal copper-rich thionein-like protein involved in metal detoxification in the digestive gland of mussels, *Comp. Biochem. Physiol.,* 93C, 389, 1989.

24. **Carpenè, E., Cortesi, P., Crisetig, G., and Serrazanetti, G. P.,** Cadmium-binding proteins from the mantle of *Mytilus edulis* (L.) after exposure to cadmium, *Thalassia Jugosl.,* 16, 317, 1980.

25. **Roesijadi, G. and Klerks, P. L.,** Kinetics analysis of cadmium binding to metallothionein and other intracellular ligands in oyster gills, *J. Exp. Zool.,* 251, 1, 1989.

26. **Chiancone, E., Boffi, A., Verzili, D., and Ascoli, F.,** Proprietà delle emoglobine del mollusco *Scapharca inaequivalvis* in rapporto con le condizioni ecologiche del mare Adriatico, in *Proc. Simp. Satellite 31 Congr. Nazionale Soc. Italiana Biochim.,* Editoriale Grasso, Bologna, 1986.

27. **Cortesi, P., Cattani, O., Vitali, G., Carpenè, E., De Zwaan, A., van den Thillart, G., Roos, J., van Lieshout, G., and Weber, R. E.**, Physiological and biochemical responses of the bivalve *Scapharca inaequivalvis* to hypoxia and cadmium exposure: erythrocytes versus other tissues, in *Proc. Mar. Coast. Eutrophication Int. Conf.*, Bologna, Vollenweider, R., Ed., Elsevier, Amsterdam, 1991, in press.

28. **Mackay, E. A.**, Polymorphism of Cadmium-Induced Mussel Metallothionein. Thesis, University of Aberdeen, 1989.

29. **Carpenè, E.**, Unpublished data, 1988.

30. **Langston, W. J., Bebianno, M. J., and Zhou, M.**, A comparison of metal-binding proteins and cadmium metabolism in the marine molluscs *Littorina littorea* (Gastropoda), *Mytilus edulis* and *Macoma balthica* (Bivalvia), *Mar. Environ. Res.*, 28, 195, 1989.

31. **Stone, H. C., Wilson, S. B., and Overnell, J.**, Cadmium binding components of scallop (*Pecten maximus*) digestive gland. Partial purification and characterization, *Comp. Biochem. Physiol.*, 85C, 259, 1986.

32. **Otvos, J. D., Olafson, R. W., and Armitage, I. M.**, Structure of an invertebrate metallothionein from *Scylla serrata*, *J. Biol. Chem.*, 257, 2427, 1982.

33. **Roesijadi, G. and Drum, A. S.**, Influence of mercaptoethanol on the isolation of mercury-binding proteins from the gills of *Mytilus edulis*, *Comp. Biochem. Physiol.*, 71B, 445, 1982.

34. **Roesijadi, G., Kielland, S., and Klerks, P.**, Purification and properties of novel molluscan metallothioneins, *Arch. Biochem. Biophys.*, 273, 403, 1989.

35. **Minkel, D. T., Poulsen, K., Wielgus, S., Shaw, C. F., III, and Petering, D. H.**, On the sensitivity of metallothioneins to oxidation during isolation, *Biochem. J.*, 191, 475, 1980.

36. **Roesijadi, G. and Fellingham, G. W.**, Influence of Cu, Cd, and Zn preexposure on Hg toxicity in the mussel *Mytilus edulis*, *Can. J. Fish. Aquat. Sci.*, 44, 680, 1987.

37. **Viarengo, A., Canesi, L., Pertica, M., and Livingstone, D. R.**, Seasonal variations in the antioxidant defence systems and lipid peroxidation of the digestive gland of mussels, *Comp. Biochem. Physiol.*, 100C, 187, 1991.

38. **Wright, C. E., Tallan, H. H., and Lin, Y. Y.**, Taurine: biological update, *Ann. Rev. Biochem.*, 55, 427, 1986.

39. **Zurburg, W. and de Zwaan, A.**, The role of amino acids and osmoregulation in bivalves, *J. Exp. Zool.*, 215, 315, 1981.

40. **Unger, M. E., Chen, T. T., and Roesijadi, G.**, Molecular cloning of the cDNA of molluscan metallothioneins (MT), in *Proc. 6th Int. Symp. Responses Mar. Organisms Pollut.*, Woods Hole Oceanographic Institution, Woods Hole, Massachusetts, 1991.

41. **Kägi, J. H. R. and Kojima, Y.**, Chemistry and biochemistry of metallothionein, in *Metallothionein II*, Kägi, J. H. R. and Kojima, Y., Eds., Birkhauser, Boston, 1987.

42. **Udom, A. O. and Brady, F. O.**, Reactivation *in vitro* of zinc-requiring apoenzymes by rat liver zinc-thionein, *Biochem. J.*, 187, 329, 1980.

43. **Frazier, J. M. and George, S. G.**, Cadmium kinetics in oysters — a comparative study of *Crassostrea gigas* and *Ostrea edulis*, *Mar. Biol.*, 76, 55, 1983.

44. **Coombs, T. L.**, The nature of zinc and copper complexes in the oyster *Ostrea edulis*, *Mar. Biol.*, 28, 1, 1974.

45. **Wong, V. W. T. and Rainbow, P. S.,** A low molecular weight Zn-binding ligand in the shore crab *Carcinus maenas, Comp. Biochem. Physiol.,* 87C, 203, 1987.
46. **Wong, V. W. T. and Rainbow, P. S.,** Apparent and real variability in the presence and metal contents of metallothioneins in the crab *Carcinus maenas* including the effects of isolation procedure and metal induction, *Comp. Biochem. Physiol.,* 83A, 157, 1986.
47. **George, S. G.,** Biochemical and cytological assessments of metal toxicity in marine animals, in *Heavy Metals in the Marine Environment,* Furness, R. W. and Rainbow, P. S., Eds., CRC Press, Boca Raton, FL, 1990, chap 8.

CHAPTER 5

Metal Concentrations in Antarctic Crustacea: The Problem of Background Levels

Gerd-Peter Zauke and Gabriele Petri

TABLE OF CONTENTS

0-87371-734-1/93/$0.00 + $.50

I. INTRODUCTION

In several countries great effort is spent on monitoring programs to assess environmental quality, for example, using organisms as monitors of heavy metals. However, the significance of detected trace metal concentrations is seldom addressed.[1-5]

One method to qualify observed chemical concentration levels is to compare them with data from ''background'' standards, obtained by analyzing samples from remote, uncontaminated areas. Regarding sediments, the average standard shale concept is widely accepted,[6] whereas for organisms no such global reference values exist as yet. In this context, analyses of organisms from the Antarctic Ocean merit consideration.

There is growing evidence that the Antarctic environment seems not to be endangered as yet by global anthropogenic input of chemicals. It has been reported, for example, that heavy metals in snow cores from Antarctica have not increased since the end of the last century.[7] However, in a recent review[8] attention is focused on the validity of such data,which may be obscured by analytical and contamination problems or local emissions.

The problem of anthropogenic hydrocarbons is addressed by Cripps,[9] indicating that there is no obvious general anthropogenic influence in the marine environment. Furthermore, waters of the Antarctic Ocean are regarded to be rather isolated, so that substances of anthropogenic origin will be transferred to that region only over large time scales.[10] However, increasing scientific and tourist activities in Antarctica focus attention on problems of local contamination.[8,11] Examples invoke PCB and PCT as well as metals and other human impacts at McMurdo Station,[12,13] or the grounding of the Bahia Paraiso.[14]

The goal of this paper is to evaluate whether metal concentrations in mesopelagic and benthic crustaceans from the Antarctic Ocean may be suitable as background levels, as an aid for interpretation of data from areas influenced by man. As yet, there is no evidence that such marine organisms from the Antarctic Ocean are influenced by anthropogenic input of chemicals, except perhaps on small local scales.

II. METHODOLOGICAL ASPECTS

A. Constraints of the Investigation

Expeditions to remote areas are normally expensive and are thus performed to fulfill multiple goals of high priority. Usually, the collection of samples for analysis of background levels of chemicals is not included among such goals, as for example in the case of cruises of the German research vessel, RV Polarstern, to the Antarctic Ocean. Consequently, such samples are often collected, not by relevant expert investigators, but (more or less by chance) by nonexpert scientists taking part in the expeditions. These circumstances clearly focus attention on problems of validation and quality control of any such investigations.

Moreover, these constraints also determine the types of samples which might be collected on such occasions. While organisms (namely, macrozoobenthos or macrozooplankton) may be collected without severe danger of contamination if a detailed sampling schedule is provided (see next section), it is not recommended, for example, to collect water samples under these circumstances. To do this, trained persons are required as well as special equipment which is usually not routinely available on board research vessels (see Mart et al.[15]). This will lead to drawbacks in the application of certain approaches. For example, in our investigation use of the concept of bioconcentration factors (an important tool for "calibration" of monitor organisms, see below) has to rely on dissolved metal concentration data from different sources, thus only allowing approximate estimations.

B. Study Area and Organisms

In the present study, we used organisms from the Antarctic Ocean (focusing on the Weddell Sea and the Antarctic Peninsula) collected during three cruises of RV Polarstern, and on the occasion of the First German Underwater Expedition (Figure 1 and Table 1).

Samples were collected according to the following guidelines:

1. Collect organisms from the center of a catch to avoid contamination by oil, metals, or paint originating from the ship itself;
2. Maintain organisms for about 24 h in clean seawater to allow for defecation (white plastic or glass aquaria);

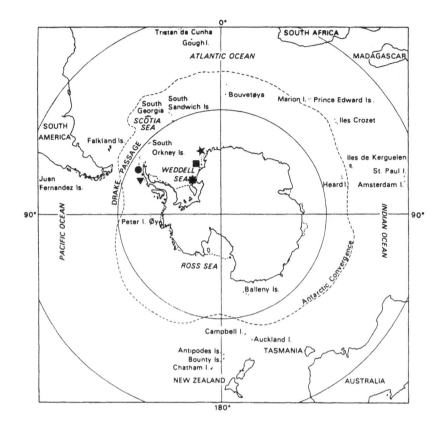

FIGURE 1. Sampling locations in the Antarctic Ocean (Weddell Sea and Antarctic Peninsula).

Symbol	Station	Range of positions
★	Vest Cap[a]	72°24′S–76°04′S to 17°35′W–28°06′W
✳	Gould Bay[b]	73°21′S–77°31′S to 21°24′W–42°12′W
●	Elephant Island[c]	60°49′S–65°45′S to 55°24′W–68°15′W
■	Haley Bay	no information available
▼	King George Island[d]	no information available

[a] RV Polarstern 'Antarktis V/3 1986'; Stat. No. 584–589.
[b] RV Polarstern 'Antarktis III 1985'; Stat. No. 23–43.
[c] RV Polarstern 'Antarktis VI/2 1987'; Stat. No. 67–212.
[d] First German Underwater Expedition 1988; near the Brazilian Research Station 'Com. Ferraz'.

Table 1. List of Species Sampled in the Antarctic Ocean for Determination of Heavy Metals

Species	Taxon[a]	Ind.[b] (n)	Method[d] of sampling	Station location
Chorismus antarcticus (Pfeffer, 1887)	D	13	AGT	✳✳
Notocrangon antarcticus (Pfeffer, 1887)	D	11	BT	✳✳●
Euphausia superba Dana, 1850	E	11	AGT/KT	★■
Aega antarctica (Hodgson, 1910)	I	27	BT	✳✳●
Ceratoserolis trilobitoides (Eights, 1833)	I	32	BT	✳✳●
Glyptonotus antarcticus Eights, 1853	I	6	BT	●●
Natatolana spp. (Vanhöffen, 1914)	I	8	BT	●
Serolis bouvieri Richardson, 1906	I	15	BT	●
Serolis pagenstecheri Pfeffer, 1887	A	3[c]	SCU	►
Bovallia gigantea Pfeffer, 1888	A	2[c]	SCU	►
Cheirimedon femoratus (Pfeffer, 1888)	A	3[c]	SCU	►
Eurymera monticulosa Pfeffer, 1888	A	3[c]	SCU	►
Eusirus properperdentatus Andres, 1978/79	A	21	RMT/BT	●
Gondogeneia antarctica Chevreux, 1906	A	3[c]	SCU	►
Maxilliphimedia longipes (Walker, 1906)	A	4	BT	●●
Paraceradocus gibber Andres, 1984	A	4	BT	●
Waldeckia obesa Chevreux, 1905	A	3[c]	SCU	►

Note: See Figure 1 for explanation of station symbols.

[a] A (Amphipoda); E (Euphausiacea); D (Decapoda); I (Isopoda).
[b] Number of individual organisms collected, processed, and analyzed as independent parallel samples.
[c] Number of pooled samples each consisting of up to 10 individuals
[d] AGT (Agassiz Trawl), BT (Bottom Trawl), RMT (Rectangular Midwater Trawl), KT (Krill Trawl), SCU (Sampling by Scuba Diving).

3.   Rinse organisms briefly in distilled water and remove surface water by drying organisms on good quality filter paper;
4.   Store large organisms individually and smaller ones in pools of 10 in a deep freezer ($-18°C$) using good quality freezer bags (polyethylene);
5.   Handle organisms with good quality stainless steel or polypropylene instruments only; and
6.   Provide information on station, date, method of sampling (trawls), depth, etc.

C. Experimental Details

Details of the analytical procedures are outlined in Zauke et al.[16] and Petri and Zauke.[17] Instead of diammonium hydrogen phosphate, a palladium nitrate-magnesium nitrate matrix modifier was applied for Cd and Pb according to Yin et al.[18]

To analyze background levels, a concise statistical data evaluation is imperative. As will be shown below, not only are mean concentrations of interest, but also variabilities on different hierarchical levels need to be considered (Figure 2). Thus, one important question to be answered is the nature of the smallest unit of investigation. By defining this unit, the observation grain is determined.[19,20] We prefer individual organisms as basic units because they may be easily collected and properly cleaned aboard ship. Pooled samples, on the other hand, offer more danger of contamination, e.g., due to residues of paints or grease which may occur with the handling of trawls on vessels. Furthermore, possible influences of biological variables can be best analyzed using individuals (even *a posteriori*), and investigation of higher hierarchical levels is possible by aggregation of data using statistical procedures. The relatively great effort of this approach is justified by the substantial gain of information as compared to analysis of pooled samples, although one has to bear in mind that an adequate experimental design depends, of course, on the specific goals of the investigation.[1]

The statistical evaluation was performed using the package BMDP 88 (IBM PC/DOS), especially programs 7d, 2d, and 1r.[21] Groups were identified by the Student-Newman-Keuls Multiple Range Test at the 95% level of significance. Results were confirmed by the Tukey Studentized Range Method and a *t* test adjusted for multiple comparisons which are not shown in this paper. After testing for equality of variances using the Levene test, equality of means was tested using either classical ANOVA or a nonclassical Welch test.

III. VALIDITY AND RELIABILITY OF RESULTS

A. Quality Control Using Reference Samples

In this investigation we found the following mean values for reference samples (SD, n in brackets; mg kg^{-1} dry weight):

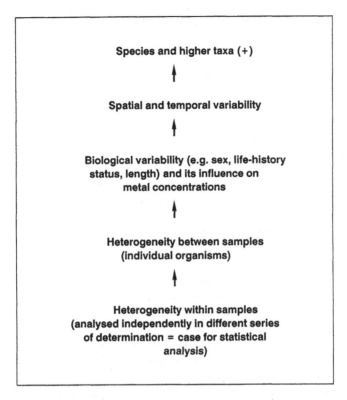

FIGURE 2. Representation of different hierarchical levels of consideration for investigation of metals in Antarctic crustaceans.

NBS SRM 1566 (oyster tissue) — Cd: analyzed 4.0 (0.5, 25), certified 3.5 (0.4); Pb: analyzed 0.45 (0.36, 24), certified 0.48 (0.04); Cu: analyzed 76 (9, 24), certified 63 (3.5); Zn: not analyzed because out of working range used, certified 852 (14);

IAEA MA-A-1 (copepod) — Cd: analyzed 0.75 (0.15, 22), certified 0.75 (0.03); Pb: analyzed 1.0 (0.6, 21), certified 2.1 (0.3); Cu: analyzed 7.2 (1.7, 21), certified 7.6 (0.2); Zn: analyzed 135 (27, 16), certified 158 (2).

The results analyzed are in most cases in good agreement with certified values, except for Pb in the copepod standard, where the low measured Pb concentrations may have resulted from inhomogeneities of the standard material. This interpretation is supported by the fact that the observed Pb value for oyster tissue proved to be valid, and that in other studies we found Pb values for the copepod standard to be in good agreement with the certified value. Limits of detection proved to be 0.7 mg kg^{-1} for Pb, 0.1 mg kg^{-1} for Cd, 4 mg kg^{-1} for Cu, and 16 mg kg^{-1} for Zn.[16]

B. *A posteriori* Quality Control of Sample Collection

As has been stated above, quality control of the analytical procedure is probably not the most crucial feature in determining background levels. In contrast to the widely accepted concept of "good laboratory practice", there are no straightforward protocols available for quality control of sampling techniques and sample preparation. Therefore we can only present arguments based more or less on plausibility. The basic idea is that the agreement of results obtained independently may be interpreted as an indication of validity, if different possible systematic errors are likely to occur. Regarding individual organisms collected on one particular cruise, it is very likely that a possible contamination will affect all samples (e.g., due to contaminated water or dust or equipment), or some individual organisms (not species) by chance (e.g., due to residues of paints or grease).

In the present study more than 15 crustacean species have been collected. In most cases, Pb concentrations were close to or below the limit of detection. For Cu and Zn many concentrations were within the range reported in the literature, whereas for Cd we found a wide range, with some species showing rather low values and others showing quite high levels. In line with arguments given above, substantial contamination is thus not very likely to have occurred.

This may be also concluded from a comparison of results for the isopod *Ceratoserolis trilobitoides,* collected on different cruises from different localities by different scientists (Table 2), where either no significant differences, or only small ones, were detected between means of metal concentrations.

The reliability of the analytical procedure (including independent digestions and determinations using a micro method and sequential multielement analysis) is given by relative standard deviations (coefficients of variance) for reference samples: 12 to 24% for Cd, Cu, and Zn, and 60 to 80% for Pb (the latter near the limit of detection). Taking the data set given in Table 2, an additional variability is obvious as compared to the reliability of the analytical procedure, focusing attention on intraspecific heterogeneities and possible influences of biological variables on incorporated metal levels (see next sections).

IV. INTRASPECIFIC HETEROGENEITIES

The problem of intraspecific heterogeneities will be treated in some detail, taking one decapod, one isopod, and one amphipod as examples. The element Pb will not be considered, since almost all concentrations detected have been close to or below the limit of detection (0.1 to 1.6 mg kg^{-1}), so that analytical reliability does not allow for further differentiation. A test of heterogeneity is provided by the Student-Newman-Keuls Multiple Range Test which allocates respective samples to different groups, if subsequent means of metal concentrations differ significantly (95% probability level) after sorting the data.[21]

**Table 2. Spatial and Temporal Variability of Heavy Metals in the Isopod
Ceratoserolis trilobitoides from the Antarctic Ocean; Mean
Concentrations in mg kg^{-1} d.w.**

	Gould Bay	Elephant Island	Tests for equality of					
			Variances		Means		Means	
			Levene		ANOVA		Welch	
	1985	1987	F	p	F	p	F	p
Pb								
Mean	0.49	0.25	0.75	0.39	14.7	0.00	14.8	0.00
SD	0.36	0.26						
n	78	88						
Cd								
Mean	1.9	2.1	15.3	0.00	1.83	0.18	2.01	0.16
SD	0.7	1.5						
n	78	90						
Cu								
Mean	38	46	4.71	0.03	15.2	0.00	15.5	0.00
SD	11	14						
n	78	85						
Zn								
Mean	40	40	0.54	0.47	0.01	0.92	0.01	0.92
SD	16	19						
n	60	90						

Note: n: number of determinations regarding 12–18 individual organisms analyzed
in 5 parallel samples in different series of determinations, respectively. AN-
OVA: assuming equality of variances. Welch test: not assuming equality of
variances. Null hypotheses are to be rejected if $p < 0.05$.

For the decapod *Notocrangon antarcticus* (Table 3), no heterogeneity is
obvious between individual organisms for Cu and Zn, whereas for Cd four groups
have been identified, with Cd concentrations ranging from about 6 to more than
18 mg kg^{-1}. Results obtained for the isopod *Ceratoserolis trilobitoides* (Table
4) differ markedly from those for *N. antarcticus*. For Cd an enormous heterogeneity is obvious, since up to 12 different groups have been identified (some
of which, however, show great overlap). Mean concentrations vary by almost
a factor of 10, from 0.7 to 6.3 mg kg^{-1}. A similar heterogeneity is depicted for
Cu, although the concentration range seems to be narrower (29 to 68 mg kg^{-1})
than for Cd. Finally, a somewhat smaller heterogeneity can also be seen for Zn.

A similar result is found for the amphipod *Eusirus propeperdentatus* (Table
5), except that for Cd and Cu the groups identified show almost no overlap.
Regarding these intraspecific heterogeneities, attention should be focused on
possible relationships between biological variables and accumulated metal concentrations which may explain these findings, at least to some degree (see next
section). On the other hand, these results may also be due to differences in gut

Table 3. Intraspecific Heterogeneities of Heavy Metals in Individual Organisms of the Decapod *Notocrangon antarcticus* From the Antarctic Ocean

Ind.	Cd Mean	SD	Groups 1 2 3 4	Ind.	Cu Mean	SD	Group 1	Ind.	Zn Mean	SD	1
124	5.9	0.5		109	43	4		124	36	9	
125	9.8	1.8		106	43	16		125	36	14	
111	10.1	4.3		107	53	9		122	44	11	
123	11.6	1.6		123	68	35		121	45	12	
110	12.7	2.2		108	70	7		110	47	11	
109	13.4	0.9		122	71	37		111	48	5	
122	14.5	2.3		110	72	5		106	48	21	
108	15.0	1.5		124	74	3		109	50	9	
121	18.1	3.1		125	75	8		123	51	9	
107	18.2	2.4		111	86	55		107	51	15	
106	18.7	3.0		121	92	7		108	54	20	

Note: n: 4–5 independent parallel samples, analyzed in different series of determinations. Ind.: code for individual organism analyzed. Bars (|) indicate groups identified by the Student-Newman-Keuls Multiple Range Test (95% level of significance).

Mean concentrations in mg kg^{-1} d.w.

Table 4. Intraspecific Heterogeneities of Heavy Metals in Individual Organisms of the Isopod *Ceratoserolis trilobitoides* From the Antarctic Ocean

	Cd		Groups
Ind.	Mean	SD	1 2 3 4 5 6 7 8 9 10 11 12
441	0.7	0.1	
412	1.0	0.1	
406	1.0	0.2	
445	1.1	0.1	
440	1.2	0.2	
413	1.2	0.1	
430	1.3	0.1	
429	1.3	0.2	
414	1.6	0.3	
439	1.6	0.3	
407	1.7	0.3	
426	1.7	0.3	
432	1.8	0.4	
443	1.9	0.4	
444	1.9	0.2	
447	2.0	0.2	
434	2.1	0.3	
446	2.1	0.4	
411	2.2	0.2	
409	2.2	0.4	
428	2.3	0.4	
425	2.4	0.6	
410	2.4	0.3	
438	2.6	0.7	
424	2.7	0.5	
408	2.8	0.2	
433	2.9	0.5	
436	3.0	0.3	
427	3.0	0.4	
448	3.3	0.2	
435	3.9	0.9	
431	6.3	0.8	

	Cu		Groups			Zn		Groups
Ind.	Mean	SD	1 2 3 4 5 6 7 8	Ind.	Mean	SD	1 2 3
430	29	1		448	13	6	
407	24	4		434	19	5	
427	30	6		426	25	7	
448	30	2		424	27	16	
411	30	3		429	27	8	
406	31	3		446	27	13	
428	33	3		445	28	8	
443	34	3		407	28	14	

Table 4 (continued). Intraspecific Heterogeneities of Heavy Metals in Individual Organisms of the Isopod *Ceratoserolis trilobitoides* From the Antarctic Ocean

Cu			Groups								Zn			Groups		
Ind.	Mean	SD	1	2	3	4	5	6	7	8	Ind.	Mean	SD	1	2	3
432	35	5									432	30	6			
409	35	1									412	31	19			
425	38	7									410	32	18			
447	40	2									447	32	27			
426	40	3									430	35	13			
435	40	6									431	36	12			
414	41	5									433	37	11			
434	42	5									443	38	9			
410	43	6									439	38	10			
440	43	3									436	38	5			
439	44	4									441	42	8			
424	45	9									413	44	11			
429	45	4									444	45	8			
412	45	5									425	45	10			
438	46	6									438	46	7			
436	46	2									440	46	13			
408	51	20									408	47	4			
413	55	8									411	47	10			
445	56	6									428	49	12			
446	58	14									435	50	7			
441	61	7									427	50	12			
444	61	3									414	51	18			
433	62	7									409	53	1			
431	68	9									406	61	7			

Notes: n: 4–5 independent parallel samples, analyzed in different series of determinations. Ind.: code for individual organism analyzed.

Mean concentrations in mg kg^{-1} d.w. Bars (|) indicate groups identified by the Student-Newman-Keuls Multiple Range Test (95% level of significance).

contents of individual organisms despite intended defecation. Whether there is a residual gut content (not avoidable under any circumstances) which is part of the sample for analysis of whole body metal concentration remains questionable.[3]

V. RELATIONSHIPS BETWEEN BIOLOGICAL VARIABLES AND ACCUMULATED METAL CONCENTRATIONS

Analysis of such relationships is a precondition for using organisms as biomonitors, since they should reflect the bioavailable portion of the environment. To do this they have to be "net accumulators" in the sense of Rainbow,[3] and not regulators.[22] If, on the other hand, biological variables do have a strong

Table 5. Intraspecific Heterogeneities of Heavy Metals in Individual Organisms of the Amphipod *Eusirus propeperdentatus* From the Antarctic Ocean

Cd Ind.	Mean	SD	Groups 1 2 3 4 5 6 7	Cu Ind.	Mean	SD	Groups 1 2 3 4
708	2.3	0.4		704	48	5	
711	2.9	0.4		708	50	4	
713	3.0	0.3		707	83	5	
705	5.7	0.9		714	85	9	
712	6.0	0.6		713	85	7	
718	6.1	0.6		715	86	4	
704	6.2	0.8		718	89	9	
714	6.4	0.6		712	95	6	
715	6.6	0.2		706	97	16	
707	7.2	0.6		716	99	11	
706	7.3	1.9		721	102	18	
716	9.4	1.4		717	106	18	
719	9.4	1.7		711	107	13	
717	11.2	1.6		719	110	28	
703	12.4	2.0		705	111	12	
701	12.6	0.8		720	131	11	
710	14.1	2.8		701	135	10	
720	14.3	2.1		702	143	6	
721	15.0	0.6		710	143	8	
709	16.0	0.7		703	152	14	
702	23.6	1.0		709	176	8	

Zn Ind.	Mean	SD	Groups 1 2 3 4 5 6
715	19	9	
717	23	7	
716	26	11	
718	27	3	
719	34	10	
708	39	10	
710	44	10	
721	44	7	
701	47	9	
713	47	8	
704	52	12	
702	55	7	
711	56	7	
720	57	9	
709	58	12	
703	64	14	
707	65	7	
706	66	13	

Table 5 (continued). Intraspecific Heterogeneities of Heavy Metals in Individual Organisms of the Amphipod *Eusirus propeperdentatus* From the Antarctic Ocean

Zn			Groups					
Ind.	Mean	SD	1	2 3	4	5 6		
705	66	6						
714	71	5						
712	72	5						

Note: n: 4–5 independent parallel samples, analyzed in different series of determinations. Ind.: Code for individual organism analyzed.

Mean concentrations in mg kg^{-1} d.w. Bars (|) indicate groups identified by the Student-Newman-Keuls Multiple Range Test (95% level of significance).

influence on accumulated metal concentrations, this would jeopardize the concept of a background metal concentration in a particular organism, unless such effects can be corrected for. It is therefore imperative to analyze such relationships quantitatively.

A conceptual model is given in Figure 3. An adequate tool for quantitative evaluation is the so-called "second generation of multivariate analysis", modeling latent variables, for example using LISREL or LVPLS.[23] Such an approach has been elaborated in a detailed study on gammarids from the Weser estuary, Germany.[24] In the present study on Antarctic crustaceans, we are only able to present preliminary results on the basis of bivariate linear correlation analysis, but more advanced investigations are encouraged in the future.

A. Body Length and Dry Weight

Individual body length and dry weight are considered as indicator variables of growth or age status. In the case of *Notocrangon antarcticus,* the observed intraspecific heterogeneity for Cd (Table 3) does not correspond to a significant correlation with the variables mentioned above. Most surprisingly, a positive correlation has been found between Cu and either body length or dry weight (Figure 4, H and I). Regarding *Ceratoserolis trilobitoides,* only a weak correlation has been found for Cd (Figure 4G) but no significant one for Cu, despite substantial heterogeneities detected by the Student-Newman-Keuls Multiple Range Test (Table 4). For both organisms mentioned there is thus no convincing evidence that intraspecific heterogeneities can be explained by indicator variables of individual growth or age.

The opposite seems to be true for *Eusirus propeperdentatus,* where intraspecific heterogeneities for Cd, Cu, and Zn (Table 5) coincide with significant correlations of these metals versus body length and dry weight (Figure 4, A to F).

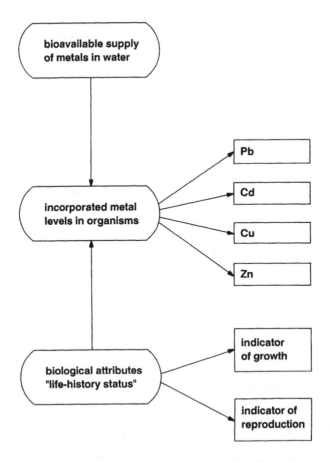

FIGURE 3. Conceptual model for biological monitoring of metals in organisms, showing latent variables on left hand side and measurable indicator variables on right hand side.

The findings outlined above for the decapod and isopod are somewhat surprising, since relationships between indicators of growth and accumulated metal concentrations are commonly reported in the literature.[1,25] On the other hand, a lack of correlation has also been shown for other Antarctic crustaceans.[26,27]

B. Sex

If reproduction has some influence on accumulated metal concentrations, one would expect that males and females would differ in their mean metal concentrations. Data obtained for *Ceratoserolis trilobitoides* allow the testing of this hypothesis (Table 6), whereas data for *Eusirus propeperdentatus* (Table 7) only allow for preliminary interpretation, there being very few males within the collection.

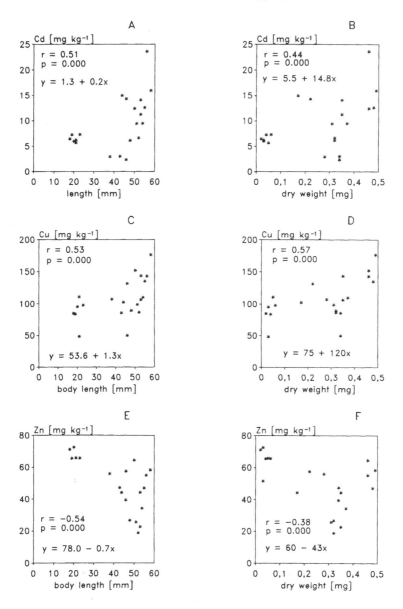

FIGURE 4. Relationship between accumulated metal concentrations in Antarctic crustaceans and their body length or individual dry weight: *Eusirus propeperdentatus* (A to F), *Ceratoserolis trilobitoides* (G), and *Notocrangon antarcticus* (H to I).

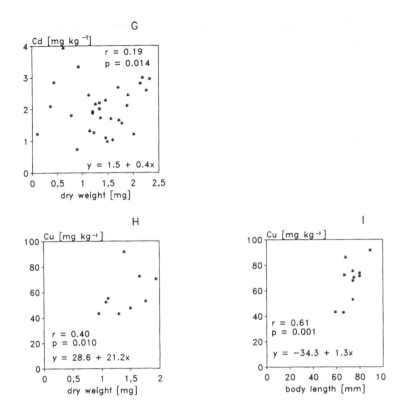

FIGURE 4 G, H, I.

In case of *C. trilobitoides,* differences between the means of Cd and Cu concentrations of males and females did coincide with substantial intraspecific heterogeneities detected by the Student-Newman-Keuls Multiple Range Test (Table 4). On the other hand, for *E. propeperdentatus* there was only little evidence for such an effect, in this case for Zn.

Preliminary results presented here encourage more detailed studies in the future. Organisms should not be collected more or less by chance, but specifically to cover a wide range of biological variables. Not only sex may be important in determining accumulated metal concentrations, but also the fecundity status of the females, as has been shown for gammarids from the Weser estuary modeling study.[24]

VI. INTERSPECIFIC HETEROGENEITIES

A. Whole Body Concentrations

In this paper we will only summarize results on whole body concentrations; details are given in Petri and Zauke.[17] In the case of Cd, substantial interspecific

Table 6. Variability of Heavy Metals in Males and Females of the Isopod
***Ceratoserolis trilobitoides* from the Antarctic Ocean**

			Tests equality of					
			Variances		Means		Means	
	Males	Females	Levene		ANOVA		Welch	
	1987	1987	F	p	F	p	F	p
Pb								
Mean	0.24	0.30	1.97	0.16	0.47	0.49	0.60	0.44
SD	0.24	0.44						
n	34	55						
Cd								
Mean	1.4	2.6	0.51	0.48	16.3	0.00	18.4	0.00
SD	1.1	1.5						
n	35	55						
Cu								
Mean	36	51	0.08	0.78	36.7	0.00	34.5	0.00
SD	13	11						
n	32	53						
Zn								
Mean	44	37	24.4	0.00	2.70	0.10	2.04	0.16
SD	25	12						
n	35	55						

Note: Mean concentrations in mg kg^{-1} d.w.

n: number of determinations regarding 7–11 individual organisms analyzed in 5 parallel samples in different series of determinations, respectively. ANOVA: assuming equality of variances. Welch test: not assuming equality of variances. Null hypotheses are to be rejected if $p < 0.05$.

heterogeneity is obvious from Figure 5, with mean concentrations ranging from 0.9 up to 13 mg kg^{-1}. Species showing relatively low concentrations are within the range reported in the literature for amphipods from the Weser estuary[2] and euphausiids from the Mediterranean[28] as well as from the northeastern Atlantic Ocean.[25] Most surprisingly, the highest Cd concentrations in Antarctic decapods coincide with those of taxonomically similar organisms from the Atlantic Ocean.[29-31] Even intermediate concentrations in some Antarctic isopods and amphipods (Figure 5) are within the range reported in the literature for amphipods from the Arctic and Mediterranean.[28,32,33] A similar coincidence has also been reported for amphipods of the genus *Themisto* from the Antarctic and Atlantic oceans,[25] showing Cd concentrations as high as 70 mg kg^{-1}.

As has been mentioned earlier, most of Pb concentrations in Antarctic crustaceans were close to or below the limit of detection, ranging from 0.1 mg kg^{-1} in the amphipod *Gondegoneia antarctica* to 2.5 mg kg^{-1} in the isopod *Glyptonotus antarcticus*.[17] For Cu and Zn most of the concentrations found for Ant-

Table 7. Variability of Heavy Metals in Males and Females of the Amphipod *Eusirus propeperdentatus* from the Antarctic Ocean

			Tests for equality of					
			Variances		Means		Means	
			Levene		ANOVA		Welch	
	Males	Females	F	p	F	p	F	p
Pb								
Mean	0.62	0.49	0.17	0.68	0.37	0.55	0.45	0.51
SD	0.56	0.64						
n	10	65						
Cd								
Mean	8.6	10.9	1.07	0.31	1.37	0.25	1.22	0.29
SD	6.2	5.7						
n	10	65						
Cu								
Mean	118	115	8.75	0.00	0.05	0.83	0.12	0.73
SD	17	36						
n	9	59						
Zn								
Mean	57	41	4.56	0.04	9.18	0.00	24.8	0.00
SD	8	16						
n	10	64						

Note: Mean concentrations in mg kg^{-1} d.w.

n: number of determinations regarding 3–13 individual organisms analyzed in 3–5 parallel samples in different series of determinations, respectively. ANOVA: assuming equality of variances. Welch test: not assuming equality of variances. Null hypotheses are to be rejected if $p < 0.05$.

arctic crustaceans are within the range reported in the literature for marine crustaceans, that is, 40 to 120 mg kg^{-1} Cu and 30 to 120 mg kg^{-1} Zn (see references mentioned for Cd). Theoretical considerations have shown that the lower end of this range may be equivalent to enzymatic requirements of these metals for decapods (about 26 mg kg^{-1} for Cu and 35 mg kg^{-1} for Zn), whereas the upper end of the observed concentration range may represent total metabolic (enzymatic plus hemocyanin component) requirements (about 83 mg kg^{-1} for Cu and 70 mg kg^{-1} for Zn).[3,34,53]

If similar essential metal requirements hold for Antarctic crustaceans (which has to be proved in future studies), then some of the species examined might be suffering some Cu or Zn deficiency, in that only enzymatic requirements appear to have been met. This may be especially true for the amphipod *Maxilliphimedia longipes* and the isopod *Aega antarctica,* which have copper concentrations as low as 6 to 8 mg kg^{-1}. Potential Cu deficiencies have also been suggested for the deep sea decapod *Systellaspis debilis*. It is speculated that these

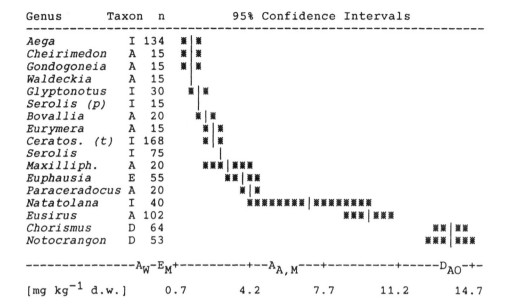

FIGURE 5. Cd in crustaceans from the Antarctic Ocean with indication of mean and 95% confidence interval (✳✳✳ | ✳✳✳) and order of magnitude of literature data; A_W: Amphipods, Weser estuary, Germany;[2] E_M: Euphausiids, Mediterranean;[28] A_A,M: Amphipods, Arctic and Mediterranean;[28,32,33] D_AO: Decapods, Atlantic Ocean;[29-31] and n = number of determinations regarding individual organisms analyzed in 3 to 5 parallel samples in different series of determinations, respectively.

organisms contain little if any hemocyanin, possibly limiting their activity level.[3] On the other hand, the amphipod *Waldeckia obesa* shows very high Zn levels of 300 mg kg⁻¹, an accumulated concentration which is regarded to be toxic for many decapods.[35,36]

B. Molar Ratios of Trace Metals

Molar ratios of trace metals give further insight into potential accumulation strategies and requirements of organisms, as we started to discuss in the previous subchapter. These are listed for Antarctic crustaceans in Table 8 (relative to Cu = 1,000) after sorting of the data according to whole body Cu concentrations. Roughly four classes may be distinguished on the basis of potential Cu requirements for decapods given above:

1. Two species apparently not reaching minimum enzymatic requirement;
2. Five species probably reaching the level of enzymatic requirement;
3. Seven species probably reaching enzymatic plus hemocyanin requirement; and
4. Three species exceeding proposed theoretical Cu requirements.

In the case of the theoretical Zn and Cu requirements of decapods mentioned above, a molar ratio of approx. 1,000:1,000 might be expected in crustacean Zn and Cu body concentrations. Measured values for the decapod *Paleamon elegans*[4] under control and metal exposure conditions are about 600:1,000 Zn:Cu. Similar ratios are found in many Antarctic crustacean species allocated to classes (2) and (3) (Table 8), whereas in the amphipod *Waldeckia obesa* Zn reaches accumulated concentrations often toxic for many decapods (see above). The implication, then, is that *W. obesa* is storing the excess zinc in a detoxified form not metabolically available to play a toxic role, whereas a higher percentage of the accumulated Zn concentration of decapods remains in a metabolically available toxic form.[3] Species allocated to class (1) may be showing Cu deficiency, and consequently Zn:Cu ratios reach relatively high values. On the other hand, species allocated to class (4) showing enhanced Cu levels have Zn:Cu ratios of approximately 500:1,000, indicating that "normal" Zn requirements have probably been met.

Regarding Pb and Cd, interpretations of molar ratios are difficult. For class (1) species, accumulated Pb and Cd concentrations occur in nearly the same ratios to Cu as those found for dissolved metals in Antarctic Ocean waters. In other cases, Pb and Cd to Cu ratios are distinctly lower in organisms than in Antarctic Ocean water, suggesting perhaps that these nonessential metals are less bioavailable. Whereas only a low variability in molar ratios is apparent for Pb, the opposite is true for Cd where the decapods *Notocrangon antarcticus* and *Chorismus antarcticus* show the highest relative ratios.

In general, the observed molar ratios of Pb to Cu in Antarctic crustaceans show good agreement with those in amphipods and isopods from southern German Bight and its estuaries, but ratios of Cd to Cu seem to be distinctly higher.[2,37] In German crustaceans Cd:Cu ratios are about 1 to 4:1,000; only values for semiterrestrial amphipods (*Talitrus* spp) reach ratios of 10 to 20:1,000. Cd:Cu ratios of 5:1,000 also seem to be characteristic of other marine crustaceans (e.g., euphausiids[25]), whereas some caridean decapods from the northeastern Atlantic Ocean[31] show ratios as high as 75:1,000, comparable to data for our Antarctic decapods and some amphipods (Table 8).

C. Bioconcentration Factors

The concept of bioconcentration factors (BCFs) is an important tool for "calibration" of biomonitor organisms, as has been emphasized by Zauke et al.[38] and, more implicitly, by Phillips and Segar[1] and Rainbow et al.[5] It implies that accumulated metal concentrations are related to the bioavailable supply of metals in the environment. Thus this concept does not apply for elements which are regulated to relatively constant body concentrations by organisms; it does apply, however, for "net accumulators".[3,5] It is also not adequate if biological variables influence accumulated metal concentrations significantly.

To compare relative bioavailabilities of metals using biomonitors from different habitats or regions, bioconcentration factors need to be rather similar, at

Table 8. Molar Ratios of Trace Metals (Relative to Cu = 1,000) and Mean Cu Concentrations [mg kg⁻¹ d.w.] in Crustaceans From the Antarctic Ocean

Species	Taxon[a]	Molar ratios				Cu (mg kg⁻¹)	Class[d]
		Pb	Cd	Zn	Cu		
Antarctic Ocean waters		20[b]	150[b]	1,100[c]	1,000[b]	—	1
Maxilliphimedia longipes	A	15	330	33,000	1,000	5.5	1
Aega antarctica	I	10	62	14,800	1,000	7.7	2
Gondogeneia antarctica	A	1	14	790	1,000	37.1	2
Serolis bouvieri	I	5	36	660	1,000	37.6	2
Ceratoserolis trilobitoides	I	3	27	925	1,000	41.8	2
Bovallia gigantea	A	7	19	840	1,000	53.0	2
Paraceradocus gibber	A	8	42	1,200	1,000	53.0	2
Eurymera monticulosa	A	2	18	650	1,000	63.7	3
Euphausia superba	E	1	29	490	1,000	65.7	3
Notocrangon antarcticus	D	4	110	650	1,000	67.2	3
Natatolana spp.	I	4	54	1,230	1,000	68.2	3
Waldeckia obesa	A	1	7	3,600	1,000	80.8	3
Cheirimedon femoratus	A	1	6	550	1,000	85.0	3
Chorismus antarcticus	D	5	80	460	1,000	92.7	3
Eusirus propeperdentatus	A	1	50	450	1,000	107	4
Serolis pagenstecheri	I	5	7	378	1,000	123	4
Glyptonotus antarcticus	I	5	5	415	1,000	149	4

a A (Amphipoda); E (Euphausiacea); D (Decapoda); I (Isopoda).
b From Mart, L., Rützel, H., Klahre, P., Sipos, L., Platzek, U., Valenta, P., and Nürnberg, H.W., *Sci. Total Environ.*, 26, 1, 1982.
c From Harris, J.E. and Fabris, G.J., *Mar. Chem.*, 8, 163, 1979.
d See text.

least for certain types of habitats. A straightforward concept to achieve such "calibration" makes use of dynamic toxicokinetic experiments under laboratory or field conditions, regarding different species or populations of organisms, and modeling BCFs at theoretical equilibrium with the aid of compartment models. Such investigations have been carried out for gammarid amphipods from the Weser estuary, leading to corresponding BCFs for Cd in the range of 500 to 1,000.[38,39]

These figures are in good agreement with field BCFs calculated for the Weser and Elbe estuary (500 to 1,500) using accumulated Cd concentrations in amphipods, and literature data of Cd in river water.[2] The latter values may be biased to some extent because it is not certain that metal concentrations reported for both estuaries are restricted to the dissolved fraction. However, even if not, field BCFs would be in the range of 1,100 to 5,600 (taking ratios of total to dissolved fraction in these estuaries from the work of Mart and Nürnberg[40]).

Regarding Antarctic crustaceans, field BCFs can only be calculated using literature data for metal concentrations in Antarctic Ocean waters. Furthermore, it is difficult to get appropriate data for the water phase measured. Concentrations differ by a factor of 5 to 25 in the literature,[15,27,41] and in our opinion the lowest values seem to be the most convincing (refer to Table 9). Of course, such differences have implications for the calculation of BCFs.

In the case of Cd, field BCFs for Antarctic crustaceans are within a range of 17,000 to 248,000 (Table 9), thus being several times higher than the values of amphipods from German estuaries. The same is true for other elements, where for German amphipods field BCFs of 8,500 to 32,000 for Cu and 1,000 to 3,000 for Zn have been reported.[2] In the case of Pb, BCFs in Antarctic crustaceans are of the same order of magnitude as those in amphipods from German estuaries, since the latter values (BCF = 50 to 630) probably have to be corrected to 3,600 to 27,000, in line with the arguments presented above.

Bioconcentration factors are not often addressed explicitly in the literature. Regarding Antarctic krill, BCFs (converted to a dry weight basis) are about 0.5 $\times$ 10^3 for Pb, 26×10^3 for Cd, 44×10^3 for Zn, and 96×10^3 for Cu,[26] and are thus not too far away from our values quoted in Table 9 (with the exception of Pb).

VII. SUMMARY AND CONCLUSIONS

It has been shown in the previous sections that accumulated trace metal concentrations in crustaceans from the Antarctic Ocean generally are not low. Only for Pb are the levels detected, in most cases, close to or below the limit of detection. For Cu and Zn most values are within the range reported in the literature for other marine and estuarine crustaceans, whereas Cd concentrations are relatively high in Antarctic caridean decapods as well as in certain amphipods and isopods — in good agreement with taxonomically similar organisms from

Table 9. Bioconcentration Factors (BCFs) Calculated for Antarctic Crustaceans on Basis of Dry Weight

Species	Taxon[a]	BCFs × 10^3			
		Pb	Cd	Cu	Zn
Antarctic Ocean waters	[ng kg^{-1}]	13[b]	54[b]	200[b]	230[c]
Maxilliphimedia longipes	A	19	54	28	740
Gondogeneia antarctica	A	11	18	186	132
Bovallia gigantea	A	93	34	265	200
Paraceradocus gibber	A	109	70	265	272
Eurymera monticulosa	A	32	36	318	188
Waldeckia obesa	A	20	20	404	1,309
Cheirimedon femoratus	A	15	17	425	210
Eusirus propeperdentatus	A	34	176	535	212
Notocrangon antarcticus	D	60	248	336	198
Chorismus antarcticus	D	122	246	464	190
Euphausia superba	E	19	64	329	145
Aega antarctica	I	21	17	39	526
Serolis bouvieri	I	43	49	188	111
Serolis pagenstecheri	I	142	29	615	203
Ceratoserolis trilobitoides	I	29	38	209	173
Natatolana spp.	I	70	123	341	372
Glyptonotus antarcticus	I	195	28	745	277

[a] A (Amphipoda); E (Euphausiacea); D (Decapoda); I (Isopoda).
[b] From Mart, L., Rützel, H., Klahre, P., Sipos, L., Platzek, U., Valenta, P., and Nürnberg, H.W., *Sci. Total Environ.*, 26, 1, 1982. With permission.
[c] From Harris, J.E. and Fabris, G.J., *Mar. Chem.*, 8, 163, 1979. With permission.

other regions. Despite substantial intraspecific heterogeneities, distinct interspecific differences are obvious for all elements, except for Pb.

We have argued in our introduction that the marine Antarctic environment seems not to be influenced by the global anthropogenic input of chemicals, especially metals. Such influences may eventually occur on a local scale near the numerous research stations or as a result of accidents. This view is supported by very low concentrations of metals in Antarctic Ocean waters, as summarized in Table 9.

However, regarding our results on heavy metals in Antarctic crustaceans, we conclude that these organisms are not suitable for deriving background levels. Some basic arguments are outlined in the following section. In general, they rely on the specific environmental conditions characteristic for the Antarctic Ocean and on the specific adaptations of the fauna to these conditions. For example, it is commonly accepted that reduction of energy metabolism in Antarctic organisms is an evolutionary response to the conditions of the habitat.[42] The evolutionary time scale for these processes is as long as 50×10^6 years,[43] in contrast, for example, to the Arctic environment.

There is some indication that several Antarctic crustacean species might show Cu deficiencies to some extent. This is inferred from knowledge of decapods from temperate regions. In order to prove such an hypothesis for Antarctic organisms, more detailed investigations are required, for example, on the biochemistry and ecology of these animals (enzymatic requirements, role of hemocyanin, detailed feeding strategies, and activity levels).

Furthermore, there is some evidence that productivity in the Southern Ocean may be limited by iron.[44,45] This may be eventually regarded as an indication of a general deficiency of trace elements in this region. In line with these arguments are the rather high bioconcentration factors observed for Antarctic crustaceans (Table 9). If there is really a general limitation of trace elements, development of very efficient mechanisms for uptake and storage of metals would be advantageous. To date, no detailed information has been available on these mechanisms in Antarctic crustaceans, in contrast to the case for organisms from temperate regions.[3,5]

If very efficient mechanisms for taking up essential elements have been evolved, it may be possible that nonessential elements are strongly accumulated too. This would explain the relatively high Cd levels observed in some Antarctic crustacean species. Moreover, it may also be speculated that nonessential metals may substitute for essential ones under conditions of deficiency, as has been shown, for example, for Cd and Zn in a marine diatom.[46]

High accumulation of potentially toxic metals requires, on the other hand, mechanisms of storage and detoxification, as summarized for other crustaceans by Rainbow[3] and Rainbow et al.[5] Again, detailed information is not available for Antarctic species, especially regarding the role of metallothionein and organ specific detoxification sites (e.g., in midgut glands). Detailed studies are required in the future, eventually combining these aspects with toxicokinetic experiments.

Finally we should like to stress that Antarctic crustaceans show specific features in their life cycles, classically referred to as those of a K-strategy. They are normally characterized by very slow growth rates, late reproduction, and high life expectancies of approximately 6 to 10 years.[47-50] These features may also contribute to high accumulated metal concentrations, for example of Cd, since it has been shown that rapid growth may lead to a dilution of accumulated metals.[5] In the case of crustaceans this effect may be enhanced due to molting, as discussed by Zauke,[51] although it has also been argued that substances may be resorbed from the cuticle to the body prior to molting.[3]

To conclude, we believe that accumulated metal concentrations in Antarctic crustaceans should not be used as global background levels in monitoring studies. In contrast to the standard shale concept for sediments mentioned above, it may not be possible to define any global background for organism trace metal concentrations. Instead, we recommend evaluation of regional background levels for appropriate groups of organisms (either defined taxonomically or ecologically in the sense of guilds), in line with the concept of "water quality" developed for freshwaters by Wuhrmann.[52]

ACKNOWLEDGMENTS

P. Lorenz, O. Coleman, and H.-G. Meurs (Oldenburg) collected the organisms on various expeditions to the Antarctic Ocean and helped with the identification of the animals. The experimental work was carried out by Gabriele Petri as part of a diploma thesis at the University of Oldenburg.

REFERENCES

1. **Phillips, D. J. H. and Segar, D. A.**, Use of bio-indicators in monitoring conservative contaminants: programme design imperatives, *Mar. Pollut. Bull.*, 17, 10, 1986.
2. **Zauke, G.-P., Meurs, H.-G., Todeskino, D., Kunze, S., Bäumer, H.-P., and Butte, W.**, Zum Monitoring von Cadmium, Blei, Nickel, Kupfer und Zink in Balaniden (Cirripedia: Crustacea), Gammariden (Amphipoda: Crustacea) und Enteromorpha (Ulvales: Chlorophyta), Umweltbundesamt, Berlin, Forschungsbericht Wasser 102 05 209, Part 3, UBA-FB 86-109, Texte 18/88, 1988.
3. **Rainbow, P. S.**, The significance of trace metal concentrations in decapods, in *Zool. Symp. No. 59*, Zoological Society, London, 1988, 291.
4. **Rainbow, P. S. and White, S. L.**, Comparative strategies of heavy metal accumulation by crustaceans: zinc, copper and cadmium in a decapod, an amphipod and a barnacle, *Hydrobiologia*, 174, 245, 1989.
5. **Rainbow, P. S., Phillips, D. J. H., and Depledge, M. H.**, The significance of trace metal concentrations in marine invertebrates, a need for laboratory investigation of accumulation strategies, *Mar. Pollut. Bull.*, 21, 321, 1990.
6. **Förstner, U. and Wittmann, G. T. W.**, *Metal Pollution in the Aquatic Environment,* Springer-Verlag, Berlin, 1979.
7. **Boutron, C.**, Respective influence of global pollution and volcanic eruptions on the past variation of the trace metals content of Antarctic snows since 1880's, *J. Geophys. Res.*, C85, 7426, 1980.
8. **Wolff, E. W.**, Signals of atmospheric pollution in polar snow and ice, *Antarctic Sci.*, 2, 189, 1990.
9. **Cripps, G. C.**, Problems in the identification of anthropogenic hydrocarbons against natural background levels in the Antarctic, *Antarctic Sci.*, 1, 307, 1989.
10. **Bonner, W. N.**, Conservation and the Antarctic, in *Antarctic Ecology*, Vol. 2, Laws, R. M., Ed., Academic Press, London, 1984, 821.
11. **Anderson, C., Coles, P., and Ewing, T.**, Research in Antarctica: exploring the still unexplored, *Nature*, 350, 287, 1991.
12. **Risebrough, R. W., de Lappe, B. W., and Younghans-Haug, C.**, PCB and PCT contamination in Winter Quarters Bay, Antarctica, *Mar. Pollut. Bull.*, 21, 523, 1990.
13. **Lenihan, H. S., Oliver, J. S., Oakden, J. M., and Stephenson, M. D.**, Intense and localized benthic marine pollution around McMurdo Station, Antarctica, *Mar. Pollut. Bull.*, 21, 422, 1990.

14. **Kennicutt, M. C., II, Sweet, S. T., Fraser, W. R., Stockton, W. L., and Culver, M.,** Grounding of the Bahia Paraiso at Arthur Harbor, Antarctica. 1. Distribution and fate of oil spill related hydrocarbons, *Environ. Sci. Technol.,* 25, 509, 1991.

15. **Mart, L., Rützel, H., Klahre, P., Sipos, L., Platzek, U., Valenta, P., and Nürnberg, H. W.,** Comparative studies on the distribution of heavy metals in the oceans and coastal waters, *Sci. Total Environ.,* 26, 1, 1982.

16. **Zauke, G.-P., Jacobi, H., Gieseke, U., Sängerlaub, G., Bäumer, H.-P., and Butte, W.,** Sequentielle Multielementbestimmung von Schwermetallen in Brackwasserorganismen, in *Fortschritte in der atomspektrometrischen Spurenanalytik,* Vol. 2, Welz, B. Ed., VCH Publishers, Weinheim, Germany, 1986, 543.

17. **Petri, G. and Zauke, G.-P.,** Tracemetals in crustaceans in the Antarctic Ocean, *Ambio,* in press.

18. **Yin, X., Schlemmer, G., and Welz, B.,** Cadmium determination in biological materials using graphite furnace atomic absorption spectrometry with palladium nitrate/ammonium nitrate modifier, *Anal. Chem.,* 59, 1462, 1987.

19. **Wiens, J. A.,** Spatial scaling in ecology, *Functional Ecol.,* 3, 385, 1989.

20. **Wiens, J. A.,** On the use of 'grain' and 'grain size' in ecology, *Functional Ecol.,* 4, 720, 1990.

21. **Dixon, W. J., Brown, M. B., Engelman, L., Hill, M. A., and Jennrich, R. I.,** *BMDP Statistical Software Manual,* University of California Press, Berkeley, 1988.

22. **Phillips, D. J. H. and Rainbow, P. S.,** Strategies of trace metal sequestration in aquatic organisms, *Mar. Environ. Res.,* 28, 207, 1989.

23. **Fornell, C., Ed.,** *A Second Generation of Multivariate Analysis,* Vol. 1, Praeger Publishers, New York, 1982.

24. **Bäumer, H.-P., van der Linde, A., and Zauke, G.-P.,** Structural equation models, applications in biological monitoring, *Biometrie und Informatik in Mediziu und Biologie,* 22, 156, 1991.

25. **Rainbow, P. S.,** Copper, cadmium and zinc concentrations in oceanic amphipod and euphasiid crustaceans, as a source of heavy metals to pelagic seabirds, *Mar. Biol.,* 103, 513, 1989.

26. **Yamamoto, Y., Honda, K., and Tatsukawa, R.,** Heavy metal accumulation in Antarctic krill *Euphausia superba,* in *Proc. NIPR Symp. Polar Biology, 1,* National Institute of Polar Research, Tokyo, 1987, 198.

27. **Honda, K., Yamamoto, Y., and Tatsukawa, R.,** Distribution of heavy metals in Antarctic marine ecosystsm, in *Proc. NIPR Symp. Polar Biology, 1,* National Institute of Polar Research, Tokyo, 1987, 184.

28. **Fowler, S. W.,** Trace metal monitoring of pelagic organisms from the open Mediterranean Sea, *Environ. Monit. Assessment,* 7, 59, 1986.

29. **Leatherland, T. M., Burton, J. D., Culkin, F., McCartney, M. J., and Morris, R. J.,** Concentrations of some trace metals in pelagic organisms and of mercury in Northeast Atlantic Ocean water, *Deep-Sea Res.,* 20, 679, 1973.

30. **Ridout, P. S., Willcocks, A. D., Morris, R. J., White, S. L., and Rainbow, P. S.,** Concentrations of Mn, Ie, Cu, Zn and Cd in the mesopelagic decapod *Systellaspis debilis* from the East Atlantic Ocean, *Mar. Biol.,* 87, 285, 1985.

31. **White, S. L. and Rainbow, P. S.,** Heavy metal concentrations and size effects in the mesopelagic decapod crustacean *Systellaspis debilis, Mar. Ecol. Prog. Ser.,* 37, 147, 1987.

32. **Hamanaka, T. and Ogi, H.,** Cadmium and zinc concentrations in the hyperiid amphipod, *Parathemisto libellula,* from the Bering Sea, *Bull. Fac. Fish. Hokkaido Univ.,* 35, 171, 1984.

33. **Macdonald, C. R. and Sprague, J. B.,** Cadmium in marine invertebrates and arctic cod in the Canadian Arctic. Distribution and ecological implications, *Mar. Ecol. Prog. Ser.,* 47, 17, 1988.

34. **White, S. L. and Rainbow, P. S.,** On the metablic requirements for copper and zinc in molluscs and crustaceans, *Mar. Environ. Res.,* 16, 215, 1985.

35. **White, S. L. and Rainbow, P. S.,** Regulation and accumulation of copper, zinc and cadmium by the shrimp *Palaemon elegans, Mar. Ecol. Prog. Ser.,* 8, 95, 1982.

36. **Rainbow, P. S.,** Accumulation of Zn, Cu and Cd by crabs and barnacles, *Estuarine Coastal Shelf Sci.,* 21, 669, 1985.

37. **Zauke, G.-P. and Meurs, H.-G.,** Unpublished data, 1991.

38. **Zauke, G.-P., von Lemm, R., Meurs, H.-G., Todeskino, D., Bäumer, H.-P., and Butte, W.,** Zum biologischen Monitoring von Schwermetallen in Ästuarien — konzeptionelle Aspekte und erste Untersuchungsergebnisse dargestellt am Beispiel von Cd in euryhalinen Gammariden, in *Umweltvorsorge Nordsee — Belastungen, Gütesituation, Maßnahmen,* Niedersächsisches Umweltministerium, Hannover, 1987, 325.

39. **Zauke, G.-P., von Lemm, R., Meurs, H.-G., Todeskino, D., Bäumer, H.-P., and Butte, W.,** Bioakkumulation und Toxizität von Cadmium bei Gammariden im statischen Labortest und dynamischen Test unter in-situ Bedingungen, Final report to Umweltbundesamt, Berlin, Forschungsbericht Wasser 102 05 209, Part 2, 1986.

40. **Mart, L. and Nürnberg, H. W.,** Cd, Pb, Cu, Ni and Co distribution in the German Bight, *Mar. Chem.,* 18, 197, 1986.

41. **Harris, J. E., and Fabris, G. J.,** Concentrations of suspended matter and particulate cadmium, copper, lead and zinc in the Indian sector of the Antarctic Ocean, *Mar. Chem.,* 8, 163, 1979.

42. **Clarke, A.,** A reappraisal of the concept of metabolic cold adaptation in polar marine invertebrates, *Biol. J. Linn. Soc.,* 14, 77, 1980.

43. **Maxwell, J. G. H.,** The breeding biology of *Chorismus antarcticus* (Pfeffer) and *Notocrangon antarcticus* (Pfeffer) (Crustacea, Decapoda) and its bearing on the problems of the impoverished Antarctic decapod fauna, in *Proc. 3rd SCAR Symp. Antarctic Biol.,* Gulf Publ. No. 6, Houston, Texas, 1977, 335.

44. **Martin, J. H., Gordon, R. M., and Fitzwater, S. E.,** Iron in Antarctic waters, *Nature,* 345, 156, 1990.

45. **de Baar, H. J. W., Buma, A. G. J., Nolting, R. F., Cadée, G. C., Jacques, G., and Tréguer, P. J.,** On iron limitation of the Southern Ocean: experimental observations in the Weddell and Scotia Seas, *Mar. Ecol. Prog. Ser.,* 65, 105, 1990.

46. **Price, N. M. and Morel, F. M. M.,** Cadmium and cobalt substitution for zinc in a marine diatom, *Nature,* 344, 658, 1990.

47. **Siegel, V.,** Age and growth of Antarctic euphausiacea (Custacea) under natural conditions, *Mar. Biol.,* 96, 483, 1987.

48. **Clarke, A.,** On living in cold water: K-strategies in Antarctic benthos, *Mar. Biol.,* 55, 111, 1979.

49. **Wägele, J.-W.**, Growth in captivity and aspects of reproductive biology of the Antarctic fish parasite *Aega antarctica* (Crustacea, Isopoda), *Polar Biol.*, 10, 521, 1990.

50. **Luxmoore, R. A.**, The reproductive biology of some serolid isopods from the Antarctic, *Polar Biol.*, 1, 3, 1982.

51. **Zauke, G.-P.**, Cadmium in Gammaridae (Amphipoda: Crustacea) of the rivers Werra and Weser. II. Seasonal variation and correlation to temperature and other environmental variables, *Water Res.*, 16, 785, 1982.

52. **Wuhrmann, K.**, Some problems and perspectives in applied limnology, *Mitt. Int. Ver. Limnol.*, 20, 324, 1974.

53. **Depledge, M. H.**, Re-evaluation of metabolic requirements for copper and zinc in decapod crustaceans, *Mar. Environ. Res.*, 27, 115, 1989.

CHAPTER 6

Effects of Cadmium on Survival, Bioaccumulation, Histopathology, and PGM Polymorphism in the Marine Isopod *Idotea baltica*

Marina de Nicola, Nicola Cardellicchio, Carmela Gambardella,
Sandro M. Guarino, and Claudio Marra

TABLE OF CONTENTS

I. INTRODUCTION

During recent years industrial and agricultural activities have resulted in a remarkable increase of Cd in marine environments, causing disturbance to animal populations.

It has been pointed out[1,2] that the measurement of pollution levels in marine environments should be integrated with an assessment of its biological effects. The stress response is the reaction of an organism to an environmental stimulus, which, by exceeding a threshold value, disturbs its normal functions. As a consequence, many biological parameters such as cell structure, body growth rate, survival, fecundity, genetic selection, and/or activity of some specific enzyme systems, may be indicative of a stress response.

The aim of the present work has been to determine the relation between cadmium accumulation and its biological effects on a crustacean isopod, *Idotea baltica*.

This species is widespread in European coastal areas and is an important link in food webs, being a detritus feeder and prey of benthic macroinvertebrates and fishes. Consequently, quantitative changes in *I. baltica* populations as a result of cadmium contamination may have an impact on higher trophic levels.

Our results indicate that although Cd has low acute toxicity, long-term contamination with this metal strongly modifies the structure and the genetic pattern of *I. baltica* populations.

II. MATERIALS AND METHODS

Several samples of *Idotea baltica* were collected in the Bay of Naples (Mar Morto, Bacoli). Males (18 to 22 mm long) and embryo-bearing females (15 to 18 mm long) were isolated and kept under controlled laboratory conditions (15°C; LD 12:12) in Plexiglass experimental vessels (22 × 16 × 8 cm). After hatching, one-day-old juveniles (1.5 mm long) were bred under the same laboratory conditions as adults. The animals were fed on the marine alga *Gracilaria* sp., which was replaced every second day.

A total of 980 adults and 500 juveniles were tested. All tests were performed on at least three replicates. Cadmium sulfate (Merck) was used. Concentrations of the metal (μg l^{-1} dissolved cadmium) were nominal (not measured analytically). In order to maintain constant concentrations a new solution was prepared for each trial, and this was replaced every second day. The organisms tested were observed daily and dead animals removed.

LT_{50} represents the time elapsing between the beginning of the experiment and the death of half the individuals tested. The mean LT_{50} were reported together with the standard error (SE). The mean value of survival and the growth rate at different treatment were analyzed by ANOVA or X^2 tests.

The long-term responses were obtained by evaluating survival, body growth, and sex ratio in juveniles exposed to 300 μg l^{-1} Cd from hatching to day 60, when external secondary sexual characteristics became visible.

The Cd effect on PGM (phosphoglucomutase) polymorphisms was studied in males treated with 1000 μg l^{-1} Cd or with a combination of Cd (90 μg l^{-1}) and Zn (130 μg l^{-1}). The exposed animals were observed daily and dead animals removed and deep frozen ($-80°C$); all survivors were also deep frozen on the 21st day. At the same time, controls were collected and frozen. The electrophoretical analysis for PGM was performed on the frozen whole animals.[3]

The genotype frequencies observed were compared statistically. Quantitative determinations of cadmium and zinc concentrations were carried out with atomic absorption spectrophotometry.

For electron microscopy observations (TEM), hepatopancreas tubules were fixed (2.5% glutaraldehyde + 1% formaldehyde in 0.1 M cacodylate buffer in 20% SW), rinsed in buffer (0.2 M cacodylate + 0.55 M sucrose) and postfixed in 1% osmium tetroxide. After dehydration the hepatopancreas tubules were embedded in epoxy resin. Ultrathin sections (0.06 μm) were stained with uranyl acetate and lead citrate and examined under a Philips 400 TEM. Semi-thin sections (1 μm) were stained with toluidine blue and examined under the light microscope. For scanning electron microscopy (SEM), fixed tubules were critically point dried after dehydration, mounted on aluminum stubs, coated with gold, and examined in a Philips 505 SEM.

III. RESULTS

Exposure of *I. baltica* males to increasing concentrations of Cd caused a net accumulation of the metal in the hepatopancreas proportional to Cd exposure levels (Figure 1).

Males of *I. baltica* exposed to 1000 μg l^{-1} Cd for 10 days (Table 1) accumulated Cd preferentially in the hepatopancreas, which accounts for about 5% of the total dry weight of the animals.

In Cd-exposed animals the structure of the hepatopancreas appeared strongly damaged. External swelling and fraying of the muscles controlling organ contraction were evident (Figure 2). Both B and S cells (Figure 3) showed a transformation of the rough endoplasmic reticulum into whorls and vesicles. Mitochondria appeared swollen and the presence of myelin bodies indicated intracellular lysis (Figure 4). Nuclei appeared picnotic and were characterized by a progressive

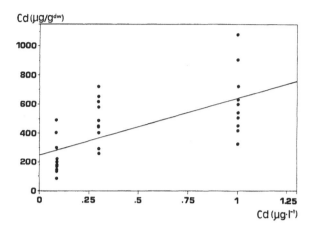

FIGURE 1. The effect of increasing Cd concentration (μg l^{-1}) on the Cd accumulation (μg/g dw) in the hepatopancreas of *Idotea baltica*. Regression line: Y = 2.48 + 3.95x.

Table 1. The Mean Concentrations of Cd in the Total Body and the Hepatopancreas of *I. baltica* Males after a 10 Day Exposure to 1000 μg l^{-1} Cd

Treatment	Cd μg/g d.w.	
	Total body $\bar{X}$ SE	Hepatopancreas $\bar{X}$ SE
Control[a]	0.68 ± 0.34	1.37 ± 0.34
Cadmium-exposed	264 ± 112	650 ± 218

[a] 1.5 μg l^{-1} Cd is the concentration in the SW where animals were collected.

transformation into irregular forms, developing small projections on their periphery (Figure 5). Crystals (Figures 3 and 5) were much more abundant in Cd-treated isopods than in the controls.[4]

In order to ascertain the lethal level of Cd accumulated in the body *in toto*, 107 animals were exposed to 1000 μg l^{-1} Cd and collected immediately after death. The data are reported in Table 2, with three different threshold lethal levels being observed in dead animals.

These different responses may be a result of individual genetic differences. For this purpose we examined the effect of 1000 μg l^{-1} Cd on the PGM isoenzyme polymorphism, which in *I. baltica* possesses 6 alleles forming 21 genotypes. Genotype frequencies in dead and surviving individuals were compared statistically. The results (Figure 6 and Table 3) indicated that the frequency of two genotypes (44 PGM and 24 PGM) was significantly higher in survivors, i.e., they were "resistant", while the frequency of two other genotypes (33 PGM and 34 PGM) was significantly higher in dead animals, i.e., they were

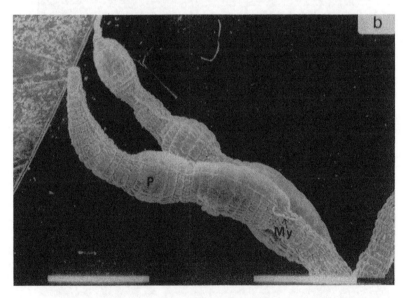

FIGURE 2. Scanning electron micrographs of the external surface of a hepato-pancreas tubule of *I. baltica* — a (control): the rows of myoepithelial cells (My) (Magnification × 245,000); b (Cd-treated): note the puffs(P) and the irregular pattern of myoepithelial cells (My) (Magnification × 37,000).

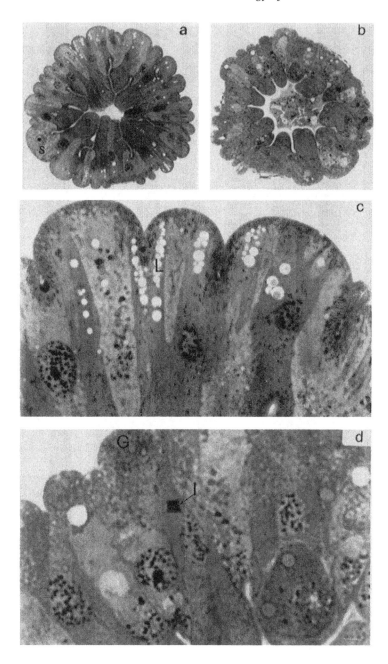

FIGURE 3. Light micrographs of transverse sections of hepatopancreas tubules
of *I. baltica* (Magnification: a,b × 230; c,d × 470); a,c = control, b,d
= Cd-treated; B cell (B) and S cell (S). Note granules in the tubule
lumen of Cd-treated isopod; L = lipid, I = crystal (probably ferritin).

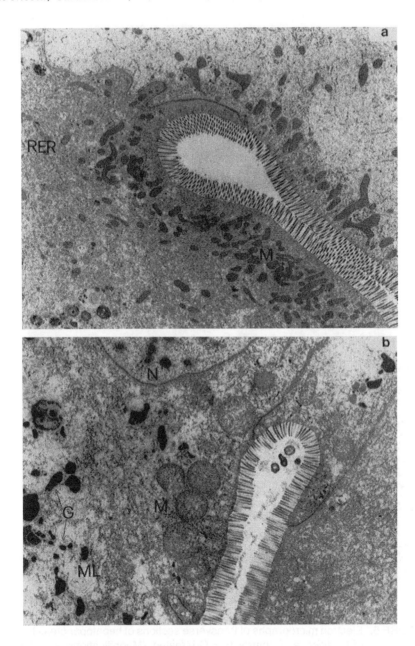

FIGURE 4. Electron micrographs of transverse section of a hepatopancreas tubule of a Cd-treated *I. baltica.* (Magnification × 7000.) Note the reduction of mitochondria (M) abundance and their swollen appearance. Rough endoplasmic reticulum (RER) is disorganized; abundance of myelin forms (ML) and granules is increased; N = nucleus.

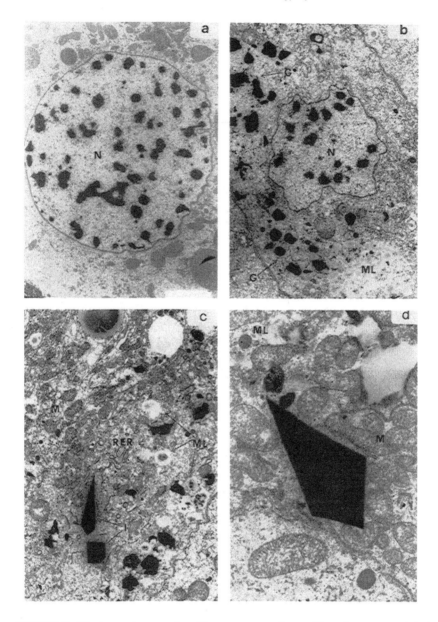

FIGURE 5. Electron micrographs of transverse sections of hepatopancreas tubule of *I. baltica;* a = control, b = Cd-treated. (Magnifications: a: × 6400; b,c: × 4900; d: × 14000.) Note pycnosis of nucleus with membrane projections and abundant granules (G) and myelin forms (ML). In micrographs c and d (Cd-treated) the rough endoplasmic reticulum (RER) is destroyed, showing myelin forms (ML), and swollen mitochondria (M). Note the crystal (I), probably of ferritin.

Table 2. Percentages of Dead Animals as a Function of Cd Accumulated in the Body

Cd (μg/g d.w.)	%
50–150	64.5
150–300	28.0
300–450	7.5

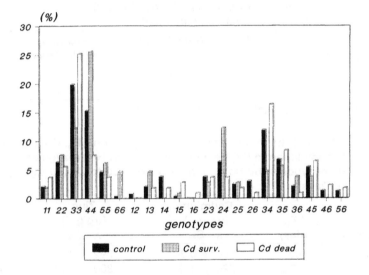

FIGURE 6. Differential survivorship of PGM genotypes of *Idotea baltica* on exposure to 1000 μg l⁻¹ Cd.

Table 3. Statistical Analysis (X² Test) of Frequencies of PGM Genotypes in *I. baltica* Males Treated With 1000 μg l⁻¹ Cd

	Frequency		Comparison	
Genotype	Dead	Survivors	X²	p
33	0.252	0.123	10.761	0.008
44	0.075	0.256	24.044	<10⁻⁶
24	0.037	0.123	9.502	0.009
34	0.164	0.047	13.040	0.006

Table 4. Statistical Analysis (X^2 Test) of the Heterozygote/Homozygote Ratio Between Dead Animals and Survivors Treated With 1000 μg l^{-1} Cd

	Dead		Survivors		Comparison	
	Freq.	He/Ho	Freq.	He/Ho	X^2	p
Het.	0.542		0.417			
		1.18		0.72	6.16	0.013
Hom.	0.458		0.583			

Table 5. Survival, Body Length and Sex Ratio in 60-day-old Juveniles of *I. baltica*, Exposed to 300 μg l^{-1} Cd During Juvenile Development

Animals	Survival (%)	Body length (mm)	Sex ratio (females/males)
Control	91	7.81	1.0
Exposed	20[a]	4.96[a]	4.0[a]

[a] All values are statistically significant at 1%.

Table 6. Effects of the Presence of Zn (130 μg l^{-1}) on LT_{50}, Cd, and Zn Accumulation in the Hepatopancreas, and Heterozygosis/Homozygosis (He/Ho) in *I. baltica* Exposed to Cd (90 μg l^{-1}) for 20 Days

Treatment	LT_{50} (day)	Cadmium ($\mu g/g^{-1}$) $\bar{X}$ SE	Zinc ($\mu g/g^{-1}$) $\bar{X}$ SE	He/Ho
Control	180	0.98 ± 0.34	292 ± 173	1.02
Cd	7.2	220 ± 107	300 ± 103	0.82
Cd + Zn	18.3	249 ± 95.0	3012 ± 1475	3.50

"sensitive". A significantly higher survival rate for the homozygote class compared with the heterozygote one was observed (Table 4).

The long-term effects of a sublethal cadmium concentration (399 μg l^{-1}) were studied in juveniles of *I. baltica* exposed during juvenile development from hatching to day 60 by determining survival, body growth, and sex ratio (Table 5).

After Cd exposure, the surviving juveniles represented only 20% of the controls and were mostly females (75%), suggesting a higher Cd sensitivity in males during development. Moreover, body length had also been greatly reduced.

Since it has been reported[5] that Zn, which is an essential metal, interacts with Cd, we experimented on the combined effects of the two metals (Table 6). These data provide evidence that the presence of Zn counteracts Cd effects by increasing survival and heterozygosis, without affecting Cd accumulation.

IV. DISCUSSION

Of a series of metals, Cd shows the lowest acute toxicity on *Idotea baltica*[6]:

$$Cd < Fe < Cu < Zn < Cr < Hg < Tributyltin\ oxide$$

A decreasing sensitivity to Cd has also been observed in *I. baltica* from females to males to juveniles.[7] Our results indicate that in this species, Cd, which is a nonessential metal, is accumulated proportionally to its exposure level, suggesting the absence of an internal regulatory mechanism, in agreement with the literature data on other crustaceans.[8-10]

The main site of Cd storage is the hepatopancreas, which accounts for 5% of the dry weight of adults. In crustaceans this gland performs several functions, e.g., secreting digestive enzymes, being the main site of food absorption, and storing metabolic reserves.[11,12] Its structure is greatly damaged by Cd exposure, causing complete degeneration of both S and B cells. Mitochondria appear swollen, the rough reticulum is disorganized, and the nuclei are picnotic. The digestive function is clearly stopped, as is also suggested by the observation that animals refuse food for several days preceding death.

Furthermore, in the hepatopancreas cells of exposed animals there appears to be a large increase in the abundance of granules, similar to the copper- and iron-rich granules reported in many crustaceans[16] and considered to be the main site of metal storage, as a detoxification mechanism. Probably, when the accumulation level of a metal exceeds the storage capacity, the animal dies.[17,18]

The observed differences in individual resistance may depend on genetic differences. This hypothesis is confirmed by the fact that Cd acts by selecting some PGM genotypes. PGM in *I. baltica* presents 21 genotypes, whose distribution showed changes according to Cd exposure. In fact, two genotypes appear to be resistant and two appear to be sensitive, due to their significant increase and decrease in survivorship, respectively. In addition, a significant increase in the total number of homozygotes was observed. As a consequence, Cd contamination may have a serious impact on the genetic structure of *I. baltica* populations. Pollution effects on the genetic structure of marine organisms have been widely reported.[19-23]

Chronic exposure to low Cd concentrations may have more dramatic consequences on *I. baltica* juveniles. In fact, exposure to 300 μg l^{-1} of Cd during early development strongly reduced survival (80%) and growth. Furthermore, the surviving population was dominated by females (75%), the resulting population structure having been noticeably modified.

The long-term effects of low Cd concentration can be counteracted by Zn. Our experiments demonstrated that addition of a low Zn concentration significantly increased survivability and heterozygosis. This antagonistic effect cannot be ascribed to a reduction of Cd uptake, Cd accumulation being the same in Cd-treated and in Cd-Zn-treated animals, suggesting that the interference of the two metals is probably at the metabolic level.[24] On the other hand, in the decapod, *Callianassa australiensis,*[5] Cd and Zn have been reported to exert toxic actions in an interactive manner, and accumulation of one metal also appeared to enhance the accumulation of the other.

In conclusion, Cd accumulates mostly in the hepatopancreas of *I. baltica.* Its accumulation was directly correlated with the ambient dissolved Cd concentration, suggesting that a regulatory mechanism is absent.

The interrelationship between Cd accumulation and death of the animals indicates that there are individual lethal concentrations that are probably genetically dependent. Cd, in fact, acted selectively on PGM genotypes.

In spite of its low acute toxicity in adults, cadmium in the long-term and in very low concentrations may cause significant alterations in population structure. In fact, if present during juvenile development, Cd reduces survivorship and growth and causes a dramatic shift in the population sex ratio. As a consequence, chronic pollution at low levels of Cd may induce a change in the population structure and the genetic pattern of *I. baltica,* thus influencing the whole marine ecosystem.

V. ABSTRACT

In *Idotea baltica,* acute Cd toxicity is higher in juveniles than in adults. Male are more sensitive than females.

The presence of sublethal Cd concentrations during embryonic and/or larval development reduces survival and body growth in the long term. Moreover, it changes the sex ratio, the population being dominated essentially by females (75%) after a two-month treatment.

The main site of Cd storage is the hepatopancreas, the ultrastructure of which appears strongly affected, showing a reduction in secretory activity.

The distribution of 21 PGM genotypes shows modifications. In fact, some genotypes appear to be either resistant or sensitive, resulting in a different genetic pattern with an increase of homozygosis in the population.

Zinc counteracts the effects of camium by increasing heterozygosis without interfering in Cd accumulation.

Our results point out the consequences of chronic contamination with cadmium on a population, and the need to take into account the modifications of biological parameters in order to monitor pollution in marine environments.

ACKNOWLEDGMENTS

We are grateful to the Zoological Station of Naples for the use of its facilities, in particular to G. Gragnaniello for his expert assistance in TEM and SEM techniques.

This research was in part financed and carried out by U.O. n.5 and n.7 of the CNR Project "Monitoraggio dell'inquinamento marino nel Mezzogiorno d'Italia".

REFERENCES

1. **Bayne, B. L.,** Responses to environmental stress: tolerance, resistance and adaptation, in *Marine Biology of Polar Regions and Effects of Stress on Marine Organisms*, Gray, J. S. and Chistiansen, M. E., Eds., John Wiley & Sons, New York, 1985, 331.
2. **Underwood, A. J. and Peterson, C. H.,** Towards an ecological framework for investigating pollution, *Mar. Ecol. Prog. Ser.*, 46, 227, 1988.
3. **Harris, H. and Hopkinson, D. A.,** *Handbook of Enzyme Electrophoresis in Human Genetics*, North-Holland, Amsterdam, 1976.
4. **de Nicola, M.,** unpublished data, 1991.
5. **Ahsanullah, M., Negilsky, D. S., and Mobley, M. C.,** Toxicity of zinc, cadmium and copper on the shrimp *Callianassa australiensis*. Effects of individual metals, *Mar. Biol.*, 64, 311, 1981.
6. **de Nicola, M., Cardellicchio, N., Guarino, S. M., Gambardella, C., Capuano, F., Scandone, L., and Marra, C.,** Heavy metal effect on the benthic crustacea *Idotea baltica*, presented at Ecotoxicologie des Sédiments, La Rochelle, France, June 5, 1991.
7. **de Nicola, M., Migliore, L., Guarino, S. M., and Gambardella, C.,** Acute and long term toxicity of cadmium on *Idotea baltica* Pall. (Crustacea, Isopoda), *Mar. Pollut. Bull.*, 18, 454, 1987.
8. **White, S. L. and Rainbow, P. S.,** Accumulation of cadmium by *Palaemon elegans* (Crustacea, Decapoda), *Mar. Ecol. Prog. Ser.*, 32, 17, 1986.
9. **Vernberg, W. B., De Corsey, P. J., and O'Hara, K.,** Multiple environmental factor effects of physiology and behaviour of the fiddler crab *Uca pugilator*, in *Pollution and Physiology of Marine Organisms*, Vernberg, F. J. and Vernberg, W. B., Eds., Academic Press, New York, 1974.
10. **Jennings, J. R. and Rainbow, P. S.,** Studies on the uptake of cadmium by the crab *Carcinus maenas* in the laboratory. I. Accumulation from seawater and a food source, *Mar. Biol.*, 50, 131, 1979.
11. **Hohn, L. and Scheer, B. T.,** *Carbohydrate Metabolism in Crustaceans*, Florkin, M. and Scheer, B. T., Eds., Vol. 5, Academic Press, New York, 1979, 147.
12. **Hopkin, S. P. and Martin, M. H.,** Distribution of zinc, cadmium, lead and copper within the hepatopancreas of a woodlouse, *Tissue & Cell*, 14, 703, 1982.
13. **Brown, B. E.,** The form and function of metal-containing "granules" in invertebrate tissues, *Biol. Rev.*, 57, 621, 1982.
14. **Mason, A. Z. and Nott, J. A.,** The role of intracellular biomineralized granules in the regulation and detoxification of metals in gastropods with special reference to the marine prosobranch *Littorina littorea*, *Aquat. Toxicol.*, 1, 239, 1981.
15. **Walker, G.,** "Copper" granules in the barnacle *Balanus balanoides*, *Mar. Biol.*, 39, 343, 1977.

16. **Moore, P. G. and Rainbow P. S.,** Ferritin in the gut caeca of *Stegocephaloides christianiensis* and other Stegocephalidae (Amphipoda: Gammarida): a functional interpretation, *Philos. Trans. R. Soc. London,* B 306, 219, 1984.

17. **Rainbow, P. S.,** Accumulation of Zn, Cu and Cd by crabs and barnacles, *Estuarine Coastal Shelf Sci.,* 21, 669, 1985.

18. **Rainbow, P. S. and White, S. L.,** Comparative strategies of heavy metal accumulation by crustaceans: zinc, copper and cadmium in a decapod, an amphipod and a barnacle, *Hydrobiologia,* 174, 245, 1989.

19. **Battaglia, B. and Bisol, P. M., Fossato, W. U., and Rodino, E.,** Studies on the genetic effects of pollution in the sea, *Rapp. P.-V. Reun. Cons. Int. Explor. Mer.,* 179, 267, 1980.

20. **Bryant, E. H.,** The adaptive significance of enzyme polymorphisms in relation to environmental variability, *Am. Nat.,* 108, 1, 1974.

21. **Ben-Shlomo, R. and Nevo, E.,** Isozyme polymorphism in monitoring of the marine environment: the interactive effect of cadmium and mercury pollution on the shrimp *Palaemon elegans, Mar. Pollut. Bull.,* 19, 314, 1988.

22. **Lavie, B. and Nevo, E.,** Heavy metal selection of phosphoglucose isomerase allozymes in marine gastropods, *Mar. Biol.,* 71, 17, 1982.

23. **Lavie, B. and Nevo, E.,** Genetic selection of homozygote allozyme genotypes in marine gastropods exposed to cadmium pollution, *Sci. Total Environ.,* 57, 91, 1986.

24. **Devineau, J. and Amiard-Triquet, C.,** Patterns of bioaccumulation of an essential trace element (zinc) and a pollutant metal (cadmium) in larvae of the prawn *Palaemon serratus, Mar. Biol.,* 86, 139, 1985.

SECTION II

Freshwater Environments

General Considerations
7. Metal Uptake, Regulation, and Excretion in Freshwater Invertebrates

8. Accumulation and Effects of Trace Metals in Freshwater Invertebrates

---

Freshwater Hydrozoa
9. Investigation of Uranium-Induced Toxicity in Freshwater Hydra

---

Freshwater Molluscs
10. Metal Regulation in Two Species of Freshwater Bivalves

Freshwater Crustacea
11. Autoradiographic Study of Zinc in *Gammarus pulex* (Amphipoda)

Freshwater Macroinvertebrate Communities
12. The Use of Freshwater Invertebrates for the Assessment of Metal Pollution in Urban Receiving Waters

13. Evolution of Resistance and Changes in Community-Composition in Metal-Polluted Environments: A Case-Study on Foundry Cove

CHAPTER 7

Metal Uptake, Regulation, and Excretion in Freshwater Invertebrates

Philip S. Rainbow and Reinhard Dallinger

TABLE OF CONTENTS

0-87371-734-1/93/$0.00 + $.50
© 1993 by Lewis Publishers

I. INTRODUCTION

There is no inherent reason to believe that the processes of metal accumulation should differ in principle between marine and freshwater invertebrates. Nevertheless the constituent processes of metal accumulation (metal uptake, sequestration, and/or excretion) may be affected to various degrees by the different physicochemistries of freshwaters in comparison to that of the relatively stable marine environment. Metal uptake rates are particularly likely to be affected by the physicochemistry of the medium, whether or not there is an additional physiological response on behalf of the invertebrate, for example, to a low concentration of an inorganic ion like calcium. It is relevant that not only do freshwater ecosystems exhibit a high natural variability in their physicochemistries, but, possibly as a consequence, they are more susceptible to anthropogenic influences than the more stable marine environments. A dramatic example is that of acidification, particularly severe in lakes with only a limited buffering capacity. Furthermore, many freshwater invertebrate communities are dominated by an animal class barely represented in marine systems — the insects — larval stages of which are particularly important in benthic freshwater habitats. The insects bring with them a novel organ with the potential to play a role in the detoxificatory sequestration of accumulated metals — that is, the Malpighian tubule.

The accumulation of metals by an invertebrate can be divided into three phases: (1) metal uptake, (2) metal transport, distribution, and sequestration within the body, and (3) metal excretion, the latter of which may be absent. The interrelationship of these processes defines the metal accumulation strategy of the invertebrate.[1-5] In addition to metal taken up into the body of the invertebrate with the potential to play a metabolic role and/or be accumulated, a further quotient of metal may be passively adsorbed onto the outside of the invertebrate, particularly in the case of insects and crustaceans with cuticular exoskeletons. The adsorbed component of the total body metal load can exchange passively with metal in the surrounding medium. It is not available to play a metabolic role and, in this account, is excluded from that metal described as accumulated by an invertebrate.

Accumulation strategies of invertebrates for metals have been discussed in some detail already in this volume,[6] and in the literature.[3,4] Accumulation strategies vary intraspecifically between metals, and interspecifically for the same metal even between quite closely related taxa. The strategies occupy a gradient from strong net accumulation through weak net accumulation (partial regulation) to regulation. Net accumulation may or may not involve excretion of metal, net accumulation being a result of the balance of absolute uptake and excretion of metals. Strong accumulators usually show no (significant) metal excretion and absolute uptake may be shown to be synonymous with net uptake of accumulation (e.g., the barnacle *Elminius modestus* and zinc[1]).

The term regulator, when used to refer to a whole animal, defines an invertebrate which shows no significant change in body metal content over time on exposure to a raised metal bioavailability. Theoretically, an invertebrate might regulate its body metal content by excluding all metal (no absolute uptake). This is difficult to conceive of as a long-term strategy, although it is certainly possible in the short term. Thus, freshwater bivalves including *Corbicula fluminea*[7] may close their shell valves for several hours, avoiding contact with the ambient water and temporarily preventing uptake of dissolved metals. A stricter definition of metal regulation at the whole body level might require a regulator to maintain body metal content approximately constant by altering the rate of metal excretion to match the rate of metal uptake. The latter varies with the bioavailability of the metal and is affected significantly by physicochemical factors beyond the physiological control of the invertebrate.

The processes of metal uptake, regulation, and excretion are considered in more detail in turn, with particular emphasis on freshwater invertebrates.

II. METAL UPTAKE

Possible mechanisms involved in the uptake of metals by aquatic organisms have been reviewed by Simkiss and Taylor.[8] Of the mechanisms described, two are considered in detail here to explain the uptake of trace metals from solution by aquatic invertebrates.

Aquatic invertebrates are bathed in a medium containing dissolved trace metals at concentrations approximately ranging from nanograms to (exceptionally) milligrams per liter. These concentrations are low in comparison to those within aquatic invertebrates, but the uptake of many trace metals by aquatic invertebrates appears to be passive, not requiring the expenditure of energy.[8-11] In contrast, the uptake of major ions like sodium, potassium, and calcium which are more 'ionic' in nature does require active pumps to cross the cell membrane.

In contrast to the major ions above, trace metals have a high affinity for proteins, bonding with sulfur and nitrogen.[12] The first model of metal uptake depends on this high affinity of trace metals for proteins, proposing that dissolved metals in the external medium bind passively onto transport proteins in the membranes of permeable surfaces of the invertebrate. By facilitated diffusion (also passive) the trace metal is transported across the membrane and into the cell, where it binds with a series of metal-binding ligands of increasing affinity. Metal uptake continues passively, apparently against a concentration gradient, for backflow is limited by the excess of nondiffusible high affinity binding sites internally, and/or cellular compartmentalization and transport of the metal out of the cell into the blood.

The binding of the trace metal to any membrane transport ligand is commonly considered to involve the free metal ion, which is therefore the bioavailable chemical species.[8,13-16] In aquatic media, dissolved trace metals are partitioned in equilibria between inorganic and organic complexing ligands. Thus, cadmium in seawater is found as chloro-complexes (e.g., $CdCl^+$, $CdCl_2^0$, $CdCl_3^-$) with only about 2.5% of the total present as the free (hydrated) cadmium ion Cd^{2+}.[17,18] For many trace metals in seawater, the free metal ion is similarly present at a relatively low equilibrium percentage of total dissolved metal.[18] In comparison with seawater, freshwaters clearly contain low concentrations of inorganic anions such as chloride, capable of complexing metal ions. Thus the proportion of total dissolved metal that is in the form of the free metal ion (the bioavailable form) is likely to be higher in freshwater than seawater, particularly in acidic freshwaters.

The relative concentrations of metal-chelating organic ligands are less predictable. Seawater contains variable concentrations of organic exudates and degradation products of phytoplankton and macrophytic seaweeds; freshwaters contain similar products of freshwater vegetation but also organic compounds (e.g., humates and fulvates) derived from terrestrial vegetation via freshwater runnoff. The presence of metal-binding ligands decreases the bioavailability of dissolved trace metals, presumably by reducing the free metal ion concentration.[13,16,19] Thus, any physicochemical change that reduces the hydrophilic complexation of a dissolved metal enhances its bioavailability by increasing the absolute concentration of the free metal ion available to bind with membrane transport ligands.

Under unusual circumstances, direct passage through the lipid membrane is probably the route of uptake of metals bound in lipophilic organometallic compounds. Such compounds may be of natural occurrence (e.g., methylmercury formed by microbial activity), or derived anthropogenically (tetraethyl lead, and tributyl tin in petrol and antifouling paint, respectively). The binding of a trace metal by a lipophilic complexing agent would have the same effect.[20]

The second possible uptake mechanism to be considered in detail involves the incorporation of dissolved trace metals into active pumps available for the major ions. As indicated above, major metal ions (Na^+, K^+, and Ca^{++}) do not have a high affinity for organic ligands and are moved across hydrophobic cell membranes against concentration gradients via active transport pumps. It is inevitable that the free ions of some trace metals will become incorporated into such pumps,[21] but the relative significance of this route of entry for trace metals will vary between organisms and with environmental conditions.

Nonmetallic ions like chloride and sulfate also require active pumps for transport across the cell membrane. In oxygenated waters certain trace metals exist as oxygenated anions (e.g., chromate, molybdate, and vanadate) rather than as the free metal ion or as a free metal ion complex. Given the lack of availability of a free metal ion to bind with a membrane transport protein for uptake, it is possible that such metals are taken up via active transport pumps. Such pumps might be metal-specific, or the metal anions may become incorporated by chance into pumps normally employed for sulfate, phosphate, etc.

Examples of trace metals with free ions potentially able to enter an active pump for the uptake of a major ion concern cadmium and calcium. The free cadmium ion has an ionic radius of 109 pm, similar to that of calcium (114 pm) (assuming a coordination number of 6 in each case[22]). It is inevitable that some cadmium ions will enter calcium pumps, the latter being under physiological control as the organism responds to changes in the requirement for calcium uptake. Wright[21] has presented evidence that cadmium accumulation by the freshwater amphipod crustacean, *Gammarus pulex,* may be at least partially accounted for by a process of 'accidental' active cadmium uptake. The question remains as to the relevant significance of the two routes of cadmium uptake — the first via facilitated diffusion, the second via calcium pumps.

The passive route of facilitated diffusion is beyond the short-term physiological control of the invertebrate. Physicochemical changes in solution (e.g., relative concentrations of inorganic and organic chelating agents) will alter the absolute concentration of free Cd ions available for uptake and change the cadmium uptake rate, all independently of the invertebrate. The only control that could be exerted by the invertebrate with respect to this route would be via some alteration to the number, accessibility, or affinity of the membrane transport ligands. Such alterations do not seem to be a feature of the physiology of most marine invertebrates, including the littoral prawn *Palaemon elegans,*[14,15] although exceptionally it does appear to occur in other euryhaline crustaceans including the talitrid amphipod *Orchestia gammarellus*[23] and the shore crab *Carcinus maenas.*[24] In these latter examples such physiological control over metal uptake rates at low external salinities may be related to changes in apparent water and electrolyte permeabilities.[23,24] It is stressed that such physiological responses are only found in certain estuarine invertebrates experiencing substantial salinity changes.

The second route, the incorporation of the free ion of a trace metal into an active pump for a major ion, as exemplified by cadmium and calcium, is potentially under physiological control as a matter of routine. The uptake of calcium requires energy and is therefore under physiological control, the random incorporation of cadmium being directly proportional to the activity of the pump. Thus, cadmium uptake will be directly proportional to calcium uptake. To provide an example: an estuarine crustacean like the crab *Carcinus maenas* suffers from osmotic uptake of water on passage up an estuary into low salinity. Water balance is restored by increased expulsion of urine from the pair of antennary glands. The urine, however, is isosmotic with the blood, the expulsion of the excess water therefore being associated with the unfortunate loss of dissolved salts, including calcium. This loss of salts is made good by the active uptake of ions in the gills. Thus the active uptake of calcium by the crab is increased when the crab is exposed to decreased salinity.

Sure enough, the uptake of cadmium is also increased on exposure of the crab to low salinity,[25] as expected by this model of uptake. The facilitated diffusion model, however, also predicts that the uptake of cadmium should increase at low salinity as a result of decreased complexation of the free cadmium ion as chloride concentrations decrease. Either model therefore fits the data.

It has been stressed that the second model involving active pumps is more routinely under the physiological control of the invertebrate, directly able to control pump activity. Nevertheless, there remains one component of even this model beyond the physiological control of the invertebrate. In the case of *C. maenas* and cadmium uptake, it is the free cadmium ion that is incorporated into the calcium pump. Thus any change in the equilibrium concentration of the free cadmium ion (e.g., by physicochemical change) will change the cadmium uptake rate via the calcium pump, even in the absence of a change in pump activity. This effect is probably much smaller and may be masked by the effects of changes in pump activity.

Is it possible to assess the relative significance of the two proposed routes of uptake? Nugegoda and Rainbow[15] designed an experiment separating changes in chloride concentration (affecting complexation — in this case of dissolved zinc) from changes in osmolality (affecting rates of uptake of major metal ions in a euryhaline crustacean), using sucrose and the decapod crustacean *Palaemon elegans*. *P. elegans* showed an increase in the rate of zinc uptake with decrease in salinity (total inorganic solute concentration, dominated by chloride) irrespective of the osmolality of the medium. This indicates, that in this case at least, incorporation of the trace metal into an active ion pump is not the major route of dissolved metal uptake.

It is tempting to generalize that for marine invertebrates, the binding of the free metal ion to a membrane transport ligand can explain the uptake of trace metals, with metal incorporation into an active pump being of minor significance. What about freshwater invertebrates? Freshwater invertebrates can be expected to have higher activities of uptake pumps for calcium, for they live in hypotonic media with associated osmotic and ionic balance problems. Pumps for calcium will be particularly well represented in freshwater crustaceans (e.g., crayfish[26] and molluscs such as unionid bivalves[27]) requiring significant concentrations of calcium for calcified cuticles and shells, respectively; indeed, such high requirements for calcium typically restrict many molluscs to calcium-rich freshwaters.

Indeed, a high requirement for calcium may increase the relative importance of cadmium entry via the calcium pump, even in a marine invertebrate. Bjerregaard and Depledge[28] have evidence that the uptake of cadmium by the intertidal winkle *Littorina littorea* is significantly related to the calcium concentration of the medium, independently of an salinity effect. In the Australian freshwater mussel *Velesunio angasi* there is a strong correlation between accumulated labelled cadmium and labelled calcium concentrations,[29] explicable by the uptake of both metals through the calcium pump. As reported above, at least a component of the uptake of cadmium by the freshwater amphipod *Gammarus pulex* can be accounted for by the 'accidental' incorporation of cadmium into a calcium pump.[21]

The above discussion has concentrated on metal uptake from solution. Freshwater invertebrates will also take up metals from particulates, essentially from food. It is difficult to make generalizations about the relative importance of these two major entry routes; the two routes are linked and their relative significance will vary between metals, and for each invertebrate, according to ambient physicochemical conditions, physiological and developmental stage, food availability and type, etc. van Hattum et al.[30] concluded that uptake of cadmium from water predominates over cadmium uptake from food in the freshwater isopod *Asellus aquaticus*. In nymphs of the burrowing mayfly *Hexagenia rigida,* however, net uptake of cadmium and zinc in the gut from sediment consumed as food was the most important uptake route.[31] Similarly, in the freshwater snail *Lymnaea palustris* and in the crayfish *Astacus leptodactylus,* the uptake of radiocobalt (^{60}Co) via contaminated food was responsible for a relatively higher accumulation of the isotope in the animals than was direct uptake of radiocobalt from solution.[32]

III. METAL REGULATION

Regulation as a metal accumulation strategy of invertebrates is much less common than net accumulation.[2,4] Constant body concentrations of metals in specimens of a species collected from many different sites of different metal bioavailabilities[33,34] do provide some indication that regulation is occurring, but strictly regulation needs to be verified in laboratory experiments.[4,35] For example, the decapod *Palaemon elegans* collected from different sites around Britain has relatively constant body concentrations of zinc and copper,[34,36] but confirmation that metal uptake is significant and is matched by excretion has only been provided in the case of zinc.[37] Such verification requires the use of labelled tracers to separately identify new and original metal accumulated, as has been carried out for zinc,[1] cadmium,[1] and cobalt.[38] Copper radiotracers with short half lives are less amenable to long-term experiments.

Depledge and Rainbow[39] discuss the concept of regulation at the whole body level of marine invertebrates. Whole body metal loads are a consequence of the summation of the metal contents of individual tissues, in some of which (or all, in the case of regulators) specific metal levels are maintained within narrow ranges by regulatory mechanisms that do not involve significant accumulation of excess metals.[6,39] Even strong net accumulators of metal will have particular tissues that regulate their metal content, net accumulation at the whole body level resulting from net accumulation in one or more detoxificatory storage tissues. Kraak et al.,[40] Hopkin,[41] and Dallinger[5] interpret body and tissue metal concentrations in freshwater mussels and terrestrial invertebrates in this way.

As stated above, regulation at the whole body level is not a common metal accumulation strategy of invertebrates, apparently being restricted to certain essential trace metals, particularly zinc and copper, and perhaps to particular invertebrate taxa especially decapod crustaceans.[2,42] Partial regulators which show little net accumulation of metal are exemplified by one group of bivalve molluscs, the mussels, in the case of zinc.[6,43] Oysters, which are also bivalves, on the other hand, are strong accumulators of zinc.[6,44]

It is of interest, therefore, that among freshwater invertebrates a possible example of a regulator is a decapod crustacean, the crayfish *Austropotamobius pallipes,* in the case of zinc.[45,46] Another malacostracan crustacean, but not a decapod, the freshwater isopod *Asellus aquaticus,* however, is a net accumulator of another essential metal, copper.[47] The freshwater mussels *Dreissena polymorpha* and *Unio pictorum* also appear to be regulators (or at least partial regulators) of zinc and copper, as investigated by Kraak et al.,[40] although verification with radiotracers is needed. Another freshwater mussel, the Australian *Velesunio ambiguus,* also appears to regulate tissue levels of zinc.[48,49] There is some evidence that the freshwater oligochaete *Lumbriculus variegatus* may also regulate body contents of zinc.[50] Another annelid, the estuarine polychaete *Hediste (Nereis) diversicolor,* has long been known as a zinc regulator.[51]

Hare et al.[31] have investigated the dynamics of zinc and cadmium through sediment-feeding larvae of the mayfly *Hexagenia rigida.* Total body concentrations of these metals did not change while new metal was taken up, but it is not possible from the concentrations quoted (dpm mg^{-1} for radioisotopes, μg g^{-1} for total metal) to determine whether uptake was significant enough to cause a rise in total body metal concentration above background variation. Thus net accumulation (rather than regulation) could still have been occurring, but at a low rate. Radioactive zinc and cadmium are excreted again but at lower rates than uptake,[31] indicating that the mayfly larvae are net accumulators, not regulators, although body metal concentrations might reach a steady-state concentration, different at each metal exposure. Shutes et al.[52] came to a similar conclusion, namely that the crustaceans *Gammarus pulex* and *Asellus aquaticus* reach equilibrium body concentrations of zinc and lead, these equilibrium concentrations increasing with metal exposure. *Asellus aquaticus* is also a net accumulator of lead and copper.[47]

Timmermans and Walker[53] have studied the accumulation of zinc by other insect larvae — in this case the chironomids (midges) *Chironomus riparius* and *Stictochironomus histrio.* There was no evidence of regulation of body levels of zinc in these larvae, body contents increasing with zinc exposure, although zinc was lost with each cast exuvium. *Chironomus riparius* larvae similarly showed no evidence for regulation of body copper contents, nor did either species for the nonessential metal cadmium.[53] Krantzberg and Stokes,[54] however, concluded that populations of larval *Chironomus* sp. could regulate zinc bioaccumulation. The apparent inconsistency here lies in the definition and use of the term regulation. The lack of a significant rise in body concentration of zinc in the two larval populations might indicate zinc regulation, though tracer work is required for an understanding of zinc turnover in the larvae.

In other studies of the accumulation of nonessential metals by freshwater invertebrates, the snail *Physa integra,*[55] larvae of the stonefly *Pteronarcys dorsata* and the caddis fly *Hydropsyche betteni,*[55] and the isopod *Asellus aquaticus*[30] are all net accumulators of cadmium. Both *H. betteni* larvae[55] and *A. aquaticus*[30] show some evidence of decreases in net cadmium accumulation at high exposures[55] or over time.[30] *P. integra,* larvae of *P. dorsata,* larvae of the caddis fly *Brachycentrus* sp., and the amphipod *Gammarus pseudolimnaeus* all accumulate lead in proportion to the dissolved lead concentration in the medium.[55]

IV. METAL EXCRETION

Metal excretion needs to balance metal uptake if regulation is to be achieved when there is significant metal uptake occurring. Many freshwater invertebrates have evolved from marine ancestors via estuaries. Decapod and amphipod crustaceans and bivalves have taken this route, which does not apply, however, to insect larvae or to many pulmonate gastropods invading freshwater from land.

One feature of the invasion of freshwaters by decapods and amphipods has been an increase in tegument impermeability in comparison to their marine relatives. Such impermeability restricts osmotic entry of water, but might also serve as a preadaptation to the adoption of regulation as a metal accumulation strategy for it will restrict trace metal uptake, at least via any facilitated diffusion pathway. The freshwater crayfish *Austropotamobius pallipes,* for example, has relatively low permeability to zinc.[45,46] Any reduction in metal uptake will increase the possibility that metal excretion can match metal uptake. The crayfish *A. pallipes* certainly excretes zinc during zinc regulation, in this case via the feces.[45]

Contrastingly, adoption of the freshwater habit by malacostracan crustaceans might also increase metal uptake by active pump routes. Relative degrees of reduction in the facilitated diffusion route and increase in the active pump routes will vary between invertebrates and determine whether metal uptake rates rise or fall, overall.

Freshwater bivalve molluscs, on the other hand, have had little scope for reducing uptake by reducing permeability of the outer surface. As lamellibranchs, these bivalves have huge gill surface areas for feeding, with the inevitable potential for osmotic uptake of water across these permeable surfaces. Indeed, such bivalves have huge osmotic throughputs of water, in spite of a reduction in the osmotic pressure of body fluids reducing the osmotic gradient across the gills. High metal uptake rates may be inevitable, but perhaps the high exit rate of osmotic water offers a vehicle for metal excretion. Anyway, freshwater mussels do appear to regulate or at least partially regulate soft tissue concentrations of zinc,[40,48] so there is presumably some balancing of zinc excretion and uptake. The unionid bivalves *Anodonta anatina* and *Unio pictorum* are certainly able to excrete aluminum accumulated during experimental exposure to the metal.[56]

Recent work has been carried out on metal excretion by freshwater insect larvae. Timmermans and Walker[53] have confirmed that significant levels of zinc, copper, and cadmium accumulated from laboratory exposure to high dissolved metal concentrations are excreted during the larval development of the chironomid midge larvae *Chironomus riparius* and *Stictochironomus histrio.* Losses of zinc and cadmium were associated with cast exuvia at molting, but net accumulation of both metals still occurred. Copper accumulated by *C. riparius* larvae was also lost at metamorphosis, but apparently not to any significant extent in exuvia.[53] There is clear interspecific variation in metal accumulation strategies even among midge larvae, for field samples of *S. histrio* showed considerable losses of zinc, copper, and cadmium upon metamorphosis, whereas those of *Chironomus anthracinus* did not.[53] Mayfly larvae are also capable of trace metal excretion. Sediment-feeding larvae of the mayfly *Hexagenia rigida* excrete zinc, cadmium, and lead accumulated originally from sediment as food.[31]

V. CONCLUSIONS

A clear feature of this viewpoint of metal uptake, regulation, and excretion in freshwater invertebrates has been the requirement to call upon examples involving marine invertebrates. There is a relative lack of available data on the biology of trace metals in freshwater invertebrates.

There is a particular need for more data on the mechanisms of uptake of trace metals by freshwater invertebrates. It might be expected that routes of uptake involving major ion active pumps should be of greater significance in freshwater invertebrates than in marine invertebrates in osmotic balance.

Also required are studies of trace metal kinetics through freshwater invertebrates using radioactive tracers. Such studies allow concrete conclusions to be drawn on metal uptake and excretion rates and on the presence of regulation as an accumulation strategy. Such data are prerequisites to the use of any freshwater invertebrate as a biomonitor of metals.

REFERENCES

1. **Rainbow, P. S. and White, S. L.,** Comparative strategies of heavy metal accumulation by crustaceans: zinc, copper and cadmium in a decapod, an amphipod and a barnacle, *Hydrobiologia*, 174, 245, 1989.
2. **Phillips, D. J. H. and Rainbow, P. S.,** Strategies of trace metal sequestration in aquatic organisms, *Mar. Environ. Res.*, 28, 207, 1989.
3. **Rainbow, P. S.,** Heavy metal levels in invertebrates, in *Heavy Metals in the Marine Environment*, Furness, R. W. and Rainbow, P. S., Eds., CRC Press, Boca Raton, FL, 1990, 67.
4. **Rainbow, P. S., Phillips, D. J. H., and Depledge, M. H.,** The significance of trace metal concentrations in marine invertebrates: a need for laboratory investigation of accumulation strategies, *Mar. Pollut. Bull.*, 21, 321, 1990.
5. **Dallinger, R.,** Strategies of metal detoxification in terrestrial invertebrates, this volume, chap. 14.

6. **Rainbow, P. S.,** The significance of trace metal concentrations in marine invertebrates, this volume, chap. 1.

7. **Doherty, F. G., Cherry, D. S., and Cairns, J., Jr.,** Valve closure responses of the Asiatic Clam *Corbicula fluminea* exposed to cadmium and zinc, *Hydrobiologia,* 153, 159, 1987.

8. **Simkiss, K. and Taylor, M. J.,** Metal fluxes across the membranes of aquatic organisms. *CRC Crit. Rev. Aquat. Sci.,* 1, 173, 1989.

9. **Williams, R. J. P.,** Physico-chemical aspects of inorganic element transfer through membranes, *Philos. Trans. R. Soc. London, Ser. B,* 294, 57, 1981.

10. **Williams, R. J. P.,** Natural selection of the chemical elements, *Proc. R. Soc. London, Ser. B,* 213, 361, 1981.

11. **Carpenè, E. and George, S. G.,** Absorption of cadmium by gills of *Mytilus edulis* (L.), *Mol. Physiol.,* 1, 23, 1981.

12. **Nieboer, E. and Richardson, D. H. S.,** The replacement of the nondescript term 'heavy metals' by a biologically and chemically significant classification of metal ions, *Environ. Pollut.,* B 1, 3, 1980.

13. **Nugegoda, D. and Rainbow, P. S.,** Effect of a chelating agent (EDTA) on zinc uptake and regulation by *Palaemon elegans* (Crustacea: Decapoda), *J. Mar. Biol. Assoc. U.K.,* 68, 25, 1988.

14. **Nugegoda, D. and Rainbow, P. S.,** Effects of salinity changes on zinc uptake and regulation by the decapod crustaceans *Palaemon elegans* and *Palaemonetes varians, Mar. Ecol. Prog. Ser.,* 51, 57, 1989.

15. **Nugegoda, D. and Rainbow, P. S.,** Salinity, osmolality and zinc uptake in *Palaemon elegans* (Crustacea: Decapoda), *Mar. Ecol. Prog. Ser.,* 55, 149, 1989.

16. **O'Brien, P., Rainbow, P. S., and Nugegoda, D.,** The effect of a chelating agent EDTA on the rate of uptake of zinc by *Palaemon elegans* (Crustacea: Decapoda), *Mar. Environ. Res.,* 30, 155, 1990.

17. **Zirino, A. and Yamamoto, S.,** A pH-dependent model for the chemical speciation of copper, zinc, cadmium and lead in seawater, *Limnol. Oceanogr.,* 17, 661, 1972.

18. **Bruland, K. W.,** Trace elements in seawater, in *Chemical Oceanography,* Vol. 8, Riley, J. P. and Chester, R., Eds., Academic Press, New York, 1983, 157.

19. **Poldoski, E. J.,** Cadmium bioaccumulation assays. Their relationship to various ionic equilibria in Lake Superior water, *Environ. Sci. Technol.,* 13, 701, 1979.

20. **Ahsanullah, M. and Florence, T. M.,** Toxicity of copper to the marine amphipod *Allorchestes compressa* in the presence of water- and lipid-soluble ligands, *Mar. Biol.,* 84, 41, 1984.

21. **Wright, D. A.,** Cadmium and calcium interactions in the freshwater *Gammarus pulex, Freshwater Biol.,* 10, 123, 1980.

22. **Huheey, J. E.,** *Inorganic Chemistry,* 3rd ed., Harper International, Cambridge, 1983.

23. **Rainbow, P. S., Malik, I., and O'Brien, P.,** Physico-chemical and physiological effects on the uptake of dissolved zinc and cadmium by the amphipod crustacean *Orchestia gammarellus, Aquat. Toxicol.,* in press.

24. **Chan, H. M., Bjerregaard, P., Rainbow, P. S., and Depledge, M. H.,** Uptake of zinc and cadmium by two populations of shore crabs *(Carcinus maenas)* at different salinities, *Mar. Ecol. Prog. Ser.,* in press.

25. **Wright, D. A.,** The effect of salinity on cadmium uptake by the tissues of the shore crab *Carcinus maenas* (L.), *J. Exp. Biol.,* 67, 137, 1977.

26. **Malley, D. F.**, Decreased survival and calcium uptake by the crayfish *Orconectes virilis* in low pH, *Can. J. Fish. Aquat. Sci.*, 37, 364, 1980.

27. **Pynnönen, K.**, Accumulation of ^{45}Ca in the freshwater unionids *Anodonta anatina* and *Unio tumidus*, as influenced by water hardness, protons and aluminium, *J. Exp. Zool.*, 260, 18, 1991.

28. **Bjerregaard, P. and Depledge, M. H.**, Unpublished data, in Depledge, M. H., interactions between heavy metals and physiological processes in estuarine invertebrates, in *Estuarine Ecotoxicology*, Chambers, P. L. and Chambers, C. M., Eds., JAPAGA, Ashford, Ireland, 1990, 89.

29. **Jeffree, R. A.**, A radioecological approach to problems of bioaccumulation, in *Proc. Bioaccumulation Workshop, Feb. 1991, Sydney*, The Water Board, Sydney, in press.

30. **van Hattum, B., de Voogt, P., van den Bosch, L., van Straalen, N. M., and Joosse, E. N. G.**, Bioaccumulation of cadmium by the freshwater isopod *Asellus aquaticus* (L.) from aqueous and dietary sources, *Environ. Pollut.*, 62, 129, 1989.

31. **Hare, L., Saouter, E., Campbell, P. G. C., Tessier, A., Ribeyre, F., and Boudon, A.**, Dynamics of cadmium, lead and zinc exchange between nymphs of the burrowing mayfly *Hexagenia rigida* (Ephemeroptera) and the environment, *Can. J. Fish. Aquat. Sci.*, 48, 39, 1991.

32. **Amiard, J. C., and Amiard-Triquet, C.**, Distribution of cobalt 60 in a mollusc, a crustacean and a freshwater teleost: variations as a function of the source of pollution and during elimination, *Environ. Pollut.*, 20, 199, 1979.

33. **Bryan, G. W.**, Concentrations of zinc and copper in the tissues of decapod crustaceans, *J. Mar. Biol. Assoc. U.K.*, 48, 303, 1968.

34. **White, S. L., and Rainbow, P. S.**, Regulation and accumulation of copper, zinc and cadmium by the shrimp *Palaemon elegans, Mar. Ecol. Prog. Ser.*, 8, 95, 1982.

35. **Timmermans, K. R.**, Accumulation and effects of trace metals in freshwater invertebrates, this volume, chap. 8.

36. **Rainbow, P. S.**, Trace metal concentrations in a Hong Kong penaeid prawn, *Metapenaeopsis palmensis* (Haswell), in *Proc. 2nd Int. Mar. Biol. Workshop: The Marine Flora and Fauna of Hong Kong and Southern China, Hong Kong, 1986*, Morton, B. Ed., Hong Kong University Press, Hong Kong, 1990, 1221.

37. **White, S. L. and Rainbow, P. S.**, Regulation of zinc concentration by *Palaemon elegans* (Crustacea: Decapoda): zinc flux and effects of temperature, zinc concentration and moulting, *Mar. Ecol. Prog. Ser.*, 16, 135, 1984.

38. **Rainbow, P. S. and White, S. L.**, Comparative accumulation of cobalt by three crustaceans: a decapod, an amphipod and a barnacle, *Aquat. Toxicol.*, 16, 113, 1990.

39. **Depledge, M. H. and Rainbow, P. S.**, Models of regulation and accumulation of trace metals in marine invertebrates, *Comp. Biochem. Physiol.*, 97C, 1, 1990.

40. **Kraak, M. H. S., Toussaint, M., Bleeker, E. A. J., and Lavy, D.**, Metal regulation in two species of freshwater bivalves, this volume, chap. 10.

41. **Hopkin, S. P.**, Deficiency and excess of copper in terrestrial isopods, this volume, chap. 18.

42. **Rainbow, P. S.**, The significance of trace metal concentrations in decapods, *Symp. Zool. Soc. London*, 59, 291, 1988.

43. **Phillips, D. J. H. and Rainbow, P. S.,** Barnacles and mussels as biomonitors of trace elements: a comparative study, *Mar. Ecol. Prog. Ser.,* 49, 83, 1988.
44. **Phillips, D. J. H. and Yim, W. W.-S.,** A comparative evaluation of oysters, mussels and sediments as indicators of trace metals in Hong Kong waters, *Mar. Ecol. Prog. Ser.,* 6, 285, 1981.
45. **Bryan, G. W.,** The metabolism of Zn and ^{65}Zn in crabs, lobsters and fresh-water crayfish, in *Radioecological Concentration Processes,* Aberg, B. and Hungate, F. P., Eds., Pergamon Press, New York, 1966, 1005.
46. **Bryan, G. W.,** Zinc regulation in the freshwater crayfish (including some comparative copper analyses), *J. Exp. Biol.,* 46, 281, 1967.
47. **Brown, B. E.,** Uptake of copper and lead by a metal-tolerant isopod *Asellus meridianus* Rac., *Freshwater Biol.,* 7, 235, 1977.
48. **Millington, P. J. and Walker, K. F.,** Australian freshwater mussel *Velesunio ambiguus* (Philippi) as a biological monitor for zinc, iron and manganese, *Aust. J. Mar. Freshwater Res.,* 34, 873, 1983.
49. **Maher, W. A. and Norris, R. H.,** Water quality assessment programs in Australia: deciding what to measure, and how and where to use bioindicators, *Environ. Monit. Assessment,* 14, 115, 1990.
50. **Bauer-Hilty, A. and Dallinger, R.,** unpublished data.
51. **Bryan, G. W. and Hummerstone, L. G.,** Adaptation of the polychaete *Nereis diversicolor* to estuarine sediments containing high concentrations of zinc and cadmium, *J. Mar. Biol. Assoc. U.K.,* 53, 839, 1973.
52. **Shutes, R. B. E., Ellis, J. B., Revitt, D. M., and Bascombe, A. D.,** The use of freshwater invertebrates for the assessment of metal pollution in urban receiving waters, this volume, chap. 12.
53. **Timmermans, K. R. and Walker, P. A.,** The fate of trace metals during the metamorphosis of chironomids (Diptera, Chironomidae), *Environ. Pollut.,* 62, 73, 1989.
54. **Krantzberg, G. and Stokes, P. M.,** Metal regulation, tolerance and body burdens in the larvae of the genus *Chironomus, Can. J. Fish. Aquat. Sci.,* 46, 389, 1989.
55. **Spehar, R. L., Anderson, R. L., and Fiandt, J. T.,** Toxicity and bioaccumulation of cadmium and lead in aquatic invertebrates, *Environ. Pollut.,* 15, 195, 1978.
56. **Pynnönen, K.,** Aluminium accumulation and distribution in the freshwater clams (Unionidae), *Comp. Biochem. Physiol.,* 97C, 111, 1990.

CHAPTER 8

Accumulation and Effects of Trace Metals in Freshwater Invertebrates

Klaas R. Timmermans

TABLE OF CONTENTS

0-87371-734-1/93/$0.00 + $.50
© 1993 by Lewis Publishers

I. INTRODUCTION

It is a generally accepted view that underwater sediments act as a sink for trace metals.[1] Freshwater invertebrates can, as a result of their preference for the benthic habitat, be confronted with an environment high in trace metal concentrations. As a consequence, trace metals will be accumulated, but there is no straightforward relationship between trace metal concentrations in sediment and those in the invertebrates inhabiting that sediment.[2-5]

In this chapter, an overview of trace metal accumulation in, and effects on, freshwater invertebrates will be given. For reasons of clarity, accumulation is defined as (trace metal) uptake minus elimination in organisms. It would be too ambitious to try to take all trace metals and all freshwater invertebrates into account, and therefore this overview will be limited to some dominant benthic freshwater macroinvertebrates and two essential metals (zinc and copper) and two nonessential metals (cadmium and lead). Throughout the chapter, the discussion will be focused on possible explanations for the fact that some organisms take up more metals than others living in the same habitat and on the effects that result from this uptake.

II. ACCUMULATION

A. Normal Versus Elevated Concentrations

It is like stating the obvious: trace metals can be accumulated by freshwater invertebrates. The ultimate result of accumulation has been determined for decades now, in the form of trace metal residue analyses in organisms. In recent years, modern analytical equipment, new clean laboratory techniques, and use of certified reference materials have greatly improved the accuracy and precision of trace metal analyses in organisms. Still, a review of the literature reveals that only a limited data set of trace metal analyses in freshwater invertebrates is available. The reasons for this may be numerous: tedious sampling, sorting, and identification of organisms, lack of interest on the (mostly) small individuals or their limited commercial value. In addition, data are often presented without possible explanations or without statements on significance of the observations.

An overview of cadmium, zinc, lead, and copper concentrations in freshwater invertebrates collected in polluted and unpolluted freshwater ecosystems is presented in Table 1. The species include organisms with different feeding habits (deposit feeders and filter feeders), different trophic levels (predators and prey organisms), and from different habitats (benthic, pelagic, lenthic, and lotic).

As the authors give indications of the extent of trace metal pollution in the habitats sampled, the data presented in Table 1 offer the opportunity to derive normal and elevated trace metal concentrations in freshwater invertebrates. The majority of invertebrates collected in unpolluted ecosystems have cadmium concentrations ranging from 0.01 to 1.0 μg g^{-1} dry weight. In polluted ecosystems much higher values are reported, for example, in organisms collected in the River Dommel: 0.3 to 147 μg Cd g^{-1} dry weight. Similar observations can be made for lead. Invertebrates collected in unpolluted habitats have concentrations varying between 0.29 and 5.0 μg Pb g^{-1} dry weight. In polluted aquatic ecosystems, as the River Irwell, lead concentrations were between 30 and 10,820 μg g^{-1} dry weight and in the River Dommel between 0.7 and 74 μg g^{-1} dry weight. Zinc and copper concentrations of invertebrates collected in unpolluted or moderately polluted ecosystems range between 75 and 200 μg g^{-1} dry weight, and between 2 and 40 μg g^{-1} dry weight, respectively. In severely polluted ecosystems, such as the River Irwell, much higher copper concentrations were reported in invertebrates.

B. Species Specific Differences

Apart from normal and elevated metal concentrations, Table 1 reports several species-specific differences in trace metal accumulation. Species collected at the same location can exhibit different metal concentrations. Several explanations for these differences can be indicated.

In the first place, biomagnification provides an explanation for differences in zinc concentration in some predators and their prey. Biomagnification is defined as the descriptive phenomenon that concentrations of pollutants can increase along trophic pyramids.[14,15] Zinc is apparently accumulated in higher trophic levels, as for example in water mites or leeches collected in the Maarsseveen Lakes and the River Dommel. Both these predator species prey upon chironomid larvae, and have zinc concentrations twice as high as their prey organisms. However, some caution is necessary, as it cannot be excluded that zinc is either directly taken up from water, or that predators have an extra need for zinc (e.g., in metalloenzymes). Secondly, physiology can play a role. Macroinvertebrates, such as crustaceans and molluscs with hemocyanin, a copper containing hemolymph pigment, have the highest copper levels. Obviously, the physiological ability of hemocyanin synthesis and storage (in the hepatopancreas) enables these organisms to maintain a high body concentration of copper, irrespective of concentrations in the environment.[16,17] Finally, body weight can play a role, especially in determination of lead and copper concentrations in ma-

Table 1. Cadmium, Zinc, Lead, and Copper Concentrations in Freshwater Benthic Invertebrates

Site and Reference	Taxon	Cadmium	Zinc	Lead	Copper	Classification
River Irwell (U.K.) (6)	*Asellus aquaticus*	N.A.	146–183	154–595	160–277	Polluted
	Erpobdella octoculata	N.A.	983–1565	30–111	15–54	
River Irwell (U.K.) (7)	Oligochaeta	N.A.	43330	10820	9000	Polluted
	Asellus aquaticus	N.A.	3590	630	5370	
	Chironomidae	N.A.	43560	9080	1130	
	Erpobdella octoculata	N.A.	22490	1260	330	
Rivers Brett and Chelmer (U.K.) (8)	*Asellus* sp.	0.105–0.202	N.A.	0.465–0.867	N.A.	Unpolluted
	Gammarus sp.	0.101	N.A.	0.234	N.A.	
	Haemopsis sp.	0.106	N.A.	0.617	N.A.	
	Limnaea sp.	0.095–0.262	N.A.	0.788–1.921	N.A.	
	Sigara sp.	0.148–0.183	N.A.	0.217–0.440	N.A.	
	Trichoptera	0.094–0.283	N.A.	0.608–1.025	N.A.	
	Ephemeroptera	0.293–0.525	N.A.	0.852–1.879	N.A.	
	Dytiscus sp.	0.043–0.124	N.A.	0.799–1.098	N.A.	
	Sialis sp.	0.050–0.275	N.A.	0.218–0.533	N.A.	
	Agriidae	0.100–0.164	N.A.	0.384–0.531	N.A.	
River Hayle (U.K.) (9)	Trichoptera	N.A.	330–774	N.A.	488–1000	Polluted
	Plecoptera	N.A.	132–410	N.A.	404–615	
	Odonata	N.A.	428–481	N.A.	448–768	
	Neuroptera	N.A.	350	N.A.	183	
	Coleoptera	N.A.	63–173	N.A.	44–47	
Fox River (U.S.) (10) Elgin	*Physa* sp.	2.44	49.06	19.05	22.77	Point sources
	Chironomidae	2.71	152.16	32.70	16.92	
	Sigara sp.	1.46	165.63	9.63	15.07	

Site	Descriptor	Species				
Algonquin	Point sources	Hydropsyche sp.	1.42	353.63	39.43	12.67
		Asellus sp.	1.63	132.51	18.61	86.60
		Physa sp.	5.66	174.30	34.81	17.79
		Sigara sp.	1.63	163.16	18.34	19.34
		Hydropsyche sp.	0.53	122.71	4.99	9.29
		Chironomidae	1.97	95.05	32.87	10.02
River Derwent (U.K.) (11)	Metal-contaminated	Ephemeroptera	6–151	812–15050	19–1625	N.A.
		Coleoptera	0.5–53	72–1060	18–504	N.A.
		Trichoptera	0.7–52	234–1879	6–372	N.A.
		Chironomus sp.	58.7	1504	305	N.A.
River Dommel (Netherlands) (12)	Polluted	Tubifex sp.	1.09–147	109–659	9.27–74.42	11–48
		Lymnaea ovata	0.32	78	0.68	14
		Physa fontinalis	0.51–3.15	70–89	1.99–9.60	79–108
		Piscicola geometra	1.15	740	13.08	51
		Glossiphonia compl.	2.74–31.45	786–1622	0.98–15.31	10–13
		G. heteroclita	5.04–13.49	1249–1723	4.69–6.45	17–21
		Haemopsis sanguisina	2.60–11.13	968–797	4.31–7.66	14–20
		Gammarus pulex	4.26	110	11.16	111
		Asellus aquaticus	3.18–12.51	201–267	10.81–15.59	99–136
		Ischnura elegans	1.66	163	15.34	29
		Sigara striata	0.73	303	2.68	47
		Cloeon dipterum	4.42	317	48.88	26
		Ilybius fuliginosus	0.83–5.43	162–426	5.48–50.68	22–32
		Glyptotendipes pallens	1.66	279	8.32	39
		Psectrotanypus varius	0.95	123	6.03	27
		Macropelopia sp.	4.59	205	6.92	26
		Prodiamesia sp.	5.25	128	16.52	19
		Chaoborus sp.	0.50	88	1.51	20

Table 1 (continued). Cadmium, Zinc, Lead, and Copper Concentrations in Freshwater Benthic Invertebrates

Site and Reference	Taxon	Cadmium	Zinc	Lead	Copper	Classification
Maarsseveen (Netherlands) (13)						
Lake I	Erpobdella octoculata	0.07–0.36	307–336	2.64–2.88	23–26	Unpolluted
	Tubifex sp.	0.02–0.24	103–267	1.7–13	3–23	
	Piscicola geometra	0.43–0.90	324–642	1.01–2.15	9–26	
	Glossiphonia compl.	0.17	553	1.6	16	
	Potamopyrgus jenkinsi	0.68–0.80	75–86	4.72–5.16	96–107	
	Dreissena polymorpha	0.08–16	106–113	1.07–1.46	11–15	
	Gammarus pulex	0.17–0.53	79–130	0.77–2.5	68–82	
	Asellus aquaticus	0.13–0.80	170–185	2.6–2.8	86–131	
	Hygrobates trigonicus	1.1–1.2	265–325	2.4–5.7	40–51	
	H. longipalpis	1.8	267	1.1	33	
	Mideopsis orbicularis	1.5	272	12	65	
	Ischnura elegans	0.12–0.46	103–107	2.2–4.0	24–25	
	Sialis lutaria	0.03	127	0.48	24	
	Limnephilus marmoratus	0.38–1.1	184–304	2.2–6.8	18–21	
	Gyrinus marinus	0.01	56	1.8	9	
	Stictochironomus histrio	0.10–0.16	102–118	2.7–5.4	22–32	
	C. anthracinus	0.64–1.56	98–135	3.1–10	8–32	
	Chironomus sp.	0.05	136	2.1	19	
	Tanytarsus sp.	0.05–3.2	87–123	13–27	42–101	
Lake II	Erpobdella octoculata	0.01–0.06	213–438	0.37–3.47	10–34	Unpolluted
	Tubifex sp.	0.03	222	2.7	7	
	Piscicola geometra	0.15–1.65	433–453	0.58–0.84	21–23	
	Glossiphonia complanata	0.36	229	3.3	15	
	Potamopyrgus jenkinsi	0.90–7.6	87–101	10.7–11.7	108–111	

Lymnaea ovata	0.10	78	0.79	24
Dreissena polymorpha	0.01–0.09	74–103	0.26–1.12	11–12
Cyclops sp.	0.21–0.47	119–154	4.4–34	29–160
Eurycercus sp.	0.29	103	13	28
Gammarus pulex	0.02–0.11	71–132	0.74–1.4	40–60
Asellus aquaticus	0.07–0.24	101–135	2.1–3.1	55–69
Hygrobates trigonicus	0.28–0.29	213–230	1.6–3.0	3–14
H. nigromaculatus	0.33	279	4.2	36
H. longipalpis	0.37–1.3	328–346	0.39–1.2	19–28
Hydrodroma despiciens	0.03–1.4	222–1202	2.7–6.3	11–184
Ischnura elegans	0.04–0.26	98–106	1.2–5.1	12–13
Micronecta minutissima	0.07–0.10	111–112	1.9–2.1	5–11
Sialis lutaria	0.10	125	0.29	28
Limnephilus marmoratus	0.14–0.20	200–402	3.7–8.3	11–19
Holocentropus picicornis	0.01–0.08	112–121	1.3–1.4	9–12
Mystacides sp.	0.17	94	2.6	20
Glyptotendipes pallens	0.04–0.06	90–111	0.44–2.5	15–16
Stictochironomus histrio	0.04–0.24	91–171	2.1–12	13–46
Chironomus muratenis	0.09–0.63	106–209	1.2–2.1	15–27
Chironomus sp.	0.43	189	4.2	11

Note: The concentrations are expressed in $\mu g\ g^{-1}$ dry weight (average or range of averages). N.A. = not analyzed.

croinvertebrates. Individuals with the lowest weight generally have the highest concentrations of these two metals. This is the case not only for single species, but also for groups of species, and was demonstrated in both polluted and unpolluted ecosystems.[12,18] This latter observation can be explained in several ways: small organisms have a relatively large surface area, which may result in an increased external adsorption of trace metals; or small organisms may eat smaller particles with relatively higher trace metal concentrations; or the metabolism of large organisms may be slower than that of small organisms.

C. Essential and Nonessential Trace Elements

Zinc and copper, two essential trace metals, are commonly assumed to be regulated in aquatic organisms.[19,20] Essential metals have many functions in animals. Zinc has been described as a co-factor in many enzymatic reactions, and copper occurs, for example, in hemocyanin. As demonstrated in Table 1, zinc and copper concentrations vary between narrow margins. Even at much higher ambient levels, zinc and copper concentrations in the organism increase only slightly. Crustaceans and molluscs clearly exhibit higher copper concentrations than other organisms and this can be related to their hemocyanin metabolism.[17,19] Under normal, unpolluted conditions cadmium and lead concentrations are low in aquatic invertebrates. Upon exposure to elevated concentrations, both metals are readily accumulated in these organisms. Aquatic invertebrates therefore provide appropriate monitor organisms of environmental cadmium and lead pollution. Cadmium and lead have no known biological function and are therefore generally classified as nonessential.

D. Uptake and Elimination Routes

Free metal ions are assumed to be the most available form of dissolved metals to aquatic organisms. By a process of facilitated diffusion, metals can enter the body. The importance of trace metal uptake via water has been stressed by several authors,[21-23] but trace metal uptake via food has been largely ignored.[15,24]

Some doubt can be raised as to whether the uptake of dissolved metal ions via water is really the most important route of trace metal absorption in freshwater invertebrates. Trace metal concentrations in water are often extremely low — below limits of detection. How can these low concentrations account for significant trace metal accumulations in invertebrates? Can the relatively higher trace metal concentrations in food be of more importance? Very few field studies have focused on trace metal concentrations in invertebrate predators and their food organisms.[12,18] Laboratory experiments in which invertebrate predators are fed with (contaminated) prey are even more scarce.[25,26] An explanation for this lack of scientific interest is not obvious, but one can think of the complexity of the experimental design, in which both consumer and food have to be kept alive.

In experimental work with water mites (*Limnesia maculata*) and caddisfly larvae (*Mystacides* spp.), which were exposed via water or food (chironomid larvae) to zinc or cadmium, clear indications were found that trace metals can be accumulated via food.[26] As shown in Figure 1, both water mites and caddisfly larvae accumulate zinc from their food (ratio exposed:control organisms >1.0), but no biomagnification *sensu stricto* is observed (ratio exposed organisms:food organisms <1.0). The same observations were made for cadmium, but here the ratio exposed:control organisms was much higher. With a first-order one-compartment model it was calculated that cadmium was mainly accumulated via food, whereas zinc was mainly accumulated via water. These experimental results led the authors to the conclusion that in addition to uptake from water, aquatic invertebrate predators can accumulate trace metals from their food. In this respect, some doubt can be raised on the strict definition of biomagnification. A significant increase in trace metal body concentrations in predators, without exceeding concentrations in the prey, cannot be classified as biomagnification, but can form a serious threat to predators.

As with uptake routes, trace metal elimination routes in freshwater invertebrates are generally poorly documented. A passive process of elimination has been described for insects with stages of aquatic larvae.[27,28] It is most likely that trace metal elimination can take place during molt or metamorphosis, when larval and pupal skins are shedded and externally and internally (gut epithelium) adsorbed metals are removed.

The reader should bear in mind that the possible explanations for differences in trace metal concentrations of freshwater invertebrates in this section on accumulation, are mainly theoretically derived. The experimental proof of uptake and elimination strategies is often lacking. How exactly essential metals are regulated in freshwater invertebrates and why nonessential metals obviously lack such regulation is poorly understood. The only way to find answers to these open questions is to perform laboratory studies. In these studies trace metal uptake and elimination can be modeled, the importance of the different uptake routes can be determined, and the physiological role of trace metals in organisms can be assessed. Only then can the significance of differences in trace metal concentrations in freshwater invertebrates be fully understood.

E. Bioavailability

As a final topic on accumulation of trace metals in freshwater invertebrates, bioavailability should be considered. Bioavailable metals are defined according to Campbell et al.[29] as metals in such a biologically available chemical state that they can be taken up by an organism and can react with its metabolic machinery. A simple relationship between external trace metal concentrations in sediment or water and internal concentrations in organisms is often absent.[2,4,29] High trace metal concentrations in sediment can be quite harmless for organisms, whereas low concentrations can be very available and result in a considerable accumulation.

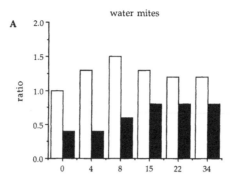

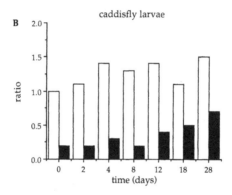

□ ratio zinc in exposed individuals : control individuals
■ ratio zinc in food : exposed individuals

FIGURE 1.

A. Water mites (*Limnesia maculata*).

Open bars: ratio of zinc concentrations in water mites feeding on contaminated chironomid larvae (1152 ± 52 μg Zn g⁻¹) and water mites feeding on control chironomid larvae (320 ± 9 μg Zn g⁻¹) over a period of 34 days. **Shaded bars:** ratio of zinc concentrations in contaminated food (chironomid larvae, 1152 ± 52 μg Zn g⁻¹) and water mites feeding on contaminated food over a period of 34 days. Ratio >1.0 indicates biomagnification, ration <1.0 indicates no biomagnification.

B. Caddisfly larvae *Mystacides* spp.

Open bars: ratio of zinc concentrations in caddisfly larvae feeding on contaminated chironomid larvae (778 ± 32 μg Zn g⁻¹) and caddisfly larvae feeding on control chironomid larvae (320 ± 32 μg Zn g⁻¹). **Shaded bars:** ratio of zinc concentrations in contaminated food (chironomid larvae, 778 ± 32 μg Zn g⁻¹) and caddisfly larvae feeding on contaminated food over a period of 28 days. Ratio >1.0 indicates biomagnification, ratio <1.0 indicates no biomagnification.

Calculations are based on metal concentrations per dry weight. Recalculated from Timmermans, K.R., Spijkerman, E., Tonkes, M., and Govers, H., *Can. J. Fish. Aquat. Sci.,* 1992. With permission.

The binding capacity of a sediment determines its trace metal availability. Sediments with high percentages of smaller grain size fractions (e.g., clay or silt), with high percentages of iron and manganese which can scavenge trace metals, or sediments with a high organic carbon content generally have a low availability of trace metals, irrespective of their total load. A widely used, although not completely accepted, method for determination of bioavailable trace metals, is the use of sequential sediment extractions.[30] Using different extractants, exchangeable, reducible, and organically bound metals are operationally determined.[2,4,29] In combination with normalization of sediment parameters (e.g., iron content and grain size) moderate to good predictions have been made of trace metal concentrations to be expected in organisms.[4,31] In most of these studies, precautions were taken to limit the survey to one or a few species, and to work in an environment with a limited number of abiotic variables. It was demonstrated that when surveys were performed in more complex ecosystems, the predictive power of only sediment- and water-related parameters was limited. A combination of abiotic and biotic parameters gave better predictions, although the total percentage of variance explained remained relatively low.[12]

III. EFFECTS

A. Lethal and Sublethal Effects

Acute lethal effects of aqueous trace metals in freshwater invertebrates are relatively well documented, especially for cadmium.[32,33]

An overview of the available data demonstrates that cadmium and copper are the most toxic metals for freshwater invertebrates, whereas zinc is hardly toxic. However, some caution is necessary, as generally relatively insensitive species (only they can survive laboratory conditions) and mostly adults have been used in toxicity tests. It has frequently been demonstrated that differences of orders of magnitude exist between the sensitivities of different life history stages, younger instars being more sensitive than older ones.[34-37]

However, the wealth of LC_{50} data cannot hide the fact that death is a rather crude index of stress, being of limited relevance in ecotoxicology. Most of the data available can merely serve as an illustration of toxicity of the trace metals studied. More information on sublethal effects is therefore required. An example for this is given by studies of Heinis et al.,[38] who determined the effects of cadmium on the behavior of filter-feeding chironomid larvae. Cadmium concentrations resulting in effects on behavior in these organisms were orders of magnitude lower than concentrations causing mortality.

B. Long-Term Toxicity Effects

In contrast to acute, lethal effects, long-term (sublethal) effects are poorly documented. These kinds of long-term-effect studies can provide very interesting

Table 2. Zinc Exposure of *Chironomus riparius*

| | No. of days stages were present[a] | | |
Larval stage	Control	0.10 mg Zn l⁻¹	1.0 mg Zn l⁻¹
First	2	5	10
Second	3	9	13
Third	33	64	65
Fourth	65	57	7

[a] Total length of this experimental period was 72 days.

From Timmermans, K.R., Peeters, W., and Tonkes, M., *Hydrobiologia*, in press.

insight in for example, delay in development and emergence of aquatic insects. Several authors have demonstrated that long-term exposure to sublethal concentrations can cause a significant delay in the growth, larval development, and reproduction of freshwater invertebrates.[36,37,39-41] As an illustration of this, Table 2 shows effects of long-term exposure to zinc on the life cycle of *Chironomus riparius*. Upon exposure, the time during which the first three instar stages are present increases, whereas the fourth stage is present during a significantly shorter period. In addition, adult emergence is retarded upon zinc exposure (data not shown). Only in a modest number of cases could the observations be related to elevated body concentrations or morphological changes in the organisms.[37,42,43]

IV. ECOTOXICOLOGICAL CONSEQUENCES

Acute effects of trace metal pollution on freshwater invertebrates are (fortunately) only seldom encountered. It is long-term exposure that is the most realistic threat for organisms. Long-term exposure has much less spectacular consequences than acute mortality, and is therefore likely to be ignored. Often only the result, in the form of an "a-biotic" environment is encountered.

Without any doubt, trace metal accumulation has its cost for invertebrates. It can result in change or retardation of life cycles or predator-prey interactions can be altered. Knowledge of these parameters for undisturbed situations is scarce, and hence the establishment of a causal relationship between trace metal pollution and its effects on organisms will be almost impossible to do. However, available data make clear that even small changes can have a profound influence on freshwater invertebrate communities. In order to assess the actual extent of the effects of trace metal pollution more studies in ecology, toxicology, and physiology are required.

V. EPILOGUE

In addition to a (limited) overview of available data on trace metal concentrations in freshwater invertebrates, the author has tried to give some background information on how these data might be explained, and what the possible consequences of metal pollution on aquatic invertebrates might be. It is a long step from field observations and acute toxicity experiments to long-term experiments determining the importance of metal uptake via water or food in predators, or relating metal uptake to effects of metal exposure. Many efforts in this field have been made successfully, but much research remains to be done. In particular, more attention should be paid to laboratory experiments elucidating uptake and elimination strategies and metal kinetics in different invertebrate species. So far, efforts have been made to document the ultimate results of trace metal accumulation in aquatic ecosystems. Now the time has come to look at the (physiological) processes governing accumulation, and thereby to open the black box of trace metal uptake and elimination in freshwater invertebrates. In combination with ecology this can lead to an assessment of the threats imposed by toxic metals to freshwater invertebrates, and thus of the conditions required to preserve the structure and function of freshwater ecosystems.

ACKNOWLEDGMENTS

This paper summarizes a four-year Ph.D. project on trace metal uptake and effects in chironomids and other benthic macroinvertebrates. During this time many people have contributed to the research in their own unique way. I am particularly indebted to Cees Davids, Bert van Hattum, Michiel Kraak, Harrie Govers, Wilma Peeters, Paddy Walker, Elly Spijkerman, and Marcel Tonkes. I wish to thank the two anonymous reviewers for their valuable comments on the manuscript.

REFERENCES

1. **Salomons, W. and Förstner, U.,** *Metals in the Hydrocycle,* Springer-Verlag, New York, 1984.
2. **Luoma, S. N.,** Can we determine the biological availability of sediment-bound trace elements?, *Hydrobiologia,* 176/177, 379, 1989.
3. **Tessier, A., Campbell, P. G. C., Auclair, J. C., and Bisson, M.,** Relationships between the partitioning of trace metals in sediments and their accumulation in the tissues of the freshwater mollusk *Elliptio complanata* in a mining area, *Can. J. Fish. Aquat. Sci.,* 41, 1463, 1984.

4. **Tessier, A. and Campbell, P. G. C.,** Partitioning of trace metals in sediments: relationships with bioavailability, *Hydrobiologia,* 149, 43, 1987.

5. **Campbell, P. G. C. and Tessier, A.,** Biological availability of metals in sediments: analytical approaches, *Proc. Int. Conf. Heavy Metals in the Environment,* Vernet, J. P., Ed., CEP Consultants, Edinburgh, 516, 1989.

6. **Eyres, J. P. and Pugh-Thomas, M.,** Heavy metal pollution of the river Irwell (Lancashire, U.K.) demonstrated by analysis of substrate material and macroinvertebrate tissue, *Environ. Pollut.,* 16, 129, 1978.

7. **Dixit, S. S. and Witcomb, D.,** Heavy metal burden in water, substrate and macroinvertebrate body tissue of a polluted River Irwell (England), *Environ. Pollut., Ser. B,* 6, 161, 1983.

8. **Barak, N. A.-E. and Mason, C. F.,** Heavy metals in water, sediment and invertebrates from rivers in eastern England, *Chemosphere,* 10/11, 1709, 1989.

9. **Brown, B. E.,** Effects of mine drainage on the river Hayle, Cornwall. A) factors affecting concentrations of copper, zinc and iron in water, sediments and dominant invertebrate fauna, *Hydrobiologia,* 52, (2-3), 221, 1977.

10. **Anderson, R. V., Vinikour, W. S., and Brower, J. E.,** The distribution of Cd, Cu, Pb, and Zn in the biota of two freshwater sites with different trace metal inputs, *Holartic Ecol.,* 1, 377, 1978.

11. **Burrows, I. G. and Whitton, B. A.,** Heavy metals in water, sediment and invertebrates from a metal-contaminated river free of organic pollution, *Hydrobiologia,* 106, 263, 1983.

12. **Van Hattum, B., Timmermans, K. R., and Govers, H.,** Abiotic and biotic factors influencing *in situ* trace metal levels in macro-invertebrates in freshwater ecosystems, *Environ. Toxicol. Chem.,* 10, 275, 1991.

13. **Timmermans, K. R., Van Hattum, B., Peeters, W., and Davids, C.,** Trace metals in the benthic habitat of the Maarsseveen Lakes System, The Netherlands, *Hydrobiol. Bull.,* 24(2), 153, 1991.

14. **Moriarty, F.,** *Ecotoxicology — The Study of Pollutants in Ecosystems,* Academic Press, London, 1983.

15. **Dallinger, R., Prosi, F., Segner, H, and Back, H.,** Contaminated food and uptake of heavy metals by fish: a review and a proposal for further research, *Oecologia (Berlin),* 73, 91, 1987.

16. **Hopkin, S. P. and Martin, M. H.,** The distribution of zinc, cadmium, lead and copper within the woodlouse *Oniscus asellus* (Crustacea, Isopoda), *Oecologia (Berlin),* 54, 227, 1982.

17. **Rainbow, P. S.,** The significance of trace metal concentrations in decapods, *Symp. Zool. Soc. London,* 59, 291, 1988.

18. **Timmermans, K. R., Van Hattum, B., Kraak, M. H. S., and Davids, C.,** Trace metals in a littoral foodweb: concentrations in organisms, sediment and water, *Sci. Total Environ.,* 87/88, 477, 1989.

19. **Phillips, D. J. H.,** *Quantitative Aquatic Biological Monitors — Their Use to Monitor Trace Metal and Organochlorine Pollution,* Applied Science Publishers, London, 1980.

20. **Luoma, S. N.,** Bioavailability of trace metals to aquatic organisms — a review, *Sci. Total Environ.,* 28, 1, 1983.

21. **Williams, D. R. and Giesy, J. P., Jr.,** Relative importance of food and water sources to cadmium uptake by *Gambusia affinis* (Poeciliidae), *Environ. Res.,* 16, 326, 1978.

22. **Taylor, D.,** The significance of the accumulation of cadmium by aquatic organisms, *Ecotoxicol. Environ. Saf.,* 7, 33, 1983.
23. **Van Hattum, B., Voogt, P. de, Bosch, L. van den, van Straalen, N. M., and Govers, H.,** Bioaccumulation of cadmium by the freshwater isopod *Asellus aquaticus* (L.) from aqueous and dietary sources, *Environ. Pollut.,* 62, 129, 1989.
24. **Leland, H. V. and Kuwabara, J. S.,** Trace metals, *Fundamentals of Aquatic Toxicology,* Rand, G. M. and Petrocelli, S. R., Eds., Hemisphere, Washington, D.C., 1985.
25. **Giesy, J. P., Bowling, J. W., and Kania, H. J.,** Cadmium and zinc accumulation and elimination by freshwater crayfish, *Arch. Environ. Contam. Toxicol.,* 9, 683, 1980.
26. **Timmermans, K. R., Spijkerman, E., Tonkes, M., and Govers, H.,** Cadmium and zinc uptake by two species of aquatic invertebrate predators from dietary or aqueous sources, *Can. J. Fish. Aquat. Sci.,* 49(4), 655, 1992.
27. **Krantzberg, G. and Stokes, P. M.,** The importance of surface adsorption and pH in metal accumulation by chironomids, *Environ. Toxicol. Chem.,* 7, 653, 1988.
28. **Timmermans, K. R. and Walker, P. A.,** The fate of trace metals during the metamorphosis of chironomids (Diptera, Chironomidae), *Environ. Pollut.,* 62, 73, 1989.
29. **Campbell, P. G. C., Lewis, A. G., Chapman, P. M., Fletcher, W. K., Imber, B. E., Luoma, S. N., Stokes, P. M., and Winfrey, M.,** Biologically Availabile Metals in Sediments, Publ. No. 27694, National Research Council of Canada, Ottawa, 1988.
30. **De Groot, A. J., Zschuppe, K. H., and Salomons, W.,** Standardization of methods of analysis for heavy metals in sediment, *Hydrobiologia,* 92, 689, 1982.
31. **Luoma, S. N. and Bryan, G. W.,** Factors controlling the availability of sediment-bound lead to the estuarine bivalve *Scrobicularia plana, J. Mar. Biol. Assoc. U.K.,* 58, 793, 1978.
32. **Williams, K. A., Green, D. W., and Pascoe, D.,** Studies on the acute toxicity of pollutants to freshwater macro-invertebrates. 1. Cadmium, *Arch. Hydrobiol.,* 102, 461, 1985.
33. **Brown, A. F. and Pascoe, D.,** Studies on the acute toxicity of pollutants to freshwater macro-invertebrates. 5. The acute toxicity of cadmium to twelve species of predatory macro-invertebrates, *Arch. Hydrobiol.,* 114(2), 311, 1988.
34. **Williams, K. A., Green, D. W., Pascoe, D., and Gower, D. E.,** The acute toxicity of cadmium to different larval stages of *Chironomus riparius* (Diptera:Chironomidae) and its ecological significance for pollution regulation, *Oecologia (Berlin),* 70, 362, 1986.
35. **Gauss, J. D., Woods, P. E., Winner, R. W., and Skillings, J. H.,** Acute toxicity of copper to three life stages of *Chironomus tentans* as affected by water hardness-alkalinity, *Environ. Pollut. Ser. A,* 37, 149, 1985.
36. **Hatakeyama, S.,** Chronic effect of Cu on reproduction of *Polypedilum nubifer* (Chironomidae) through water and food, *Ecotoxicol. Environ. Saf.,* 16, 1, 1988.
37. **Timmermans, K. R., Peeters, W., and Tonkes, M.,** Cadmium, zinc, lead and copper in larvae of *Chironomus riparius* (Meigen) (Diptera: Chironomidae): uptake and effects, *Hydrobiologia,* 241, 119, 1992.
38. **Heinis, F., Timmermans, K. R., and Swain, W. R.,** Short-term sublethal effects of cadmium on the filterfeeding chironomid larva *Glyptotendipes pallens* (Meigen Diptera), *Aquat. Toxicol.,* 16, 73, 1989.

39. **De Nicola Guidici, M., Migliore, L., Gambardella, C., and Marotta, A.,** Effects of chronic exposure to cadmium and copper in *Asellus aquaticus* (L.) (Crustacea, Isopoda), *Hydrobiologia,* 157, 265, 1988.

40. **Kosalwat, P. and Knight, A. W.,** Chronic toxicity of copper to partial life cycle of the midge *Chironomus decorus, Arch. Environ. Contam. Toxicol.,* 16, 283, 1987.

41. **Pascoe, D., Williams, K. A., and Green, D. W. J.,** Chronic toxicity of cadmium to *Chironomus riparius* (Meigen) — effects upon larval development and adult emergence, *Hydrobiologia,* 175, 109, 1989.

42. **Seidman, L., Bergtrom, G., and Remsen, C. C.,** Structure of the larval midgut of the fly *Chironomus thummi* and it relationship to sites of cadmium sequestration, *Tissue & Cell,* 18, 40, 1986.

43. **Janssens de Bisthoven, L. and Timmermans, K. R.,** The concentration of cadmium, lead, copper and zinc in *Chironomus* gr. *thummi* larvae (Diptera, Chironomidae) with deformed versus normal menta, *Hydrobiologia,* in press.

CHAPTER 9

Investigation of Uranium-Induced Toxicity in Freshwater Hydra

Ross Hyne, Greg D. Rippon, and Graham Ellender

TABLE OF CONTENTS

0-87371-734-1/93/$0.00 + $.50

I. INTRODUCTION

The Office of the Supervising Scientist was established to ensure that the environment in the Alligator Rivers Region was protected from the effects of uranium mining operations. Mining for uranium in the region has so far been restricted to two sites. One of the mines, Ranger Uranium Mine, is situated on a leased site surrounded by Kakadu National Park in tropical northern Australia (Figure 1). The Park is considered to be an important conservation area and has been entered onto the World Heritage List and the Convention of Wetlands of International Importance.[1] The Park is fully contained within a larger area of conservation significance known as the Alligator Rivers Region (ARR) (Figure 1). Uranium concentrations in the surface waters of streams and rivers within Kakadu National Park are typically less than 1 μg/l.[2] Since uranium is nonessential for biological processes and is generally toxic at elevated concentrations,[3,4] concerns for the protection of aquatic biota exposed to contaminated waters must be considered.

Like many other metals, uranium is taken up by various aquatic organisms, particularly by fungi, algae, bacteria, and invertebrates.[5-8] Uranium has a high affinity for proteins and lipids at the pH found intracellulary, thus preventing desorption. Concentration factors for algae in the range 10^2 to 10^3 based on wet weight have been reported.[9,10] Uranium can be taken up and accumulated in fungi, algae, and bacteria to relatively high concentrations without apparent toxic consequences. A wide spectrum of results has been obtained in studies concerned with the site of uptake of uranium in microorganisms. Strandberg et al.[5] demonstrated this variability in studies on a species of yeast and of bacteria. In the yeast, uranium accumulated extracellularly on the surface, while in the bacteria there was a rapid and massive intracellular accumulation of uranium. Further, Galan et al.[8] showed that in fungi, uranium can accumulate to high levels through both intracellular and extracellular mechanisms, but that this depended on ex-

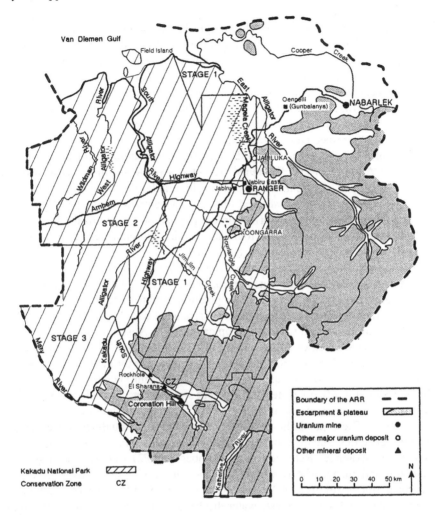

FIGURE 1. Map of the Alligator Rivers Region (ARR) showing Ranger Uranium Mine and the East Alligator River system.

perimental conditions. Intracellular accumulation of uranium has also been documented in the lysosomes, macrophages, gill epithelia, hindgut epithelia, and hepatopancreas cells of molluscs and crustaceans exposed to uranium-contaminated sea water or food.[11,12]

When the capacity of a living organism to detoxify uranium (or any metal which is in excess of the organism's requirements, if any) is exceeded, toxicological effects will begin. There will, therefore, exist within the organism a fine balance between detoxification processes, such as the formation of uranium phosphate microgranules as seen in a crab,[11] and the point at which uranium interferes with life processes, for instance when it binds to macromolecules such as proteins, thereby resulting in physiological or behavioral dysfunctions. One enzyme that uranium has been shown to inhibit is ATPase,[13,14] which is located on the plasma membrane of cells and is involved in osmoregulation and other active transport processes. As for the toxicity of uranium to aquatic organisms, there has been very little information published. Tarzwell and Henderson,[15] using fathead minnows, reported 96-h LC_{50}s for uranyl ion, with uranyl nitrate or acetate as the source in soft water (pH 7.4, hardness 20 mg/l as $CaCO_3$), as 3.1 and 3.7 mg/l, respectively. Poston et al.[16] obtained significant reduction in survival of *Daphnia magna* after 5 days exposure to a uranium concentration of 520 μg/l. Ahsanullah and Williams[17] reported a reduction in the growth of a marine amphipod, *Allorchestes compressa,* by uranium at 2 mg/l after 4 weeks.

In a previous study we showed that low concentrations of uranium (≥ 150 μg/l) have a direct inhibitory effect on the population growth of the freshwater hydra, *Hydra viridissima,* after four days.[18] This inhibitory effect of uranium was shown to be pH-dependent in the presence of bicarbonate and was not due to associated changes in the conductivity of the test water. Thus, at pH 8.5, uranium, even at 900 μg/l but in bicarbonate buffer, was found to have no significant effect on *H. viridissima* population growth over 4 days. However, TRIS-buffered water with added uranium remained toxic at pH 8.5.

There are two broad seasons in tropical northern Australia; a Wet and a Dry season.[19] The Wet season of six months or less (from about November to April) typically has local convection storms at first, then mostly monsoonal rains. In this period, most of the annual rainfall (about 1500 mm) occurs. As part of the Ranger Uranium Mine's water management strategy for this high seasonal rainfall, there is a series of retention ponds which act as sediment traps and runoff catchments. These ponds have differing chemical water qualities, differing retention capacities, and differing regulatory criteria governing release of contained water. For example, Retention Pond Number 2 (RP2) water has a very high uranium and manganese content,[20] and its release into the Magela Creek (a seasonal tributary of the East Alligator River) is only permitted in an exceptionally high rainfall Wet season. Waste water discharged from Ranger would flow into the river systems of Kakadu National Park (Figure 1) and therefore any release needs to be carefully controlled to avoid environmental damage in an important conservation area. This laboratory has developed a number of biological toxicity tests which can be used to establish dilution rates that will ensure a very high level of environmental protection.[21-24] One of the toxicity tests, using hydra asexual population growth as an end point, was found to be particularly sensitive for evaluating waste water containing uranium.

In order to investigate how uranium affects hydra, this study examined the ultrastructure of *H. viridissima* exposed to various water treatments containing uranium. Experimental waters were obtained from RP2 at Ranger or by addition of uranium to the control water obtained from Magela Creek. The structure of hydra was then studied by histology (light microscopy), transmission electron microscopy (TEM), and scanning electron microscopy (SEM). The elemental distribution of uranium was examined by energy dispersive X-ray microanalysis (EDAX) in the transmission mode.

II. MATERIALS AND METHODS

A. Control and Test Waters

The control water used in most experiments was Magela Creek water[2] and during the Wet season was collected near Georgetown Billabong,[25] upstream of the Ranger Uranium Mine waste water discharge pipe outlet. The pH of the Magela Creek water during the experimental period ranged from 6.1 to 6.7, and the conductivity varied from 12 to 20 μS/cm. For Dry season experiments the control water was collected from Buffalo Billabong, a permanent water-body seasonally connected with Magela Creek. Buffalo Billabong water had a pH range of 6.0 to 6.4 and conductivity varied from 25 to 33 μS/cm. All control waters were collected in clean, acid-washed plastic bottles on the morning of the day during which the test was to commence. Water was passed through a coarse filter (Whatman No. 1 filter paper) to remove zooplankton.

Test waters were either RP2 water (containing 2900 to 3900 μg/l of uranium, pH 7.5 to 8.0) or uranium added as uranyl sulfate to control water. RP2 water was collected in a clean, acid-washed plastic bottle from the mine site as close as practicable to the start of the test. The water was filtered through a coarse filter. A stock solution of uranyl sulfate was made by adding 40 mg of $UO_2SO_4.3H_2O$ (Ajax Chemicals, Sydney) to 2 liters of the control water. A dilution of this stock solution was made with filtered control water to give the desired concentration of added uranium. All waters were stored at 4°C until needed. The concentrations of uranium in RP2 water or uranium added to the test waters were measured by Scintrex Time Delay Fluorimetry.

B. Hydra Toxicity Test

The test species was *Hydra viridissima* (Cnidaria, Hydrozoa), commonly known as "green hydra" because of the presence of symbiotic algae in the gastrodermal cells, giving the whole animal a green coloration. The test animals were obtained from laboratory stocks reared in the same water that would be used as control water in the toxicity test. All animals selected were mature and asexually reproducing by budding with a tentacled bud. In this paper, 'hydroid' will be used to describe an individual animal with or without buds. Five hydroids were covered with 30 ml of test solution in acid-washed plastic petri dishes of 90 mm diameter. Each treatment was replicated three times. The treatments were 100% RP2 water, water containing uranium at a concentration of 200 μg/l or 350 μg/l, and control water.

The test solution was changed each day for a fresh test solution of the same concentration and temperature. Live brine shrimp nauplii (*Artemia salina*), hatched from commercially supplied cysts (San Francisco Brine Shrimp, Inc.) and suspended in the appropriate test water, were individually fed to each test hydroid using a glass Pasteur pipette. Hydra were allowed to feed for 30 min after which the petri dishes were cleaned and test solutions changed.[18] Daily observations were made of the degree of tentacle contraction (not contracted, partially contracted, or fully contracted) and changes in the number of hydroids in each treatment. The conductivity and pH of each test solution were also recorded daily. Test animals were fed daily and maintained at 30°C with a 12-h photoperiod. Hydra were not fed on the morning of the last day of the experiment, but just counted and then pooled. Pooled animals were randomly selected and transferred to either of three 50 mm glass petri dishes with 5 ml of treatment water. They were then fixed for either TEM, SEM, or for EDAX.

C. Transmission Electron Microscopy

TEM was undertaken to study the cellular and matrix ultrastructure of the control hydra and hydra exposed to uranium.

Hydra were fixed at room temperature in 0.75% glutaraldehyde and 0.75% paraformaldehyde in 0.05 M phosphate buffer at pH 7.4 for 30 min by adding a double-strength solution of fixative to an equal volume of hydra test solution. They were stored at 4°C until further processing (within 1 week). They were then washed in three changes of 0.05 M phosphate buffer, each of 15 min duration. For the morphological examination, the hydra were postfixed in 1% OsO_4, stained *en bloc* in 0.5% uranyl acetate in 70% acetone, dehydrated through graded acetones, infiltrated with Spurr's resin,[26] and polymerized overnight at 70°C.

Survey sections of 1 μm were stained with toluidine blue and mesas defined. Ultrathin sections were cut on an Ultracut-E microtome (Reichert-Jung, Austria) at 80 nm and mounted on uncoated copper grids. Sections were poststained with 3% uranyl acetate in 70% ethanol and Reynold's lead citrate,[27] each for 20 min. The sections were examined using a Philips EM300 at 60 kV.

D. Energy-Dispersive X-Ray Microanalysis

EDAX (in transmission mode) was used to obtain information about cellular and subcellular elemental distribution, with particular emphasis on the distribution of uranium.

Hydra intended for elemental analysis were fixed in 1% glutaraldehyde in 0.1 M phosphate buffer at pH 7.4, by adding a double-strength solution of fixative to an equal volume of hydra test solution, and then stored at 4°C. They were subsequently processed through an acetone series to Spurr's resin as previously described for TEM, but with the omission of postfixation with OsO_4 and *en bloc* staining with uranyl acetate. A range of thin sections (80 to 200 nm) was cut using a diamond knife and collected mostly on beryllium grids.

These sections were not poststained. Sections in the range of 180 to 200 nm were found to be the most informative.

Areas of interest were examined in a Philips EM300 (Philips Industries, Eindhoven, Holland) fitted with a goniometer and minilens. X-rays generated from the specimen were collected in an Edax 9100 detector (EDAX, Prairie View, U.S.). Representative spectra were plotted for presentation. Microscope operating conditions were maintained constant throughout, with an accelerating voltage of 60 kV, probe diameter approaching 1 μm, a count live time of 200 sec, and a goniometer angle of 35°. To minimize extraneous sources of background readings a specimen holder with a beryllium ring insert was used. Raw data were stored for future analysis.

E. Histology (Light Microscopy)

Histology was undertaken to initially review the overall structure of hydra. Hydra were fixed and transported in Bouin's fluid. They were processed into paraffin through graded alcohols and cleared in xylol. Sections were cut at 4 μm and stained with hematoxylin and eosin, Masson's trichrome stain,[28] picrosirius red,[29] or Miller's elastic stain.[30]

F. Scanning Electron Microscopy

SEM was undertaken to examine the surface features of hydra with particular reference to the tentacles.

Hydra were fixed, stored, and washed in phosphate buffer as for TEM. The hydra were then dehydrated through graded acetones and into 1:1 absolute acetone-camphene for 30 min. The hydra were infiltrated with molten absolute camphene (Merck, Darmstadt, Germany) (44°C) for 30 min[31] and then transferred carefully in molten camphene, with a warmed pipette, onto a specimen stub. Excess camphene was drained off and the specimens were allowed to solidify at room temperature. The stubs were placed under a moderate vacuum (~600 mm Hg) overnight to sublime the remaining camphene from the hydra surface, thus critically point drying the hydra and minimizing the morphologically disruptive liquid-vapor interface during drying. The majority of specimens were sputter coated with gold in a Polaron SEM Coating Unit PS3 (Polaron, Watford, U.K.) and examined in a Siemens Autoscan SEM (Siemens, Germany) in secondary electron mode. Remaining hydra were left uncoated for the detection of surface-bound elements with high atomic numbers in backscattered mode. Areas of interest were recorded on Kodak Graphic Arts film (Eastman Kodak, Rochester, U.S.A.).

G. Statistical Analysis

The population growth of hydra is expected to be exponential under the given experimental conditions. Consequently, the logarithm of the number of animals should be linear in time with random, but correlated, deviations over the times of observation. The final population size at the end of the test period is used to analyze the relative growth rates (K) for each concentration group, where K is defined as:[32]

$$K = \frac{\ln (n_3) - \ln (n_0)}{T \ (= 3 \text{ days})}$$

where ln (n_0) and ln (n_3) refer to the natural logarithms of the number of hydroids at the start and end of the 3-day test periods, respectively. A convenient method of checking the assumptions for analysis is to compute 'errors' from the three replicates.[21] These are chosen to be uncorrelated and may, consequently, be tested for normality of the distribution. These errors are also used to calculate the variance required for statistical testing and for the presence of outlier observations. If a single outlier was found and it was attributable to a single replicate observation, that observation was deleted and the variance recalculated. If, however, more than one outlier occurred or if there was other evidence of non-normality as judged from the plotted error values, then the analysis was abandoned.

The replicate data for each treatment were then pooled to give a mean population growth rate for each treatment. A modified Dunnett's procedure was then used to compare each treatment with a common control.[21]

III. RESULTS

A. Histology

Light microscopy of hydra tissue clearly demonstrated its diploblastic character with gastrodermis and epidermis separated by mesoglea. Fine connective tissue septa, which became more numerous and complex towards the peduncle, originated in the mesoglea and branched out into the gastrodermis. The septa contrasted clearly with the cellular components with Masson's trichrome (aniline blue for collagen), picro-sirius red (PSR) (red for collagen, reticulin, and basement membrane) and Miller's elastin stain (blue/black). The nematocyst capsule stained blue with Masson's stain. This structure, as well as the symbiotic algae in the gastrodermis, were better visualized by the greater resolution of the transmission electron microscopy (Figure 2). At the apex of the epitheliomuscular cells there was usually a large vacuole, along with others at varying depths, and an external coating of secretory product. Cnidoblasts and interstitial cells were found between the epitheliomuscular cells.

B. Toxicological Response of Hydra to Uranium

H. viridissima were exposed to uranium (200 μg/l) added to water collected from Magela Creek or to 100% RP2 water from Ranger uranium mine through

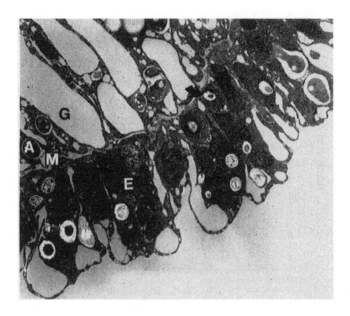

FIGURE 2. Section of *Hydra viridissima* through the body showing the characteristic layers — epidermis [E] and the gastrodermis [G] separated by the mesoglea [M]. Numerous developing nematocysts [arrow] lie adjacent to the mesoglea in the epidermis. Algae [A] are contained within cells of the gastrodermis. Connective tissue septa are seen (*). Stained with uranyl acetate and lead citrate. (Magnification × 2400, TEM).

the 1989–90 Wet season. Significant inhibition in the population growth was obtained with uranium (200 µg/l) or RP2 water, relative to the Magela Creek control on day 3 (Figure 3). Conductivity and pH were not significantly affected by the addition of uranium to Magela Creek water. Hydra exposed to the uranium treatment for greater than 24 h were observed to have a reduced ability to capture the live brine shrimp, unless the brine shrimp were placed adjacent to the hypostome (mouth).

C. Subcellular Distribution of Uranium

In initial experiments, hydra processed for routine TEM examination were treated with 1% OsO_4 postfixation and uranyl acetate *en bloc* staining. These steps were abandoned in subsequent experiments as the OsO_4 removed the vast majority of uranium crystals. The uranyl acetate *en bloc* step caused the precipitation of uranium crystals along the gastrodermal side of the mesoglea and also along the connective tissue dividing the groups of gastrodermal cells (Figure 4). Energy dispersive X-ray analysis of the background, sampled from the section immediately exterior to the tissue, showed peaks of copper and chromium arising

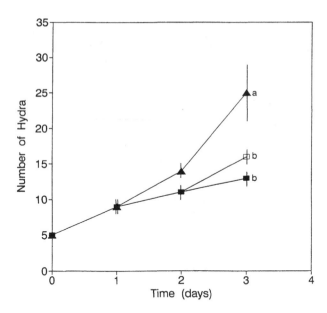

FIGURE 3. *Hydra viridissima* population growth over 3 days when exposed to 200 μg/l uranium added to Magela Creek water (■), RP2 water (□) or Magela Creek water control (▲). Each point is the mean ± SEM of three replicates. Points with a common alphabetical superscript (within each daily measurement only) are not significantly different ($p < 0.05$) from the control.

from the instrument (Figure 5). Similarly, the tissue exterior to the region containing the crystals showed no uranium peaks (Figure 6). The crystals bearing uranium seen after *en bloc* staining were not found in that location in any of the tissue processed for EDAX (stained neither with OsO_4 nor uranyl acetate).

The test hydra exposed to either 200 μg/l or 350 μg/l uranium, or RP2 water (containing 2900 to 3900 μg/l uranium), and processed for microanalysis showed electron dense accumulations with a flake-like or crystal-like appearance (Figure 7); control hydra showed no such accumulations. The electron dense accumulations were subsequently identified, using EDAX, as containing uranium with or without a small quantity of phosphorus (Figures 6 and 8). These discrete accumulations of often tightly packed crystals of uranium were located at all times within epidermal tissue in rounded groups and coexisted with a less electron dense amorphous material. Crystal lattice patterns could not be elicited from electron diffraction. The location of these uranium crystals coincided with the normal placement of nematocysts. Each dense accumulation of crystals was consistent in appearance whether the hydra were treated with uranium added to control water or to RP2 water, although more accumulations were obvious in the RP2-treated hydra. When few crystals were observed, they were situated either at the periphery along the constraining membrane, or dotted around in-

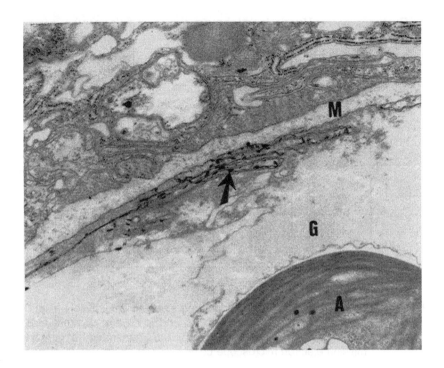

FIGURE 4. Crystals (arrows) were seen along the gastrodermal side (G) of the mesoglea (M) only after uranyl acetate *en bloc* treatment. No post-staining with either uranyl acetate or lead citrate. (Magnification × 14,300, TEM.)

dividual accumulations of amorphous material (Figure 9). The uranium crystals seen did not appear to have any connection with the external surface. Developing nematocysts of various types and free of crystals were seen in adjacent and similar regions in all hydra examined (Figure 10). The ultrastructure of the hydra treated with uranium appeared to be otherwise unaffected.

Dense inclusions were noted in the symbiotic algae under all experimental conditions. These small, electron dense, usually circular inclusions (Figure 11), showed the presence of Ca and P using EDAX (Figure 12). Occasionally, small quantities of Zn, Al, and Mg were also present but the inclusions were less regular in shape and density. Electron dense accumulations were also observed on some occasions with TEM in the external surface coat (Figure 13). These crystals contained no uranium but considerable amounts of other elements, including Fe, Al, P, and Si (Figure 14).

D. Scanning Electron Microscopy

SEM studies on hydra from these 3-day tests showed no crystalline arrangements on the surface of hydra exposed to uranium, either in RP2 water or when

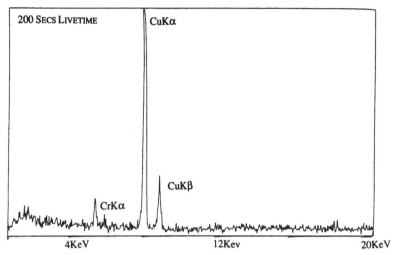

FIGURE 5. EDAX spectrum of the background sampled immediately exterior to the tissue. Peaks of CuKα, CuKβ and CrKα are apparent at 8.04, 8.90, and 5.41 KeV, respectively, and are generated from the transmission electron microscope column.

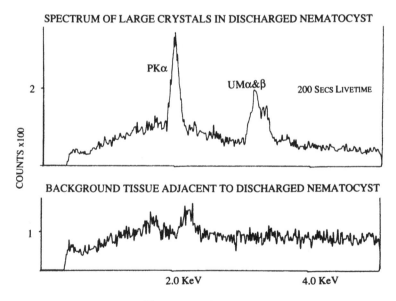

FIGURE 6. EDAX spectrum of large crystals in sites similar to Figure 7, compared with a spectrum of the epidermal tissue sampled adjacent to the region containing uranium crystals.

FIGURE 7. Unstained section of hydra processed for X-ray microanalysis showing an accumulation of flake-like crystals in the sites believed to have been occupied by a nematocyst. (The spectrum is shown in Figure 8.) (Magnification × 28,000, TEM.)

added to diluent water. Spent nematocyst capsules and numerous cnidocils (i.e., the ciliary derivatives which project from the surface of the nematocyst-bearing tentacles) were present though on the surface of these hydra. The cnidocils were distributed ubiquitously on the tentacles of all hydra, with some individual variation without any correlation to the type of treatment. Moreover, SEM showed small dome-shaped nematocyst caps along the tentacles (Figure 15), which agreed in shape with the operculum seen with TEM at the external surface of mature stenotele nematocysts (Figure 16). These caps were raised slightly from the surrounding cells. Hydra subjected to an extended 7-day period of exposure to 350 μg/l uranium and examined using SEM were seen to have large numbers of expended desmoneme nematocysts with their characteristic coiled threads (Figures 17 and 18).

IV. DISCUSSION

Hydra are complex organisms consisting of many differentiated cell types. There exists a highly organized substructure whereby groups of cells are divided

Needle Crystals

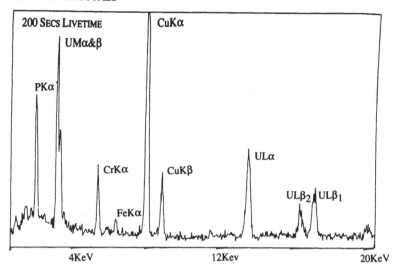

FIGURE 8. EDAX spectrum of the needle-like crystals shown in Figure 7. Peaks for uranium are present as UMα&β at 3.17 KeV, ULα at 13.6 KeV, and U Lβ$_{1\&2}$ at 17.2 and 16.42 KeV, respectively.

into segments, particularly on the gastrodermal side of the mesoglea, by fine septa of connective tissue. Many cells were found between epitheliomuscular cells. These were mainly cnidoblasts or their precursor cell — the interstitial cell. Cnidoblasts (nematoblasts) include all cells intermediate between the stages of interstitial cell and mature nematocyte.[33] A secretary product, probably a type of mucous, was seen on the epidermis which may prevent diffusion of contaminants into the intracellular environment.[34]

Nematocyte differentiation involves the formation of an intracellular nematocyst. There are four types of nematocyst found in *H. viridissima:* (1) stenotele — the largest nematocyst with a pear-shaped capsule that almost fills the nematocyte; (2) holotrichous isorhiza — a large glutinant having a long adhesive thread bearing minute spines; (3) atrichous isorhiza — a small glutinant with a straight unarmed thread; and (4) the desmoneme — a small pear-shaped nematocyst which contains a thread that when discharged coils tightly around the prey.[35] The density of the various nematocysts varies along the length of the hydra body and tentacles.[36] As the nematocyst approaches maturity, it migrates to the surface, where it becomes incorporated into the ectoderm epithelium. In the tentacles, the nematocysts are embedded in pockets of ectodermal epithelial cells known as 'battery cells' in a distinct circular arrangement.[37]

Sequestration of uranium crystals was seen clearly in epidermal regions in tightly aggregated crystals coexisting with a less electron dense material. The shape and region in which the crystals were found corresponded to the regions

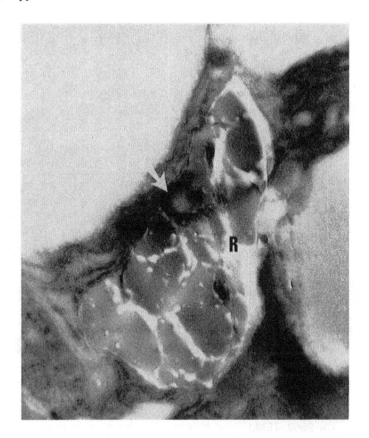

FIGURE 9. The site of a discharged nematocyst in an RP2-treated hydra containing amorphous remnants of organic debris (R) with foci of dense accretions containing uranium (arrow). (Magnification × 17,500, TEM.)

containing nematocysts. This leads us to believe that the crystals accumulate in the space after nematocyst discharge, with the remaining nematocyst capsule wall dissociating at some stage to form the amorphous material. This material is then observed in close association with the crystals. The capsule wall is reported to be collagenous in nature[38] and like many other collagens may have an affinity for uranium.[39]

Postfixation in OsO_4 was abandoned after it was noted that the majority of the crystals present in tissue not postfixed were missing after OsO_4 treatment. This occurred also in unstained sections. Vanderputte et al.[40] have shown that accumulations of lead were also substantially removed after postfixation with OsO_4. This suggests that postfixation with OsO_4 should be avoided or used with caution in any procedure likely to require elemental analysis.

Sparse crystals, identical to those previously identified as uranium, were seen along the gastrodermal side of the mesoglea in any samples (control or

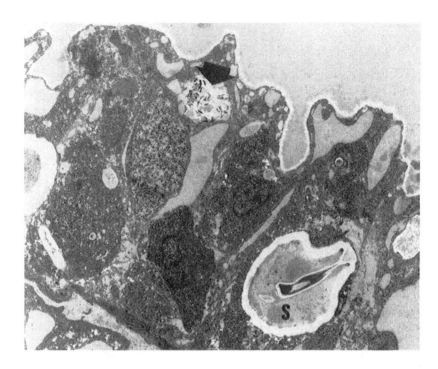

FIGURE 10. Crystals (arrow) below the external surface of an RP2-treated hydra
showing their usual placement. A maturing stenotele (S) (see text)
is present. Stained with uranyl acetate and lead citrate. (Magnification
× 2500, TEM.)

treatments) that were stained *en bloc* with uranium acetate, irrespectively of any
poststaining. These crystals were not seen when *en bloc* staining was omitted.
It is interesting to note the change in location of these crystals as compared to
those formed "naturally" with environmental uranium. It would appear that a
different mechanism of deposition was involved in each case.

Uranium was found to be the only metal detected in what appeared to be
discharged nematocysts by EDAX. This suggests that the uptake of the uranium
was by specific accumulation at this site, rather than nonspecific adsorption of
metals from the surrounding water immediately following nematocyst discharge.
During exposure to the uranium treatments, hydra were observed to have reduced
ability to capture live prey and were force-fed by placing the brine shrimp at
their hypostome (mouth). Whether the uranium-containing nematocysts inter-
fered with the cellular mechanisms involving removal of discharged nematocysts
and recrutiment of new mature nematocysts needs to be investigated.

During normal feeding of hydra, it has been reported that only the desmonemes and stenoteles are discharged.[41,42] The former anchor the prey by coiling
around the bristles; the latter penetrate the cuticle or epidermis and inject their

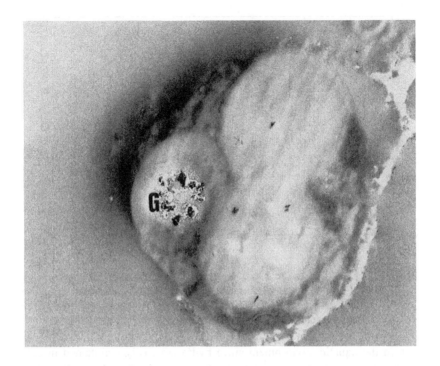

FIGURE 11. Unstained algae in hydra gastrodermis with dense granules (G) exposed to RP2 water. (Photograph after collection of spectrum). (Magnification × 15,000, TEM.)

toxic contents, paralyzing the prey. After nematocyte maturation, about 97% of these cells, including all those having desmonemes and 60% of those having stenoteles, migrate through the ectoderm into the tentacles where they are mounted in the ectodermal epithelial cells, or battery cells.[36] The turnover time for the nematocyte population on the tentacles of hydra fed daily is about 8 days, with almost a quarter of the nematocyte population in the tentacles being lost each day by sloughing off at the tips of the tenacles.[36] Exposure of hydra to uranium for 7 days caused the appearance of numerous discharged desmonemes over the surface of these hydra, even though they had not been fed for 24 h. In the absence of food stimuli, stenotele nematocysts are normally inactive.[43] It would seem that hydra were responding to the prolonged uranium treatment by failing to replace discharged desmoneme nematocysts. This could explain, in part, why hydra were observed to have some difficulty in ingesting food. Stenoteles may paralyze the prey with toxin but there may be too few desmonemes remaining to aid the hydra in successfully bringing the prey to the mouth, the normal function of desmonemes.

Freshwater hydra appear to be an excellent species to use in an ecotoxicological laboratory. This is especially true for toxicity testing for metal contami-

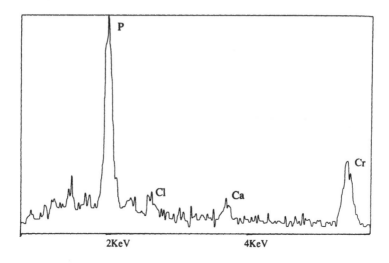

FIGURE 12. EDAX spectrum of representative electron dense deposits in algae (Figure 11) demonstrate the presence of calcium and phosphorus as CaKα and PKα at 3.40 and 2.01 KeV, respectively. Uranium was not detected, chromium is instrument derived.

nation in the aquatic environment since hydra are easily maintained in culture throughout the year, and they are sensitive to a number of metals including uranium,[18] copper,[32] and lead.[44] Freshwater hydra are perhaps particularly sensitive to metals because they lack metallothioneins or other metal-binding proteins, which are involved in the sequestration and detoxification of metals.[45] In order to prevent toxicosis by metal pollution, an aquatic organism must deal with any metal uptake by an integration of the processes of excretion and storage. The degree of integration of these processes will vary markedly between organisms. Uptake is very much dependent on the bioavailability of the metal, which is influenced markedly by environmental parameters such as pH, complexation with organics, or absorption to inorganic particles.[46] Some organisms are able to regulate their body levels of a particular metal independent of the environmental metal level, whereas others accumulate increasing body concentrations as bioavailability increases, with any necessary detoxification of the accumulated metal.[47] It is possible that marine cnidarians, including sea anemones and corals, are able to regulate heavy metal concentrations when, like *H. viridissima*, they are host to zooxanthellae or symbiotic algae.[48,49] These symbiotic algae can accumulate metal when the host is exposed to elevated metals and can then be shed from the host tissue.[49] In this study the symbiotic algae of *H. viridissima* were also observed to have small electron dense inclusions. X-ray microanalysis showed the presence of calcium and phosphorus and occasionally trace amounts of other metals (Zn, Al, and Mg), but no uranium. This mechanism of metal immobilization may be important for the hydra when exposed to higher levels of these metals, as can occur in billabongs in the early Wet Season.[50]

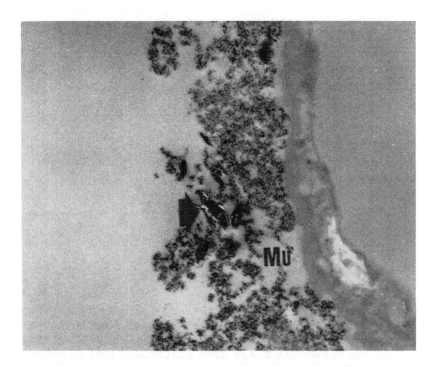

FIGURE 13. Hydra subjected to 8 h treatment with 350 μg/l uranium. The only crystals (arrow) present were in the external mucous coat (Mu) coat. These deposits contained significant quantities of Al, P, Si, and Fe. No uranium was detected. (Magnification × 7000, TEM.)

Due to accumulation of uranium in the nematocysts of hydra when exposed to abnormal concentrations of uranium, and its correlation with reduced population growth and a feeding dysfunction, hydra can be used in a post-release monitoring program to evaluate the impact of an RP2 water release on the environment. The use of hydra would be of great benefit in allowing the distinction to be made between effects on the environment due specifically to uranium from those due to other causes, e.g., sewage treatment discharge. The uranium specific end point to be monitored would therefore be the pathological accumulation of uranium in nematocysts, since other end points that could potentially be monitored, such as the feeding dysfunction and decreased population growth, are not uranium specific and would be difficult to quantify under field conditions.

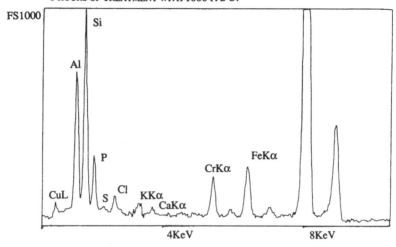

CRYSTALS WITHIN MUCUS ALONG EXTERIOR OF
HYDRA
8 HOURS OF TREATMENT WITH 1000 PPB U.

FIGURE 14. EDAX spectrum of crystals within mucous along the exterior of hydra treated for 8 h with 350 μg/l uranium showed aluminum as AlKα at 1.49 KeV, silicone as SiKα at 1.74 KeV, P as PKα at 2.01 KeV, and iron as FeKα at 6.40 KeV. (CrKα and CuKα&β peaks are generated by the instrument.)

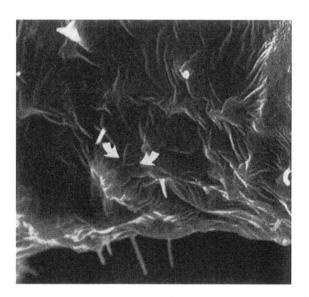

FIGURE 15. Small dome-shaped caps [arrows] are raised slightly from the tentacle surface. Numerous cnidocils are seen protruding from the tentacle taken from an RP2-exposed hydra. No crystals were seen (by backscatter mode) of SEM. (Magnification × 2700.)

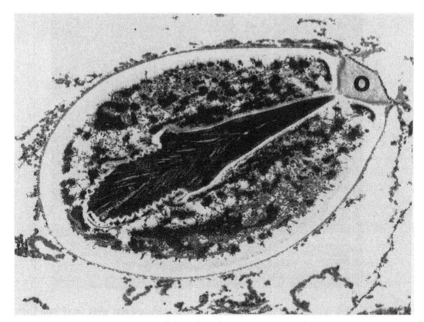

FIGURE 16. A hydra exposed to 350 μg/l U showing a mature stenotele with operculum [O]. No uranium crystals were seen associated with the undischarged nematocyst. Stained with uranyl acetate and lead citrate. (Magnification × 14,500, TEM.)

FIGURE 17. SEM of a hydra exposed to 350 μg/l uranium for 7 days showing numerous desmonemes [D] (see text) on the tentacle surface. (Magnification × 4500.)

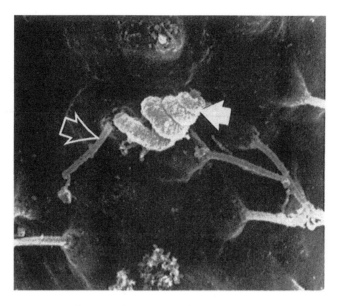

FIGURE 18. Desmonemes [arrow] were clearly associated with cnidocils [arrow-head]. (Magnification × 14,500, SEM.)

ACKNOWLEDGMENTS

This project was funded by the Office of the Supervising Scientist through a cooperative research consultancy agreement with the University of Melbourne. The technical assistance of Ms. Jacinta White is most gratefully recognized.

REFERENCES

1. **Michaelis, F. B. and O'Brien, M. O.,** Preservation of Australia's wetlands: a Commonwealth approach, in *The Conservation of Australian Wetlands,* McComb, A. J. and Lake, P. S., Eds., Surrey Beatty & Sons, Sydney, 1988, chap. 8.
2. **Hart, B. T., Ottaway, E. M., and Noller, B. N.,** Magela Creek systems, Northern Australia. 1. 1982-3 Wet season water quality, *Aust. J. Mar. Freshwater Res.,* 38, 261, 1987.
3. **Berlin, M. and Rudell, B.,** in *Handbook on the Toxicology of Metals,* Fridberg, L., Nordberg, G. F., and Vouk, V. B., Eds., Elsevier, Amsterdam, 1979, 647.
4. **Leggett, R. W.,** The behaviour of chemical toxicity of uranium in the kidney: a reassessment, *Health Phys.,* 57, 365, 1989.
5. **Strandberg, G. W., Shumate, S. E., and Parrott, J. R.,** Microbial cells as biosorbents for heavy metals: accumulation of uranium by *Saccharomyces cerevisiae* and *Pseudomonas aeruginosa. Appl. Environ. Microbiol.,* 41, 237, 1981.

6. **Nakajima, A., Horikoshi, T., and Sakaguchi, T.,** Distribution and chemical state of heavy metal ions absorbed by *Chlorella* cells, *Agric. Biol. Chem.,* 45, 903, 1981.

7. **Ellis, W. R. and Ahsanullah, M.,** The use of nuclear techniques to investigate the levels of uranium in marine waters and its uptake and distribution by marine biota, *Nucl. Tracks Radiat. Meas.,* 8, 437, 1984.

8. **Galun, M., Siegel, S. M., Cannon, M. I., Siegel, B. Z., and Galun, E.,** Ultrastructural localization of uranium biosorption in *Penicillium digetitum* by STEM X-ray microanalysis, *Environ. Pollut.,* 43, 209, 1987.

9. **Pribil, S. and Marvan, P.,** Accumulation of uranium by the chlorococcal alga *Scenedesmus quadricauda, Arch. Hydrobiol. Suppl. Algolog. Stud.,* (15)49, 214, 1976.

10. **Anonymous,** *Guidelines for Surface Water Quality, Vol. 1, Inorganic Chemical Substances,* Inland Water Directorate, Ottawa, Canada, 1983.

11. **Chassard-Bouchard, C.,** Cellular and subcellular localization of uranium in the crab *Carcinus maenas:* a microanalytical study, *Mar. Pollut. Bull.,* 14, 133, 1983.

12. **Chassard-Bouchard, C.,** Lysosomes and pollution, *Biol. Cell,* 51, 15A, 1984.

13. **Nechay, B. R., Thompson, J. D., and Saunders, J. P.,** Inhibition by uranyl nitrate of adenosine triphosphatases derived from animal and human tissues, *Toxicol. Appl. Pharmacol.,* 53, 410, 1980.

14. **Kramer, H. J., Gonick, H. C., and Lu, E.,** *In vitro* inhibition of Na-K-ATPase by trace metals: relation to renal and cardiovascular damage, *Nephron,* 44, 329, 1986.

15. **Tarzwell, C. M. and Henderson, C.,** Toxicity of less common metals to fishes, *Ind. Wastes,* 5, 12, 1960.

16. **Poston, T. M., Hanf, R. W., and Simmons, M. A.,** Toxicity of uranium to *Daphnia magna, Water, Air, Soil Pollut.,* 22, 289, 1984.

17. **Ahsanullah, M. and Williams, A. R.,** Effect of uranium on growth and reproduction of the marine amphipod *Allorchestes compressa, Mar. Biol.,* 93, 459, 1986.

18. **Hyne, R. V., Rippon, G. D., and Ellender, G.,** pH-dependent uranium toxicity to freshwater hydra, *Sci. Total Environ.,* in press, 1992.

19. **Christian, C. S. and Aldrick, J. M.,** Alligator Rivers Study. A review report of the Alligator Rivers Region Environmental Fact-finding Study, Australian Government Publishing Service, Canberra, 1977.

20. **Hyne, R. V.,** Approach to the identification and removal of the toxic compound in Retention Pond 4 at Ranger uranium mine, in *Workshop on Biological Toxicity Testing as a Regulatory Mechanism,* McGill, R. A. and Malatt, K. A., Eds., Northern Territory Department of Mines and Energy, Darwin, 1990, 81.

21. **Allison, H. E., Holdway, D. A., Hyne, R. V., and Rippon, G. D.,** OSS Procedures for the testing of waste waters for release into Magela Creek. XII. Hydra test *(Hydra viridissima* and *Hydra vulgaris),* Office of the Supervising Scientist Open File Record 72, Bondi Junction, Sydney, 1991.

22. **Holdway, D. A., Wiecek, M. M., Hyne, R. V., and Rippon, G. D.,** OSS Procedures for the testing of waste waters for release into Magela Creek. I. Embryo Gudgeon Test *(Mogurnda mogurnda),* Office of the Supervising Scientist Open File Record 69, Bondi Junction, Sydney, 1991.

23. **Hyne, R. V., Miller, K., Hunt, S., and Mannion, M.,** OSS Procedures for the testing of waste waters for release into Magela Creek. XI. Cladoceran survival test *(Pseudosida bidentata* or *Moinodaphnia macleayi),* Office of the Supervising Scientist Open File Record 17, Bondi Junction, Sydney, 1991.

24. **McBride, P., Allison, H. E., Hyne, R. V., and Rippon, G. D.,** OSS Procedures for the testing of waste waters for release into Magela Creek. X. Cladoceran reproduction test *(Moinodaphnia macleayi),* Office of the Supervising Scientist Open File Record 70, Bondi Junction, Syndey, 1991.

25. **Humphrey, C. L., Bishop, K. A., and Brown, V. M.,** Use of biological monitoring in the assessment of effects of mining wastes on aquatic ecosystems of the Alligator Rivers Region, tropical northern Australia, *Environ. Monit. Assess.,* 14, 139, 1990.

26. **Spurr, A. R.,** A low-viscosity epoxy resin embedding medium for electron microscopy, *J. Ultrastruct. Res.,* 26, 31, 1969.

27. **Reynolds, E. S.,** The use of lead citrate at high pH as an electron opaque stain in electron microscopy, *J. Cell Biol.,* 17, 208, 1963.

28. **Masson, P.,** Some histological methods. Trichrome staining and their preliminary technique, *J. Tech. Methods,* 12, 75, 1929.

29. **Constantine, V. S. and Mowry, R. W.,** Selective staining of human dermal collagen. II. The use of picro-sirius red F3BA with polarizing microscopy, *J. Invest. Dermatol.,* 50, 414, 1968.

30. **Miller, P. J.,** An elastic stain, *Med. Lab. Technol.,* 28, 148, 1971.

31. **Watters, W. B. and Buck, R. C.,** An improved simple method of specimen preparation for replicas or scanning electron microscopy, *J. Microsc.,* 94, 185, 1971.

32. **Stebbing, A. R. D. and Pomroy, A. J.,** A sublethal technique for assessing the effects of contaminants using *Hydra littoralis, Water Res.,* 12, 631, 1978.

33. **Campbell, R. D. and Bode, H. R.,** Terminology and morphology and cell types, in *Hydra Research Methods,* Lenhoff, L. M., Ed., Plenum Press, New York, 1983, 5.

34. **Haynes, J. F.,** Epithelial-muscle cells. in *Biology of Hydra,* Burnett, A. L., Ed., Academic Press, New York, 1973, 233.

35. **Boolootians, R. A. and Stiles, K. A.,** *College Zoology,* 10th ed., Macmillan, New York, 1981, 135.

36. **Bode, H. R. and Flick, K. M.,** Distribution and dynamics of nematocyte populations in *Hydra attenuata, J. Cell Sci.,* 21, 15, 1976.

37. **Hufnagel, L. A., Kass-Simon, G., and Lyon, M. K.,** Functional organization of battery cell complexes in tentacles of *Hydra attenuata, J. Morphol.,* 184, 323, 1985.

38. **Blanquet, R. and Lenhoff, H. M.,** A disulphide-linked collagenous protein of nematocyst capsules, *Science,* 154, 152, 1966.

39. **Anselme, K., Julliard, K., and Blaineau, S.,** Degradation of metal-labelled collagen implants: ultrastructural and X-ray microanalysis, *Tissue & Cell,* 22, 81, 1990.

40. **Vandeputte, D. F., Jacob, W. A.,and Van Grieken, R. E.,** Influence of fixation procedures on the microanalysis of lead-induced intranuclear inclusions in rat kidney, *J. Histochem. Cytochem.,* 38, 331, 1990.

41. **Ewer, R. F.,** On the functions and mode of action of the nematocysts of hydra, *Proc. Zool. Soc. London,* 117, 365, 1947.

42. **Picken, L. E. R. and Skaer, R. J.,** A review of researches on nematocysts, in *Symp. Zool. Soc. London,* No. 16, *The Cnidaria and their Evolution,* Rees, W. J., Ed., Academic Press, London, 1966, 19.

43. **Ruch, R. J. and Cook, C. B.,** Nematocyst inactivation during feeding in *Hydra littoralis, J. Exp. Biol.,* 111, 31, 1984.

44. **Browne, C. L. and Davis, L. E.,** Cellular mechanisms of stimulation of bud production in hydra by low levels of inorganic lead compounds, *Cell Tissue Res.,* 177, 555, 1977.

45. **Andersen, R. A., Wiger, R., Daae, H. L., and Eriksen, K. D. H.,** Is the metal binding protein metallothionein present in the coelenterate *Hydra attenuata, Comp. Biochem. Physiol.,* 91C, 553, 1988.

46. **Giesy, J. P., Geiger, R. A., Kevern, N. R., and Alberts, J. J.,** UO_2^{2+}-humate interactions in soft, acid, humate-rich waters, *J. Environ. Radioact.,* 4, 39, 1986.

47. **Depledge, M. H. and Rainbow, P. S.,** Models of regulation and accumulation of trace metals in marine invertebrates, *Comp. Biochem. Physiol.,* 97C, 1, 1990.

48. **Harland, A. D. and Brown, B. E.,** Metal tolerance in the scleractinian coral *Porites lutea, Mar. Pollut. Bull.,* 20, 353, 1989.

49. **Harland, A. D. and Nganro, N. R.,** Copper uptake by the sea anemone *Anemonia viridis* and the role of zooxanthellae in metal regulation, *Mar. Biol.,* 104, 297, 1990.

50. **Brown, T. E., Morley, A. W., Sanderson, N. T., and Tait, R. D.,** Report of a large fish kill resulting from natural acid water conditions in Australia, *J. Fish. Biol.,* 22, 335, 1983.

CHAPTER 10

Metal Regulation in Two Species of Freshwater Bivalves

Michiel H.S. Kraak, Merel Toussaint, Eric A.J. Bleeker, and Daphna Lavy

TABLE OF CONTENTS

0-87371-734-1/93/$0.00 + $.50
© 1993 by Lewis Publishers

I. INTRODUCTION

The freshwater mussels *Dreissena polymorpha* (Dreissenidae) and *Unio pictorum* (Unionidae) are commonly found in the Netherlands. These filter feeders play an important role in freshwater ecosystems: they are able to reduce phytoplankton abundances by their high filtration activity[1-3] and are the main source of food for diving ducks and some benthivorous fish.

In the presence of raised concentrations of dissolved heavy metals bivalve molluscs keep their valves closed for longer periods of time,[4-6] produce fewer byssus threads,[7] and the heart rate is lowered.[8-10] For some marine bivalves, it has been demonstrated that heavy metals reduce their filtration rate.[10-12] Both marine and freshwater bivalves exhibit high bioconcentration factors for heavy metals. At low external Zn concentrations, however, most bivalves appear to be able to regulate the body concentration of this essential metal, and no increase in the body Zn concentration occurs.[13,14] The No Observed Effect Concentration for accumulation (NOEC$_{accumulation}$) can be defined as the highest metal concentration in the water which does not result in a significant increase of the metal concentration in the organism. Once this NOEC$_{accumulation}$ is exceeded, regulation breaks down and net accumulation begins. Cu is also an essential element, and some but not all bivalves appear to be able to regulate their internal Cu concentration. *Mytilus edulis,* for instance, was capable of regulating the body concentration of Cu, but *Scrobicularia plana* was not.[14] Bivalves are incapable of regulating the body concentration of the nonessential metal Cd.[14]

The aim of this study was to examine whether and to what extent the freshwater mussels *D. polymorpha* and *U. pictorum* were able to regulate their internal Cu concentration. The size of *U. pictorum* also provides an opportunity to study organ specific regulation and accumulation. In addition, the regulation and accumulation of Zn and Cd in *D. polymorpha* were studied, in the expectation that Zn might be regulated over a certain range of external concentrations and Cd might not.[14]

II. MATERIALS AND METHODS

D. polymorpha and water were collected from Lake Markermeer, *U. pictorum* was collected from the river Gein (both in the Netherlands). The experiments were carried out in temperature controlled aquaria (15°C). The water in the aquaria was aerated and was always oxygen saturated. To prevent evaporation, the aquaria were covered with sheets of glass. The light/dark regime was synchronized with the actual day/night rhythm, but was kept constant during each experiment. For *D. polymorpha*, each experimental treatment consisted of 25 mussels of 1.6 to 2.0 cm in length, placed in a plastic aquarium containing 3 l of filtered (0.45 μm) water from Lake Markermeer (pH 7.9; hardness 15°D where 1°D = 10 mg CaO/l). The experiments were carried out twice, so there were two replicates per experimental treatment. For *U. pictorum*, each experimental treatment consisted of three mussels (7.9 ± 0.5 cm in length), placed in aquaria containing 3 l of filtered (0.45 μm) water from Lake Maarsseveen (The Netherlands).

Metals were added to the water in the aquaria the day after collection of the mussels, and both water and metals were renewed after 24 and 48 h. Metal concentrations in the animals were determined after 48 h of exposure. Water samples were taken at 1, 24, 25, and 48 h, in triplicate, and analyzed for the added metal by flame or furnace AAS. The actual dissolved concentrations of metals to which the animals were exposed during the experiments were determined from these values by integral calculus.

In the experiments with *D. polymorpha* the following concentrations of metals were studied — Zn: 0.009, 0.2, 0.5, 1, 2, 5, and 10 mg/l; Cd: 0.0002, 0.1, 0.2, 0.5, 0.75, 1, 2, and 5 mg/l; and Cu: 3, 10, 20, 50, 100, 200, 500, and 1000 μg/l. In the experiments with *U. pictorum* the following concentrations of Cu were studied: 0, 10, 20, 50, 100, and 200 μg/l. Stock solutions of 1000 mg/kg $CuCl_2$, $ZnCl_2$, and $CdCl_2$ were applied.

For *D. polymorpha*, the soft tissues of five randomly chosen mussels from each experimental treatment were placed individually in 2.2 ml polyethylene tubes, freeze dried, weighed, and dissolved by wet digestion with nitric acid and hydrogen peroxide, following Timmermans et al.[15] In the experiments with *U. pictorum*, the soft tissues of each mussel were divided into digestive gland, kidney, gill, mantle, and gonads and digested in the same way. The soft tissues of *D. polymorpha* were not divided into these anatomical parts because of the size of the animals. The samples were analyzed for metals by flame or furnace AAS, following Timmermans et al.[15] To test for differences in metal concentrations between exposed mussels and controls, Bartlett's test for homogeneity of variances, one-way ANOVA, and Sheffé's test for *a posteriori* comparison of means were used, as appropriate.

III. RESULTS

In the experiments, *D. polymorpha* was exposed to a wide range of Cu, Zn, and Cd concentrations. As an example, in Figure 1 the Cu concentration in the mussels is plotted against the Cu concentration in the water. It can be seen that over a certain, but limited, range of external concentrations the Cu concentrations in the mussels did not differ significantly ($p < 0.05$) from the controls. The No Observed Effect Concentration for accumulation ($NOEC_{accumulation}$) can be defined

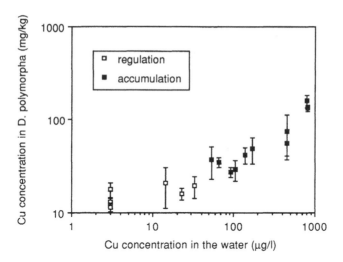

FIGURE 1. Cu concentrations in *D. polymorpha* (mg/kg dw) after 48 h of exposure to Cu in the water (μg/l). Open squares indicate the Cu concentrations in the controls and those in exposed mussels, which did not differ significantly ($p < 0.05$) from the controls (regulation). Filled squares indicate Cu concentrations in exposed mussels which differed significantly ($p < 0.05$) from the controls (accumulation).

as the highest metal concentration in the water which did not result in a significant ($p < 0.05$) increase of the metal concentration in the organism. For Cu and Zn, the $NOEC_{accumulation}$ values were 28 μg/l and 191 μg/l, respectively, suggesting that *D. polymorpha* was able to regulate its internal Cu and Zn concentration, and that Zn could be regulated up to a higher external concentration than Cu. Once the $NOEC_{accumulation}$ is exceeded, regulation breaks down and net accumulation begins (Figure 1). In contrast to Cu and Zn, every Cd concentration in the water resulted in a significant ($p < 0.05$) increase of the Cd concentration in the mussels, so that no $NOEC_{accumulation}$ could be determined, suggesting that Cd could not be regulated at all by *D. polymorpha*.

In Figures 2A to 2E the results for the exposure of *U. pictorum* to Cu are presented. Accumulation of Cu took place in the gills, the mantle, and the digestive gland. For the kidney and the gonads no significant ($p < 0.05$) difference compared to the controls was observed at the highest concentration tested (121 μg/l). The $NOEC_{accumulation}$ values in increasing order for the different organs were: gill (34 μg/l) < digestive gland (74 μg/l) = mantle (74 μg/l) < kidney (>121 μg/l) and gonads (>121 μg/l), indicating that *U. pictorum* was able to regulate Cu, and that this regulation capacity was organ-specific. These results also suggest that on the organism level the $NOEC_{accumulation}$ for *U. pictorum* was 34 μg/l.

In Table 1 the metal concentrations in control *D. polymorpha* and *U. pictorum* and in animals exposed to a nominal concentration of 200 μg/l are listed. For the essential metals Cu and Zn the background values may well represent the physiological demands of the organisms/organs. The background concentration of Cd in *D. polymorpha* represents the accumulation of Cd in the mussels from the relatively clean water of Lake Markermeer. The high Cd concentration in exposed mussels (37.1 mg/kg dry weight) compared to the low Cd concentration in the controls (1.2 mg/kg dry weight) clearly demonstrates the incapability of *D. polymorpha* to regulate Cd. Since the $NOEC_{accumulation}$ for Zn was determined as 191 µg/l, the Zn concentration in *D. polymorpha* exposed to a nominal concentration of 200 µg/l did not differ significantly from the Zn concentration in the controls. The highest Cu concentrations in exposed *U. pictorum* were observed in the gill and the digestive gland, the lowest in the gonads and the kidney (Table 1).

IV. DISCUSSION

Both vertebrates and invertebrates are unable to regulate the body concentration of the nonessential metal, Cd. In contrast, many animals are able to regulate their internal concentration of the essential metal Zn.[13,14,16,17] This is confirmed by our experiments with *D. polymorpha*. For Cu, the literature is less unambiguous. Fish and decapod crustaceans appear to be able to regulate Cu, whereas other arthropods (amphipods and barnacles) and animals like the polychaete worm *Nereis diversicolor,* were incapable of Cu regulation.[14,17] For bivalves, it has been demonstrated that *Mytilus edulis* could regulate Cu, but *Scrobicularia plana,* in contrast, was incapable of regulating Cu.[14] The marine mussel *Perna viridis* could not regulate at an external concentration of 25 µg/l Cu after 1 week of exposure.[18] Our results indicate that the freshwater mussel *D. polymorpha* could regulate its internal Cu concentration up to 28 µg/l and *U. pictorum* up to 34 µg/l or higher, depending on the organ, indicating that the regulation capacity for Cu in these two species is about the same. During these acute experiments *D. polymorpha* could regulate Cu up to lower external concentrations than Zn, which is in agreement with the results for *M. edulis.*[14]

The exposure time strongly influenced the concentration up to which Cu could be regulated: the $NOEC_{accumulation}$ for *M. edulis* decreased from 100 µg/l Cu after 4 days of exposure to 10 µg/l Cu after 16 days of exposure. Our results indicated that after 48 h of exposure the freshwater mussel *D. polymorpha* could regulate Cu up to 28 µg/l, although it is questionable whether 48 h is long enough to distinguish between regulation and low net accumulation. Chronic experiments,[19] however, did indicate that after 10 weeks of exposure *D. polymorpha* was still able to regulate at an external concentration of 13 µg/l Cu.

Rainbow and White[17] proved that Zn regulation in the decapod *Palaemon elegans* is an active process: an increased rate of Zn uptake was matched by an increase of Zn excretion. Above the $NOEC_{accumulation}$ the excretion could no longer compensate for the uptake of the metal, resulting in net accumulation of the

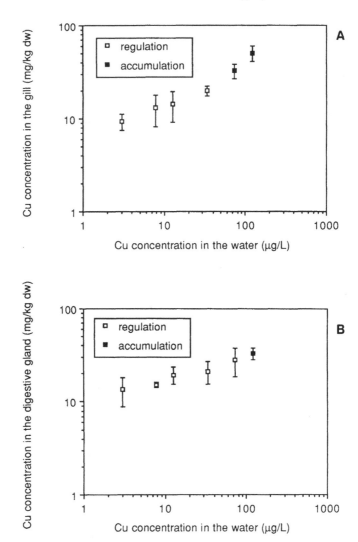

FIGURE 2. Cu concentrations in (A) the gill tissue, (B) digestive gland, (C) the mantle, (D) the kidney, and (E) the gonads of *U. pictorum* (mg/kg dw) after 48 h exposure to Cu in the water (μg/l). Open squares indicate the Cu concentrations in the controls and those in exposed mussels, which did not differ significantly ($p < 0.05$) from the controls (regulation). Filled squares indicate Cu concentrations in exposed mussels which differed significantly ($p < 0.05$) from the controls (accumulation).

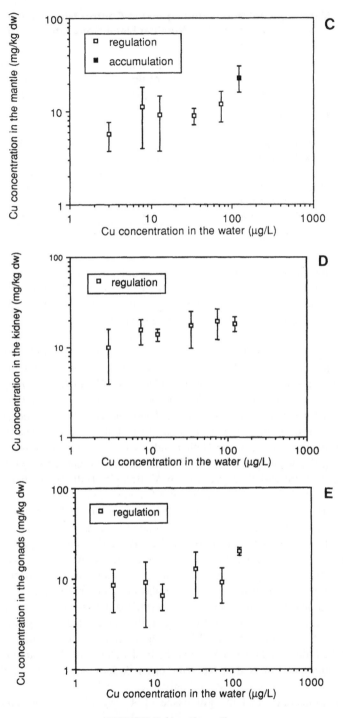

FIGURE 2 (continued).

Table 1. Metal Concentrations (mg/kg dw) in Control Mussels and in Mussels Exposed to a Nominal Metal Concentration of 200 μg/l for 48 h

Metal	Tissue	Concentration in control tissue (mg/kg dw)	Concentration in exposed tissue (mg/kg dw)
Cd	*D. polymorpha*	1.2	37.1
Zn	*D. polymorpha*	108	115
Cu	*D. polymorpha*	13.6	40.6
Cu	*U. pictorum* gill	9.4	50.0
Cu	*U. pictorum* mantle	5.7	23.1
Cu	*U. pictorum* d.g.[a]	13.4	32.7
Cu	*U. pictorum* gonad	8.5	19.8
Cu	*U. pictorum* kidney	9.9	18.4

[a] d.g. = digestive gland.

metal.[16] Thus net accumulation is the result of uptake minus excretion. Assuming that regulation is an active process, it will cost energy which cannot be used for other metabolic processes. This may explain the decrease of the regulation capacity for essential metals during prolonged exposure.[14]

Background values for Cu, Zn, and Cd in *D. polymorpha* and Cu in the different organs of *U. pictorum* (Table 1) could be compared with literature data. The 108 mg/kg (dry weight) Zn in *D. polymorpha* is in agreement with the results of Karbe et al.[20] who measured 113 mg/kg Zn in *D polymorpha* from Lake Schaalsee, which was considered to be the uncontaminated reference area. The values for Cu, Zn, and Cd in *D. polymorpha* are also in good agreement with concentrations measured in animals from Lake Heerhugowaard, another relatively clean location in the Netherlands.[21] The Cu concentration in the gill (9.4 mg/kg dry weight) and mantle (5.7 mg/kg dry weight) of *U. pictorum* are in good agreement with those reported by V.-Balogh[22] (gill 10 mg/kg dry weight) and Salanki et al.[23] (gill 14.8 and mantle 6.7 mg/kg dry weight).

The different anatomical parts of *U. pictorum* and *D. polymorpha* as a whole exhibited differential regulation and accumulation patterns. These differences are caused by differences in the original concentration in the organism/organ, the $NOEC_{accumulation}$, the accumulation rate once the $NOEC_{accumulation}$ is exceeded, and the exposure time. The physiological explanation for these differences can be found in the different role each tissue plays in metal regulation, as described by Depledge and Rainbow.[24] In these experiments the greatest Cu accumulation took place in the gills, followed by those in the mantle and the digestive gland. After 48 h of exposure to 121 μg/l Cu no significant Cd accumulation had taken place in the kidney and the gonads. For the freshwater mussel (*Anodonta cygnea*) Cu accumulation increased in the same order: kidney < mantle < gill. Accumulation of Cu in the kidney started only after 3 weeks of exposure.[5] Tallandini et al.[25] demonstrated that the gill tissue of *A. cygnea* and *Unio elongatulus* exhibited the quickest and highest Cu accumulation. Partly in contrast and partly in agreement with the above are the experimental results for *M. edulis:* the highest Cu concentration was measured in the gill, but the kidney accumulated more Cu than the mantle.[26] An explanation may be that the species mentioned above all belong to the same family (Unionidae) to which *M. edulis* does not belong. A second explanation may be that *M. edulis* is a marine species, in contrast to the Unionidae.

Comparisons can also be made with field data. In agreement with experimental data, the gills of *U. pictorum* from a polluted part of Lake Balaton (Hungary) had accumulated more Cu (203 mg/kg dry weight) than the mantle (*C.* 40 mg/kg dry weight[27]). These values are higher than those observed in this acute laboratory study. Manly and George[28] reported the following increasing order of Cu concentrations in the different organs of *A. anatina* from the river Thames (England): gonad < digestive gland < mantle = kidney < gill. In these acute laboratory experiments with *U. pictorum* the following order was observed: gonad = kidney < mantle < digestive gland < gill. Both series are partly in agreement and partly in contradiction with each other. In both studies the gill exhibited the highest and the gonads the lowest Cu accumulation. This can be explained by the fact that due to the filtration activity of the mussels much contaminated water passes through the gill tissue, which has a relatively large surface. The gonads may be protected against accumulation by the other organs. In that case, the Cu concentration in the gonads is not regulated, but accumulation is avoided. The digestive gland and the kidney differed markedly in their place in the orders mentioned above. The low accumulation in the kidney in acute laboratory experiments was also observed for *Anodonta cygnea*[5,25] and for *Unio elongatulus*.[25] For *A. cygnea* this could be explained by the fact that the Cu accumulation started only after 3 weeks of exposure. Since organisms in the field are exposed during their entire lifetime, this could very well explain the high Cu concentrations in the kidney of animals from the field compared to animals exposed in the laboratory for 48 h. Cu concentrations in the digestive gland are higher in animals from the acute experiments than in animals from the field. A redistribution of Cu from the digestive gland to the kidney may have taken place during chronic exposure in the field.

V. CONCLUSIONS

- *D. polymorpha* and *U. pictorum* could regulate their internal concentration of the essential metal Cu within limited ranges of Cu concentrations in the water.
- *D. polymorpha* could regulate Zn up to higher external concentrations than Cu. Cd could not be regulated.
- The different organs of *U. pictorum* exhibited different regulation capacities for Cu: gill < digestive gland = mantle < kidney and gonads.

• The different organs of *U. pictorum* exhibited differences in Cu accumulation, in decreasing order: gill > digestive gland > mantle > kidney = gonads.

VI. ABSTRACT

To examine whether and to what extent the freshwater mussels *Dreissena polymorpha* and *Unio pictorum* are able to regulate their body and/or internal tissue Cu concentrations these species were exposed to a range of dissolved sublethal Cu concentrations. In addition, the regulation and accumulation of Zn and Cd in *D. polymorpha* were studied. For *D. polymorpha,* total soft tissues were analyzed; for *Unio pictorum* the soft tissues of each mussel were divided into digestive gland, kidney, gill, mantle, and gonads before analyzing for Cu. Both species were able to regulate their internal concentration of the essential metal Cu. *D. polymorpha* was able to regulate the essential metal Zn up to a higher external concentration (191 μg/l) than Cu (28 μg/l), while the nonessential metal Cd could not be regulated.

The different organs of *U. pictorum* exhibited different No Observed Effect Concentrations ($NOEC_{accumulation}$) for copper: gill (34 μg/l) < digestive gland (74 μg/l) = mantle (74 μg/l) < kidney (>121 μg/l) and gonads (>121 μg/l). After exposure to the highest Cu concentration tested (nominal 200 μg/l), the highest internal Cu concentrations were observed in the gill and the digestive gland. No significant accumulation took place in the kidney and gonads in these acute experiments. The concentration in an animal or tissue is not only determined by the exposure level, but also by the regulation capability ($NOEC_{accumulation}$), the accumulation rate once the $NOEC_{accumulation}$ is exceeded, and the exposure time.

ACKNOWLEDGMENTS

We would like to thank Dr. C. Davids and Prof. Dr. Nico M. van Straalen for their comments. Dr. John Parsons corrected the English text.

REFERENCES

1. **Stanczykowska, A., Lawacz, W., and Mattice, J.,** Use of field measurements of consumption and assimilation in evaluating the role of *Dreissena polymorpha* Pall. in a lake ecosystem, *Pol. Arch. Hydrobiol.,* 22(4), 509, 1975.
2. **Lewandowski, K.,** Occurrence and filtration capacity of plant-dwelling *Dreissena polymorpha* (Pall.) in Majcz Wielki Lake, *Pol. Arch. Hydrobiol.,* 30(3), 255, 1983.

3. **Reeders, H. H., Bij de Vaate, A., and Slim, F. J.,** The filtration rate of *Dreissena polymorpha* (Bivalvia) in three Dutch lakes with reference to biological water quality management, *Freshwater Biol.,* 22, 133, 1989.

4. **Slooff, W., de Zwart, D., and Marquenie, J. M.,** Detection limits of a biological monitoring system for chemical water pollution based on mussel activity, *Bull. Environ. Contam. Toxicol.,* 30, 400, 1983.

5. **Salanki, J. and V.-Balogh, K.,** Physiological background for using freshwater mussels in monitoring copper and lead pollution, *Hydrobiologia,* 188/189, 445, 1989.

6. **Kramer, K. J. M., Jenner, H. A., and de Zwart, D.,** The valve movement response of mussels, a tool in biological monitoring, *Hydrobiologia,* 188/189, 433, 1989.

7. **Martin, J. M., Piltz, F. M., and Reish, D. J.,** Studies on *Mytilus edulis* community in Alamitos Bay, California. V. The effects of heavy metals on byssal thread production, *Veliger,* 18(2), 182, 1975.

8. **Davenport, J.,** A study of the effects of copper applied continuously and discontinuously to specimens of *Mytilus edulis* (L.) exposed to steady and fluctuating salinity levels, *J. Mar. Biol. Assoc. U.K.,* 57, 63, 1977.

9. **Akberali, H. B. and Black, J. E.,** Behavioural responses of the bivalve *Scrobicularia plana* (Da Costa) subjected to short-term copper (Cu II) concentrations, *Mar. Environ. Res.,* 4, 97, 1980.

10. **Grace, A. L. and Gainey, L. F., Jr.,** The effects of copper on the heart rate and filtration rate of *Mytilus edulis, Mar. Pollut. Bull.,* 18(2), 87, 1987.

11. **Watling, H.,** The effects of metals on mollusc filtering rates, *Trans. R. Soc. S. Afr.,* 44(3), 441, 1981.

12. **Redpath, K. J. and Davenport, J.,** The effect of copper, zinc and cadmium on the pumping rate of *Mytilus edulis* L., *Aquat. Toxicol.,* 13, 217, 1988.

13. **Simkiss, K., Taylor, M., and Mason, A. Z.,** Metal detoxification and bioaccumulation in molluscs, *Mar. Biol. Lett.,* 3, 187, 1982.

14. **Amiard, J. C., Amiard-Triquet, C., Berthet, B., and Metayer, C.,** Comparative study of the patterns of bioaccumulation of essential (Cu, Zn) and non-essential (Cd, Pb) trace metals in various estuarine and coastal organisms, *J. Exp. Mar. Biol. Ecol.,* 106, 73, 1987.

15. **Timmermans, K. R., van Hattum, B., Kraak, M. H. S., and Davids, C.,** Trace metals in a littoral foodweb: concentrations in organisms, sediment and water, *Sci. Total Environ.,* 87/88, 477, 1989.

16. **White, S. L. and Rainbow, P. S.,** Regulation and accumulation of copper, zinc and cadmium by the shrimp *Palaemon elegans, Mar. Ecol. Prog. Ser.,* 8, 95, 1982.

17. **Rainbow, P. S. and White, S. L.,** Comparative strategies of heavy metal accumulation by crustaceans, zinc, copper and cadmium in a decapod, an amphipod and a barnacle, *Hydrobiologia,* 174, 245, 1989.

18. **Krishnakumar, P. K., Asokan, P. K., and Pillai, V. K.,** Physiological and cellular responses to copper and mercury in the green mussel *Perna viridis* (Linnaeus), *Aquat. Toxicol.,* 18, 163, 1990.

19. **Kraak, M. H. S.,** Unpublished results, 1991.

20. **Karbe, L., Antonacopoulos, N., and Schnier, C.,** The influence of water quality on accumulation of heavy metals in aquatic organisms, *Verh. Int. Verein Limnol.,* 19, 2094, 1975.

21. **Kraak, M. H. S., Scholten, M. C. T., Peeters, W. H. M., and de Kock, W. C.,** Biomonitoring of heavy metals in the Western European rivers Rhine and Meuse using the freshwater mussel *Dreissena polymorpha, Environ. Pollut.,* 74, 101, 1991.

22. **V.-Balogh, K.,** Comparison of mussels and crustacean plankton to monitor heavy metal pollution, *Water, Air, Soil Pollut.,* 37, 281, 1988.

23. **Salanki, J., V.-Balogh, K., and Berta, E.,** Heavy metals in animals from Lake Valaton, *Water Res.,* 16, 1147, 1982.

24. **Depledge, M. H. and Rainbow, P. S.,** Models of regulation and accumulation of trace metals in marine invertebrates, *Comp. Biochem. Physiol.,* 97C(1), 1, 1990.

25. **Tallandini, L., Cassini, A., Favero, N., and Albergoni, V.,** Regulation and subcellular distribution of copper in the freshwater molluscs *Anadonta cygnea* (L.) and *Unio elongatulus* (Pf.), *Comp. Biochem. Physiol.,* 84C(1), 43, 1986.

26. **Sutherland, J. and Major, C. W.,** Internal heavy metal changes as a consequence of exposure of *Mytilus edulis,* the blue mussel, to elevated external copper(II) levels, *Comp. Biochem. Physiol.,* 68C, 63, 1981.

27. **V.-Balogh, K.,** Heavy metal pollution from a point source demonstrated by mussel (*Unio pictorum* L.) at Lake Balaton, Hungary, *Bull. Environ. Contam. Toxicol.,* 41, 910, 1988.

28. **Manly, R. and George, W. O.,** The occurrence of some heavy metals in populations of the freshwater mussel *Anodonta anatina* (L.) from the river Thames, *Environ. Pollut.,* 14, 139, 1977.

CHAPTER 11

Autoradiographic Study of Zinc in *Gammarus Pulex* (Amphipoda)

Qin Xu and David Pascoe

TABLE OF CONTENTS

0-87371-734-1/93/$0.00 + $.50
© 1993 by Lewis Publishers

I. INTRODUCTION

The uptake and distribution of zinc in different species of animals have been investigated using histochemical, morphological, and spectrophotometric analysis by different researchers. Although zinc has been found to be widely distributed in both soft and hard tissues of exposed animals,[1-11] the body fluid, alimentary canal, and gills of aquatic species are the main target organs, possibly in association with the existence of metallothioneins in these tissues.[12-15]

In previous studies, analyses of changes in the amino acid pool have indicated that zinc penetrated the tissues and affected the process of intracellular metabolism of the freshwater amphipod crustacean *Gammarus pulex*.[16] A study on zinc bioaccumulation suggests that *G. pulex* may be able to regulate its total body concentration of zinc to an approximately constant level.[17] With the aid of a first-order kinetic model, this study has also revealed that the concentration of zinc in the animal first increased with the time of zinc exposure and finally reached a plateau.[17]

A valuable technique for studying metal localization in tissues is autoradiography, which demonstrates the presence of radioactive isotopes in tissue sections by means of their ability to reduce silver salts in a photographic plate, film, or emulsion.[18-20] In recent years, autoradiography has been used in the study of heavy metal pollution and bioaccumulation. For instance, Carney et al.[21] have employed ^{109}Cd to illuminate the adsorption of the metal by the exoskeleton of the daphnid *Daphnia magna*.

In the present study, the distribution of zinc in tissues of *G. pulex* was investigated by the use of the radioisotope zinc-65. Equivalent doses of zinc (based upon the concentration per unit time, [exposure concentration × exposure time]) were given to two batches of animals, and the effect of short-term exposure (2 days) to an acute lethal concentration (1.5 mg l^{-1}) was compared with the effect of long-term exposure (10 days) to a low concentration (0.3 mg l^{-1}).

II. MATERIALS AND METHODS

A. Exposure of *Gammarus pulex* to Zinc

The radioisotope ^{65}Zn was purchased from Amersham International plc. (ZnCl$_2$ in 0.1 M HCl), and mixed with ZnSO$_4$ solutions to make up 2 l test solutions of nominal zinc concentrations of 1.5 and 0.3 mg l⁻¹ with dechlorinated tap water of hardness 104.79 $\pm$ 7.09 mg l⁻¹ as $CaCO_3$, and conductivity 283 $\pm$ 19 µS cm⁻¹. The pH of the solutions was indicated as pH 7.0 by pH test paper. The 4 l plastic test containers were equilibrated with 1.5 and 0.3 mg l⁻¹ zinc for 48 h before the start of the test. Total nominal activities of 30 µCi and 2.5 µCi were presented in 2 l solutions for nominal concentrations of 1.5 and 0.3 mg l⁻¹ zinc, respectively. A total of 30 male and female *G. pulex* of approximately 8 mm body length were chosen for the experiment by randomly distributing 20 into the 1.5 mg l⁻¹ and 10 into the 0.3 mg l⁻¹ test solutions. At the beginning of the experiment, 4 and 2 horse chestnut (*Aesculus hippocastanum* L.) leaf discs (diameter = 8 mm) were provided as food and cover to the 1.5 mg l⁻¹ and 0.3 mg l⁻¹ test containers, respectively. The test was performed under static (aerated) conditions at room temperature (20°C) in a fume cupboard. The 1.5 mg l⁻¹ exposure was terminated after 2 days when 11 animals had died, while the 0.3 mg l⁻¹ exposure was terminated after 10 days when 1 death had occurred. The 0.3 mg l⁻¹ solution was renewed twice during the exposure.

B. Histological Processing of *Gammarus pulex* Tissues

Survivors were sampled at the end of each exposure period; 1.5 mm transversely cut tissue blocks were fixed with cooled, modified, formaldehyde-alcohol fixative.[22] The tissue blocks were cleared in amyl acetate before embedding in paraffin wax. Then 5 µm transverse sections were cut on a Cambridge rocking microtome. Since Zn^{2+} is soluble in water, it was necessary to avoid contact of tissues with water[23,24] and, therefore, the tissue sections were floated on absolute alcohol (40°C) before mounting onto subbed slides.[19]

C. Autoradiographic Processing of *Gammarus pulex* Tissues

Sections were dewaxed in toluene and air-dried before dipping in diluted Ilford K_2 liquid emulsion (at 48°C) in the dark. The emulsion-covered slides were then air dried and packed in the dark in black boxes. In order to obtain the optimal result from the emulsion exposure, slide samples were examined at intervals for 51 days, after which all remaining slides were developed in Kodak D19 (18 to 21°C) and fixed with Amfix.

D. Staining of Autoradiographs

The fixed autoradiographs were stained with Mayer's hematoxylin and eosin, then dried and mounted on glass cover slips and examined under an Olympus BHA light microscope.

III. RESULTS

The orientation of organs of *Gammarus pulex* in transverse sections as seen by light microscopy is demonstrated in Figure 1.[25-28] The whole body of *G. pulex* is covered with cuticle produced by the underlying epidermis, next to which is a thick layer of smooth muscles (Figures 1A, 1B, and 1C). Beneath the dorsal midline of the thoracic somite lies a long slender heart with a single muscle layer enclosed in a hemocoel (Figure 1A). Paired posterior aortas leave the heart and extend along the length of the abdomen, then branch into the ventral hemocoel of the telson (Figures 1B and 1C). In the center of the body lies the midgut, accompanied by two pairs of hepatopancreatic caeca (Figures 1A, 1B, and 1C). Beneath the heart and lateral to the midgut above the hepatopancreatic caeca are the paired gonads — ovaries in female specimens (Figures 1A and 1C) and testes in male specimens. The paired ventral nerve cords have paired neuroganglia in each somite (Figures 1A, 1B, and 1C). Gills associated with appendages are seen between the thoracic coxal plates (Figure 1A). Gills are composed of two layers of epithelial cells which are separated by hemolymph spaces (Figure 1A), and the opposite epithelial cells form regular junctions across the hemolymph spaces. Marginal canals and proximal canals are observed at the two ends of the gills (Figure 1A).

The 51st day autoradiographs from two different treatments are presented in Figures 2 to 5. Although little zinc adsorption was observed on the outside of the cuticles (Figures 2A and 2B), zinc had penetrated into almost all soft tissues in both exposures (Figures 2 to 5) and radioactivity was detected within the cuticle and underlying epidermis (Figures 2A and 2B). The most obvious difference between the two exposures was seen in the gills. Although the two batches had been exposed to the same total amount of zinc (3.0 mg l^{-1} · days), *G. pulex* exposed to the 0.3 mg l^{-1} solution showed little zinc had precipitated onto the gill surface or had penetrated into the epithelial cells (Figure 3A), while the animals exposed to 1.5 mg l^{-1} zinc solution had zinc apparently associated with mucus precipitated onto the surface of the gills and zinc present in the gill epithelia (Figure 3B). After the high concentration exposure, the epithelial cells of the gills were swollen and some cell junctions in the hemolymph space had disintegrated (Figure 3B).

In the lumen of the midgut, zinc brought in with highly contaminated food was found on the apical surface of the midgut epithelium in *G. pulex* from the high concentration exposure (Figure 4B), but it was absent in the low concentration exposure (Figure 4A). However, in both exposures, very little zinc was present in the midgut epithelium itself (Figures 4A and 4B). Some radioactivity was observed in the underlying connective tissue sheath of the gut (Figures 4A and 4B). The form of the hepatopancreatic caeca of *G. pulex* (Figure 1) is similar to that described for the terrestrial isopods *Oniscus asellus*.[30-33] B cells (big cell) are larger than S cells (small cell) and project into the lumen. The space observed in a B cell could be the space left by a large lipid droplet dissolved out during preparation (Figure 5A). Zinc was found on the apical surface of both B and S cells, and had penetrated into B cells in low concentration exposure (Figure 5A). The metal was located in both B and S cells in the hepatopancreatic caeca of amphipods exposed to the higher zinc concentration (Figure 5B).

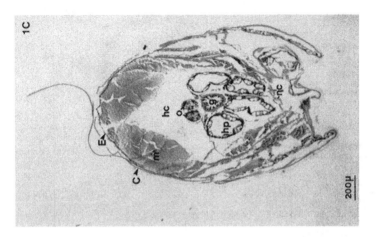

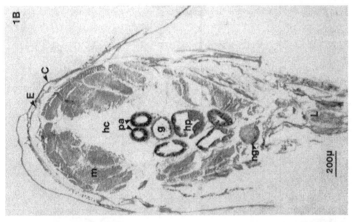

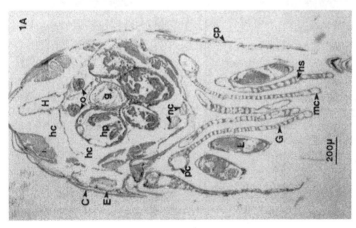

FIGURE 1. Light micrographs of transverse sections of *Gammarus pulex*. 1A: thoracic somite; 1B: abdominal somite; 1C: telson — C: cuticle; E: epidermis; m: smooth muscles; H: heart; hc: hemocoel; pa: paired posterior aortas; g: midgut; hp: hepatopancreatic caeca; o: ovaries; nc: ventral nerve cords; ng: neuroganglia; G: gills; L: appendages; cp: thoracic coxal plate; hs: hemolymph spaces in gills; mc: marginal canals of the gill; and pc: proximal canals of the gill.

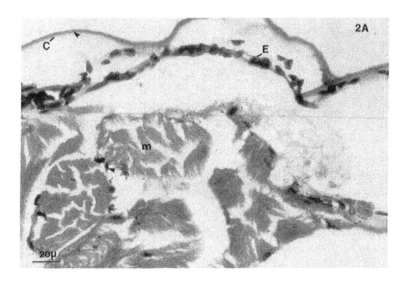

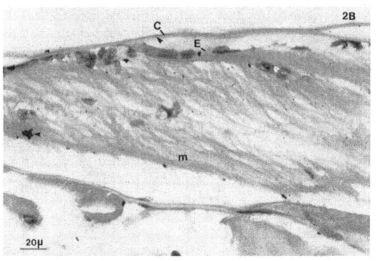

FIGURE 2. Autoradiographs of the cuticle and underlying muscles of *G. pulex.*
Figure 2A: exposed to 0.3 mg l^{-1} zinc for 10 days; Figure 2B: exposed
to 1.5 mg l^{-1} zinc for 2 days — C: cuticle; E: epidermis; m: muscle;
arrow: ^{65}Zn activity.

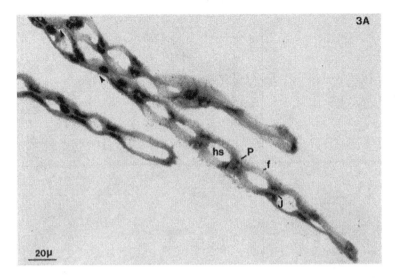

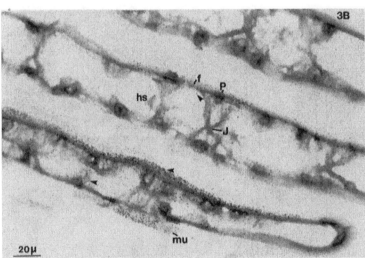

FIGURE 3. Autoradiographs of gills of *G. pulex*. Figure 3A: exposed to 0.3 mg l⁻¹ zinc for 10 days; Figure 3B: exposed to 1.5 mg l⁻¹ zinc for 2 days — P: epithelial cell; f: epithelial cell flange; hs: hemolymph space; J: junction between opposite epithelial cells; mu: mucus; arrow: ⁶⁵Zn activity.

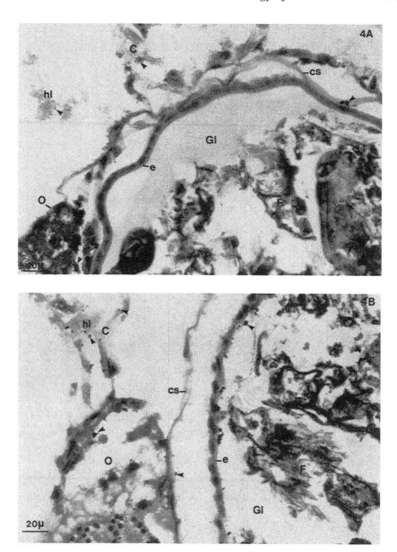

FIGURE 4. Autoradiographs of midguts and ovaries of *G. pulex*. Figure 4A: exposed to 0.3 mg l⁻¹ zinc for 10 days; Figure 4B: exposed to 1.5 mg l⁻¹ zinc for 10 days — Gl: midgut lumen; e: epithelium; cs: connective tissue sheath; F: food; O: ovary; C: connective tissues; hl: hemolymph; arrow: ⁶⁵Zn activity.

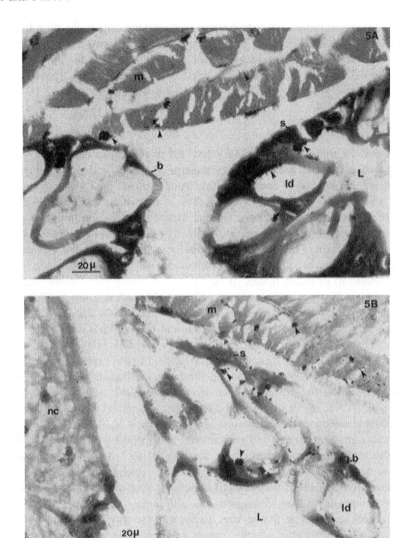

FIGURE 5. Autoradiographs of muscles and hepatopancreatic caeca of *G. pulex*. Figure 5A: exposed to 0.3 mg l^{-1} zinc for 10 days; Figure 5B: exposed to 1.5 mg l^{-1} for 2 days — m: muscle; b: big cell; s: small cell; ld: lipid droplet; L: lumen; nc: nerve cord; arrow: ^{65}Zn activity.

In other soft tissues, radioactive zinc was seen not only in the muscle fibers but also at the edge of the fiber bundles where the sinuses may be located (Figures 2 and 5). The metal was also detected in ovaries of female specimens (Figures 4A and 4B) but was not observed in testes of male specimens.

IV. DISCUSSION

There are three possible primary sites for metal uptake in aquatic animals: body surfaces, gills, and alimentary canals.[2] Zinc adsorption and absorption across the exoskeleton of crustaceans, shells of molluscs, and scales of fish have been widely reported.[1,2,11,12] In his studies of regulation and accumulation of zinc in marine and freshwater crustaceans, Bryan[3] found adsorption from a wide range of zinc concentrations by exoskeletons of the freshwater crayfish *Austropotamobius pallipes*, and absorption across the body surface in the lobster *Homarus vulgaris* exposed to sea water containing 100 μg l^{-1} zinc.[1] Analysis of zinc content in different tissues of barnacles has shown that the shell contains only a small fraction of the accumulated zinc although it comprises large proportions of dry weight (e.g., 96% in *Balanus balanoides* and 66% in *Lepas anatifera*).[12] In this research, the exoskeleton of *G. pulex* has not been found to be the major site of zinc bioaccumulation. Adsorption of the metal was not observed although some ^{65}Zn activity was present in the cuticle and epidermis.

A possible route of zinc entering the bodies of aquatic animals is that it penetrates across the gill epithelium or other permeable surfaces into the hemolymph.[3,4] As zinc can tightly bind to proteins, especially hemocyanin in the body fluid, unbound zinc may enter the hemolymph against an apparent concentration gradient, even from very diluted zinc concentrations.[1,3,4,9] With the circulation of the hemolymph inside the body, zinc is transported to internal organs. De March[15] suggested that Cu^{2+}, Zn^{2+}, and Cd^{2+}, at acutely lethal concentrations, inhibited processes of ionic regulation and led to suffocation by precipitating and causing damage to the gill surface in *Gammarus lacustris*. Similar conclusions were drawn from studies with the crayfish *Austropotamobius pallipes*[3] exposed to zinc, and the isopod *Jaera nordmanni* acutely exposed to copper, mercury, and cadmium.[34]

In the present study, 55% of the *Gammarus pulex* in the 1.5 mg l^{-1} zinc solution died within 48 h exposure at room temperature. Precipitation of zinc and damage of the gill architecture were observed in survivors and the precipitation may be associated with the production of gill mucus. Bubel[34] reported that large numbers of prominent hemocytes were present in the hemolymph spaces after the isopod *Jaera nordmanni* had been exposed to cadmium. The gill structure of *G. pulex* was not changed after 10 days exposure to 0.3 mg l^{-1} zinc. In the low exposure concentration, zinc is apparently efficiently transported from the primary sites of uptake — gills, to the internal organs via the hemolymph.

Another important route of zinc entering aquatic animals is the absorption of the metal from the alimentary canal, mostly via the hepatopancreas or equivalent caeca in crustaceans. Acting as the site for secretion of digestive enzymes, food absorption, and storage of metabolic reserves,[29] the hepatopancreatic caeca absorb the majority of zinc taken in from food, possibly binding it in metallothioneins or in granular form. Since the hepatopancreas is believed also to reabsorb and store excess zinc from the hemolymph,[1,3,29] many authors have suggested that it is the prime site for metal storage, regulation, and detoxification in crustaceans and molluscs.[1-4,12,14,29-32,36] Zinc entering *G. pulex* with food may be transported along the alimentary canal to the hepatopancreatic caeca or have a similar fate as that entering from the gills, and some of the excess zinc could be excreted in the form of feces.

The distribution of zinc in other soft tissues, i.e., muscles, ovaries, and hemolymph of *G. pulex* observed in this study was similar to that in other invertebrates examined. Accumulation of zinc in muscles of other invertebrates such as molluscs[6,8,12] and crustaceans[1-4,10] has been reported. Khan et al.[10] also suggested that the reproductive organs of crustaceans situated in the cephalothorax could accumulate high concentrations of zinc, copper, and mercury. Zinc was also reported to be present in the body fluid of many other zinc-treated animals.[1-4,8,9] Because zinc is essential and has important roles in enzyme function, protein synthesis, and nucleic acid metabolism,[37,38] metabolic processes involving zinc may occur in all soft tissues.

In *G. pulex*, ^{65}Zn activity is detected in almost all the tissues and organs, but is most concentrated in hemocoel spaces in the organs and in the hepatopancreatic caeca. Gills and the alimentary canal are two important routes of zinc entry into the body, and zinc is transported to other tissues by the hemolymph. Adsorption of zinc is not found on the exoskeleton and it is unclear whether crossing the body surface is a major route of zinc absorption.

ACKNOWLEDGMENTS

The author is grateful to Dr. R. Williams and Mr. G. Lewis in the University of Wales College of Cardiff for their professional assistance on this work. Thanks also go to Dr. X. Gao for many valuable suggestions on the manuscript and to the British Council and the Chinese Education Ministry for the financial support of the research.

REFERENCES

1. **Bryan, G. W.,** Zinc regulation in the lobster *Homarus vulgaris, J. Mar. Biol. Assoc. U.K.,* 44, 549, 1964.

2. **Bryan, G. W.,** The metabolism of Zn and ⁶⁵Zn in crabs, lobsters and fresh-water crayfish, in *Radioecological Concentration Processes,* Aberg, B. and Hungate, F. P., Eds., Pergamon Press, Elmsford, NY, 1966, 1005.

3. **Bryan, G. W.,** Zinc regulation in the freshwater crayfish (including some comparative copper analyses), *J. Exp. Biol.,* 46, 281, 1967.

4. **Bryan, G. W.,** Concentrations of zinc and copper in the tissues of decapod crustaceans, *J. Mar. Biol. Assoc. U.K.,* 48, 303, 1968.

5. **Coombs, T. L.,** The distribution of zinc in the oyster *Ostrea edulis* and its relation to enzymic activity and to other metals, *Mar. Biol.,* 12, 170, 1972.

6. **George, S. G., Pirie, B. J. S., Cheyne, A. R., Coombs, T. L., and Grant, P. T.,** Detoxication of metals by marine bivalves: an ultrastructural study of the compartmentation of copper and zinc in the oyster *Ostrea edulis, Mar. Biol.,* 45, 145, 1978.

7. **Lowe, D. M. and Moore, M. N.,** The cytochemical distributions of zinc (Zn II) and iron (Fe III) in the common mussel, *Mytilus edulis,* and their relationship with lysosomes, *J. Mar. Biol. Assoc. U.K.,* 59, 851, 1979.

8. **George, S. G. and Pirie, B. J. S.,** Metabolism of zinc in the mussel, *Mytilus edulis* (L.): a combined ultrastructural and biochemical study, *J. Mar. Biol. Assoc. U.K.,* 60, 575, 1980.

9. **Mirenda, R. J.,** Acute toxicity and accumulation of zinc in crayfish *Orconectes virilis* (Hagen), *Bull. Environ. Contam. Toxicol.,* 37, 387, 1980.

10. **Khan, A. T., Weis, J. S., and D'Andrea, L.,** Bioaccumulation of four heavy metals in two populations of grass shrimp, *Palaemonetes pugio, Bull. Environ. Contam. Toxicol.,* 42, 339, 1989.

11. **Sauer, G. R. and Watable, N.,** Temporal and metal-specific patterns in the accumulation of heavy metals by the scales of *Fundulus heterotitus, Aquat. Toxicol.,* 14, 233, 1989.

12. **Walker, G., Rainbow, P. S., Foster, P., and Crisp, D. J.,** Barnacles: possible indicators of zinc pollution?, *Mar. Biol.,* 30, 57, 1975.

13. **Olafson, R. W., Kearns, A., and Sim, R. G.,** Heavy metal induction of metallothionein synthesis in the hepatopancreas of the crab *Scylla serrata, Comp. Biochem. Physiol.,* 62B, 417, 1979.

14. **Lyon, R., Taylor, M., and Simkiss, K.,** Metal-binding proteins in the hepatopancreas of the crayfish *(Austropotamobius pallipes), Comp. Biochem. Physiol.,* 74C, 51, 1983.

15. **De March, B. G. E.,** Acute toxicity of a binary mixtures of five cations (Cu^{2+}, Cd^{2+}, Zn^{2+}, Mg^{2+} and K^+) to the freshwater amphipod *Gammarus lacustris* (Sars.): alternative descriptive models, *Can. J. Fish. Aquat. Sci.,* 45, 625, 1988.

16. **Xu, Q.,** The effects of exposure to zinc and cadmium separately and jointly on the free amino acid (FAA) pool of *Gammarus pulex,* in Toxicity of Heavy Metals to Freshwater Peracarid Crustaceans, Ph.D. thesis, University of Wales, Cardiff, 1991, chap. 3.

17. **Xu, Q.,** Bioaccumulation of zinc by *Gammarus pulex* from water and food, in Toxicity of Heavy Metals to Freshwater Peracarid Crustaceans, Ph.D. thesis, University of Wales, Cardiff, 1991, chap. 6.

18. **Gahan, P. B.,** *Autoradiography for Biologists,* Academic Press, London, 1972.

19. **Rogers, A. W.,** *Techniques of Autoradiography,* Elsevier, Amsterdam, 1973.

20. **Pearse, A. G. E.,** Autoradiography and its application, in *Histochemistry, Theoretical and Applied, Vol I,* Pearse, A. G. E., Ed., Churchill Livingstone, Edinburgh, 1980, 303.

21. **Carney, G. C., Shore, P., and Chandra, H.,** The uptake of cadmium from a dietary and soluble source by the crustacean *Daphnia magna, Environ. Res.,* 39, 290, 1986.

22. **Humason, G. L.,** *Animal Tissue Techniques,* W. H. Freeman, San Francisco, 1972.

23. **Holt, M. W., Cowing, R. F., and Warren, S.,** Preparation of radioautographs of tissues without loss of water-soluble P^{32}, *Science,* 110, 328, 1949.

24. **Canny, M. J.,** High-resolution autoradiography of water-soluble substances, *Nature,* 175, 857, 1955.

25. **Mellanby, H.,** *Animal Life in Freshwater,* Methuen & Co., Ltd., London, 1963, 113.

26. **Mclaughlin, P. A.,** *Comparative Morphology of Recent Crustacea,* W. H. Freeman, San Francisco, 1980, 112.

27. **Barra, J. A., Pequeux, A., and Humbert, W.,** A morphological study on gills of a crab acclimated to fresh water, *Tissue & Cell,* 15, 583, 1983.

28. **Bauchau, A. G.,** Crustaceans, in *Invertebrate Blood Cells,* Vol. II, Ratcliffe, N. A. and Rowley, A. F., Eds., Academic Press, London, 1981, 385.

29. **Hopkin, S. P. and Martin, M. H.,** The distributions of zinc, cadmium, lead and copper within the hepatopancreas of a woodlouse, *Tissue & Cell,* 14, 703, 1982.

30. **Hopkin, S. P. and Martin, M. H.,** The distributions of zinc, cadmium, lead and copper within the woodlouse *Oniscus asellus* (Crustacea, Isopoda), *Oecologia,* 54, 227, 1982.

31. **Hopkin, S. P. and Martin, M. H.,** Heavy metals in woodlice, *Symp. Zool. Soc. London,* 53, 143, 1984.

32. **Morgan, A. J., Gregory, Z. D. E., and Winters, C.,** Responses of the hepatopancreatic "B" cells of a terrestrial isopod, *Oniscus asellus,* to metals accumulated from a contaminated habitat: a morphometric analysis, *Bull. Environ.,* 44, 363, 1990.

33. **Hames, C. A. C. and Hopkin, S. P.,** The structure and function of the digestive system of terrestrial isopods, *J. Zool. Soc. London,* 217, 599, 1989.

34. **Bubel, A.,** Histological and electron microscopical observations on the effects of different salinities and heavy metal ions, on the gills of *Jaera nordmanni* (Rathke) (Crustacea, Isopoda), *Cell Tissue Res.,* 167, 65, 1976.

35. **McCahon, C. P., Pascoe, D., and McKavanagh, C.,** Histochemical observations on the salmonids *Salmo salar* L. and *Ecdyonurus venosus* (Fabr.) following a simulated episode of acidity in an upland stream, *Hydrobiology,* 153, 3, 1987.

36. **Brown, B. E.,** The form and function of metal-containing "granules" in invertebrate tissues, *Biol. Rev.,* 57, 621, 1982.

37. **Casey, C. E. and Hambridge, K. M.,** Epidemiological aspects of human zinc deficiency, in *Zinc in the Environment, Part II,* Nriagu, J. D., Ed., Wiley-Interscience, New York, 1980, 1.

38. **Vallee, B. L. and Falchuk, K. H.,** Zinc and gene expression, *Philos. Trans. R. Soc. London,* B294, 185, 1981.

CHAPTER 12

The Use of Freshwater Invertebrates for the Assessment of Metal Pollution in Urban Receiving Waters

Brian Shutes, Bryan Ellis, Michael Revitt, and Andrew Bascombe

TABLE OF CONTENTS

0-87371-734-1/93/$0.00 + $.50

201

I. INTRODUCTION

Environmental legislation throughout the world is primarily based on a toxicity analysis approach, with ambient pollutant concentrations for specified exposure periods forming the basis for the development of water quality criteria. The methods used to identify such standards for environmental protection are essentially based on fixed dose-response rates determined from conventional bioassays. The importance of establishing links between (laboratory and field derived) toxicological and ecological data as a prerequisite for environmental management applications has been emphasized by many workers including Williams et al.[1]

In the U.K., intermittent pulse discharges have become targets for water quality upgrading and are viewed as priority objectives in terms of cost effective sewer performance criteria and receiving water impact analysis.[2] The influence of storm sewer overflow pollutant concentrations and loadings on receiving water ecology often implies a lower quality classification for urban rivers than that derived from conventional chemical monitoring[3] and the ecotoxicological basis for establishing short-term ecological criteria in these situations is insufficiently developed.[4]

This paper provides an assessment of the effect of increased metal (Cu, Pb, and Zn) loadings due to urban surface discharges and combined sewer overflows on selected macroinvertebrate species in an urban river in northeast London, and compares this field experience with conventional laboratory toxicity procedures.

The criteria for the selection of organisms as monitoring agents of metal pollution are now well established.[5] Benthic macroinvertebrate species represent important base components of the food web in aquatic organisms[6] and the detritivore species *Gammarus pulex* (L) and *Asellus aquaticus* (L) were selected as being appropriate test organisms for this study of in-stream metal uptake. *G. pulex* has been shown to have a higher metabolic rate than *A. aquaticus*[7] and a more rapid metal biouptake rate has been confirmed in comparative laboratory toxicity tests.[8]

A. aquaticus is found naturally at the most highly polluted river sites whereas *G. pulex* is only present on colonization samplers which provide artificial substrates.[9] This suggests that its absence may be related to a combination of loss of habitat from channelization with either acute stresses imposed by extreme

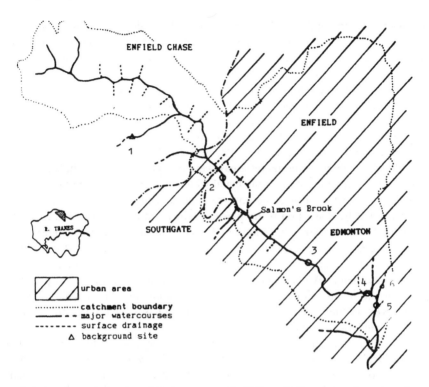

FIGURE 1. Location of sampling sites on the Salmon's Brook, northeastern London, U.K.

episodic events, and/or to chronic exposure following storm disturbance of benthic sediments which have been contaminated by the intermittent polluted discharges.

II. EXPERIMENTAL PROCEDURES

To investigate the site-specific behavior of macroinvertebrates to increasing levels of metal exposure, *G. pulex* and *A. aquaticus* were collected from unpolluted sites including an upstream tributary of the Salmon's Brook, a source of the River Lea in northeast London (site 1; Figure 1). The downstream sites on the Salmon's Brook progress from the outer fringes of the urban area (site 2), which has few contributing point sources, to locations immediately below discharges from a busy trunk road (site 3), and an industrial area (site 4). The most polluted river site however is located adjacent to a combined sewer overflow (site 5), where the impacts of undiluted storm overflows have also been separately monitored (site 6).

Fifty mature individuals, in similar instar, of each species from the background sites were transferred to two separate cages consisting of small plastic containers (1 l internal volume with nylon mesh windows of 1 mm pore size) to allow throughflow of water. The cages were firmly secured by clips and plastic coated wire to a basal concrete slab which was firmly anchored to the river bed at Sites 1 to 5. The artificial channel at site 6 required the cages to be suspended within the outfall discharge of the combined sewer overflow.

At weekly intervals (sites 1 to 5), and at appropriate intervals following a storm event (site 6), the cages were examined to determine organism mortalities. Simultaneously, live individuals of each species were removed from the cages and subsequently analyzed for Cu, Pb, and Zn. In addition to the total body metal burdens, three tissue target areas were analyzed. These included the adsorbed surface fraction of available metal which was removed by an initial dilute acid (2 M HNO$_3$) wash and the absorbed metal within the body tissue which was separated by dissection using plastic scalpels into soft tissue and hard exoskeleton fractions. After digestion in a 9:1 (v/v) nitric:perchloric acid mixture, samples were taken up in dilute nitric acid, filtered, and analyzed for Cu, Pb, and Zn by anodic stripping voltammetry or graphite furnace atomic absorption spectrophotometry. Water samples were collected automatically during storm events at site 6 and as manual grab samples on a monthly basis at sites 1 to 5. Metal levels in the soluble and suspended solid phases were determined using the same analytical techniques.

In addition to the in-stream experiments, laboratory toxicity tests were carried out in which 50 mature *Gammarus pulex* and *Asellus aquaticus* individuals from the background sites were exposed to fixed metal concentrations. A minimum number of 20 animals per test concentration is recommended for small organisms such as invertebrates.[10] A continuous-flow toxicity testing apparatus was designed for this purpose, based on that described by Green and Williams.[11] Invertebrates were simultaneously exposed in parallel tests to concentrations of 0, 50, 100, and 500 µg l^{-1} of each of the metals as solutions of their nitrate salts in water collected from a background site. When mortality in the control reached 5% of the test organisms, the associated series of tests was terminated. At least one duplicate of each test run was carried out. After 0.5, 1, 2, 4, 8, 12, and 24 h, and subsequently at 24 hourly intervals, any dead individuals were counted and removed, and tissue metal concentrations were determined in living individuals at 24-h intervals. The accumulation behavior of the metals concerned was then assessed in relation to the mortality of each species.

III. RESULTS AND DISCUSSION

A. In-Stream Metal Accumulation

The changes in mean total tissue concentrations of Pb and Zn in caged organisms over 7-week exposure periods are shown in Figures 2a to 2d. The coefficients of variation are 7.2, 14.2, 7.6, 9.3%, respectively in Figures 2a to 2d. Lead concentrations (25 to 30 µg g^{-1}) remain low at the control site and at

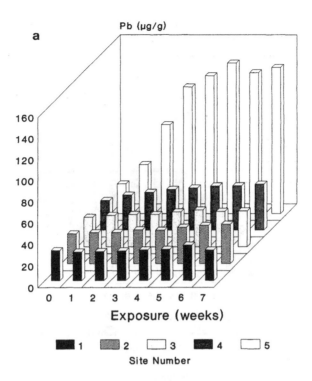

a

Pb (µg/g)

160
140
120
100
80
60
40
20
0

0 1 2 3 4 5 6 7

Exposure (weeks)

■ 1   ▨ 2   □ 3   ■ 4   □ 5

Site Number

FIGURE 2. Metal accumulation (dry weight) by caged (a) and (b) *Gammarus pulex* and (c) and (d) *Asellus aquaticus* at sites of increasing metal pollution.

sites 2 and 3 in both species. At site 4, some bioaccumulation is apparent in *G. pulex* (Figure 2a) but a marked progressive elevation of tissue Pb concentration occurs in *A. aquaticus* (Figure 2c), reaching equilibrium levels in excess of 90 µg g^{-1} over 5-week exposure periods. At site 5 there is a rapid initial metal accumulation rate to establish equilibrium concentrations of ~130 µg g^{-1} and ~110 µg g^{-1} in *G. pulex* and *A. aquaticus* after 4 and 5 weeks, respectively.

Tissue Zn concentrations in both species show a pronounced elevation in the first week of exposure at the two most polluted downstream sites followed by a slower linear uptake to equilibrium values of ~100 to ~190 µg g^{-1} (site 4) and ~200 to ~240 µg g^{-1} (site 5) in *G. pulex* and *A. aquaticus*, respectively (Figures 2b and 2d). Broadly similar trends to those observed for Zn were also obtained for Cu, with *G. pulex* and *A. aquaticus* attaining equilibrium concentrations of ~100 µg g^{-1} and ~90 µg g^{-1}, respectively. At the downstream site there is an overall increase in maximum equilibrium tissue concentrations of between four and five times compared to the background levels for Pb, Cu, and Zn.

The enhancement of ambient water and sediment metal levels at site 5 (Table 1) clearly produce biomagnifications in caged *G. pulex* as indicated by the total

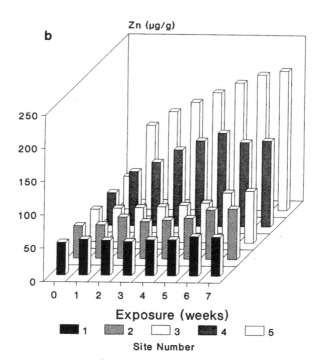

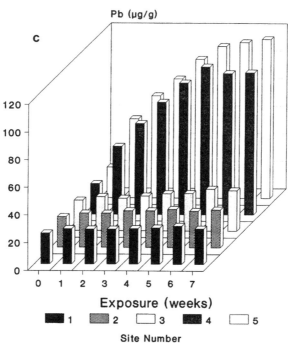

FIGURE 2b, c.

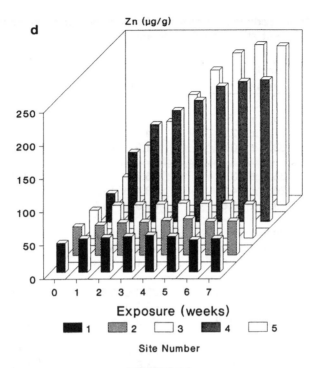

FIGURE 2d.

tissue metal levels (Figures 2a and 2b). *G. pulex* species have been shown to filter some 0.16 l of water per day to satisfy their oxygen demands.[12] This throughput is compatible with the observed rapid bioaccumulation of Zn at site 5, as concentrations of this metal are correspondingly higher for Pb or Cu in the aqueous phase (Table 1). A *G. pulex* individual of 10 mg wet weight has been estimated to require about 0.15 mg of food per day,[12] although there would be variation with food quality. Hence the high Pb bioaccumulation rate can be explained by the observed elevation in metal concentrations of the particulate phase (Table 1).

All the metals, but particularly Zn (Figure 2d), demonstrate a higher and equivalent tissue level in caged *A. aquaticus* at site 4 compared to site 3, despite only marginally elevated sediment, particulate, and dissolved metal concentrations at the former site (Table 1). This may be explained by a change in substrate composition from gravel at site 3 to unconsolidated silts downstream which facilitate metal uptake through improved feeding rates. However, only a small elevation occurs in caged *A. aquaticus* at site 5, whereas free-living organism tissue levels double as compared to site 4. This discrepancy can be explained by the increased flow rates at site 5 and the consequent reduction of settled particles in the cages which are consumed by this substrate feeding detritivore. Furthermore, the nylon mesh of the cage windows limits access to particles with a diameter of <1 mm, whereas natural populations have access to larger detrital

Table 1. Mean Sediment Suspended Particulate and Water Soluble Concentrations for Cu, Pb, and Zn

Sample source	Sediment[a] (μg g^{-1})	Suspended particulate[a] (μg l^{-1})	Soluble (<0.45 μm) (μg l^{-1})
Site 1			
Pb	19.7 ± 4.5	15.1 ± 7.8	33.8 ± 11.1
Cu	37.9 ± 12.2	34.3 ± 5.5	13.3 ± 2.3
Zn	83.6 ± 40.4	61.9 ± 9.9	38.5 ± 12.0
Site 2			
Pb	35.4 ± 13.7	16.7 ± 7.5	31.3 ± 11.3
Cu	29.5 ± 14.3	33.4 ± 10.0	17.4 ± 6.5
Zn	79.3 ± 20.8	69.0 ± 10.6	46.8 ± 16.2
Site 3			
Pb	130.3 ± 70	22.1 ± 10.4	40.0 ± 11.9
Cu	79.5 ± 35.2	40.9 ± 9.6	19.0 ± 6.2
Zn	161.6 ± 97.6	81.1 ± 17.8	61.3 ± 20.3
Site 4			
Pb	188.5 ± 109.4	23.6 ± 9.6	49.0 ± 10.8
Cu	95.8 ± 27.9	49.5 ± 11.1	20.4 ± 4.7
Zn	200.9 ± 106.9	96.0 ± 21.9	66.5 ± 34.9
Site 5			
Pb	219.5 ± 133.6	50.8 ± 22.4	88.6 ± 28.9
Cu	247.0 ± 65.0	83.9 ± 20.6	40.9 ± 20.5
Zn	445.3 ± 234.3	266.6 ± 88.4	155.4 ± 62.1
Site 6			
Pb	—	21.8 ± 4.4	7.3 ± 3.2
Cu	—	17.9 ± 4.1	22.1 ± 2.6
Zn	—	124.0 ± 40.1	343.0 ± 35.6

[a] Concentrations are expressed for dry weights.

particles. It is clearly desirable to compare tissue metal levels where possible in both captive and free-living populations to satisfactorily interpret metal accumulation behavior. It must be realized, however, that both river channelization and intermittent hydraulic shocks are also comparable in importance to ambient pollution levels in determining the availability of habitats and the distribution of species in the natural environment. Such physical factors are likely to have a significant effect on spatial variation in metal uptake rates. These parameters have been effectively ignored in the experimental technique and further work is needed to separately identify the nature and scale of such geometry and hydraulic controls.

B. Separate Tissue Metal Accumulation

The differential accumulations of Pb and Zn by soft tissue, exoskeleton, and surface adsorption at site 5 are shown for *G. pulex* and *A. aquaticus* in Figures 3a to 3d. Equilibrium concentrations in caged *G. pulex* and *A. aquaticus* for

a

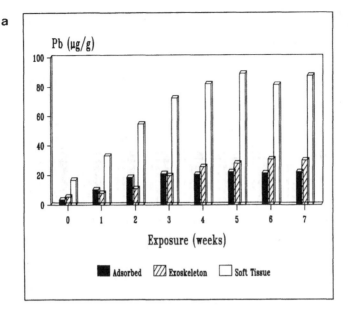

Standard Deviations								
Week	0	1	2	3	4	5	6	7
Adsorbed	0.6	1.3	1.6	1.0	1.5	1.4	1.4	0.7
Exosekeleton	1.7	1.3	1.4	0.6	1.3	0.6	0.7	0
Soft Tissue	4.7	4.3	5.3	6.1	4.2	7.9	2.8	2.8

FIGURE 3. Distribution of metal accumulation (dry weight) in caged (a) and (b) *Gammarus pulex* and (c) and (d) *Asellus aquaticus* at site 5.

adsorbed Pb ($\sim$22 and $\sim$25 μg g^{-1}), Zn ($\sim$42 and $\sim$50 μg g^{-1}), and Cu ($\sim$13 and $\sim$15 μg g^{-1}) are reached after 2 to 4 weeks, suggesting that the exoskeleton surface area is the limiting factor. The exoskeleton tissue metal concentrations, which normally exceed adsorbed concentrations, appear to reach equilibrium after 5 to 6 weeks, attaining values of 32 and 35 μg Pb g^{-1}, 54 and 62 μg Zn g^{-1}, and 13 and 15 μg Cu g^{-1} in *G. pulex* and *A. aquaticus*, respectively. A progressive and rapid bioaccumulation of Pb and Zn is apparent in soft tissue for *G. pulex* and *A. aquaticus* (Figures 3a to 3d) with equilibrium concentrations in *G. pulex* of 90, 70, and 110 μg g^{-1} and in *A. aquaticus* of 55, 50, and 130 μg g^{-1} for Pb, Cu, and Zn, respectively.

The equilibrium metal distributions in the different tissue components shown in Table 2 indicate the greater affinity of Pb and Cu for adsorbed and exoskeleton fractions in *A. aquaticus* as compared to *G. pulex*. However, the high proportion and rapid rate of soft tissue metal bioaccumulation, particularly of Pb and Cu in *Gammarus pulex*, confirms the suitability of this target area for monitoring metal toxicity over relatively short exposure periods. It is possible that metal

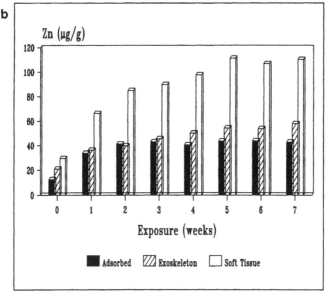

Standard Deviations								
Week	0	1	2	3	4	5	6	7
Adsorbed	2.1	3.3	4.5	2.6	2.2	5.3	2.8	1.4
Exosekeleton	5.3	5.6	6.4	9.5	9.0	8.3	2.8	4.2
Soft Tissue	7.5	7.2	18.4	7.3	6.9	4.9	2.5	3.5

FIGURE 3b.

diffusion across the exoskeleton to the soft tissue may be complementing contaminated particle ingestion and uptake through the gills.

C. Combined Sewer Overflow Metal Accumulation

Three separate overflow events have been monitored at site 6. Each of the macroinvertebrate species exposed to the undiluted overflow waters demonstrates a distinct increase in total tissue metal levels during the monitored post-storm periods, although the highest and most rapid metal accumulation to attained equilibrium concentrations in the adsorbed soft tissue and exoskeleton fractions is exhibited by Zn (Table 3; Figures 4a and 4b). In the 40-day period following overflow 1, the *G. pulex* total tissue Cu, Pb, and Zn concentrations achieved levels of 83, 86, and 254 μg g^{-1}, respectively. The increased bioaccumulation of Zn is consistent with the higher soluble levels of this metal (Table 1) and the high water throughput of *G. pulex* species to satisfy their oxygen demands. The cumulative effect of overflows 2 and 3, which occurred within the same monitoring period and are therefore considered together, produces an even higher total tissue Zn accumulation of 329 μg g^{-1} (Table 3). A similar enhancement for Zn between overflow 1 (298 μg g^{-1}) and overflows 2 and 3 (337 μg g^{-1})

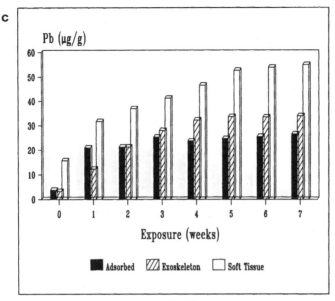

Standard Deviations								
Week	0	1	2	3	4	5	6	7
Adsorbed	2.5	2.0	3.2	2.1	2.5	1.5	2.1	2.1
Exosekeleton	0.6	1.5	3.5	4.0	4.4	1.5	3.5	1.1
Soft Tissue	2.1	1.5	1.7	3.1	3.1	2.5	2.8	1.4

FIGURE 3c.

is observed for *Asellus aquaticus*. However, the discrepancy between the total Zn levels between the two monitoring periods can be clearly seen to correspond to elevated adsorbed metal concentrations which existed when this experiment was terminated immediately following the overflow 3. The unique behavior of Zn at the overflow sites is demonstrated by comparisons with the results obtained at the downstream river location (site 5), where maximum total tissue metal levels are always lower for Zn but are consistently higher for Cu and Pb.

The patterns of Zn bioaccumulation rates in each species at site 6 (Figures 4a and 4b) are generally typical of those demonstrated by the other monitored metals, with total tissue levels exhibiting a relatively rapid post-storm increase followed by the attainment of an equilibrium level within a 3- to 5-week period. The individual tissue fractions, however, clearly exhibit different uptake behaviors. Thus, the surface-adsorbed *G. pulex* Zn fraction shows a pronounced enhancement following overflow 1 to a maximum level of 110 μg g^{-1} after 5 days. The subsequent decrease to a value of 42 μg g^{-1} Zn after 6 weeks is possibly a result of ionic diffusion into the exoskeleton and particularly the soft tissue, which both exhibit increasing accumulation rates to establish equilibrium levels after 4 to 5 weeks. The immediate post-storm response of soft tissue to Zn bioaccumulation indicates its importance as an early indicator of toxic metal

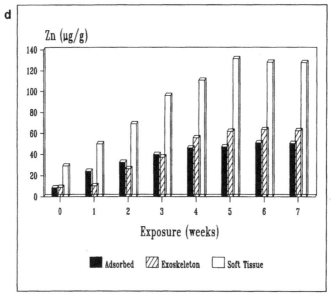

Standard Deviations

Week	0	1	2	3	4	5	6	7
Adsorbed	0.6	3.6	1.5	2.1	3.5	4.2	2.8	3.5
Exosekeleton	2.9	2.5	3.8	5.5	7.0	5.3	0.7	0.7
Soft Tissue	2.0	2.0	12.2	6.2	14.6	22.0	14.1	9.2

FIGURE 3d.

Table 2. Percentage Equilibrium Metal Tissue Distribution in Caged Organisms at Site 5 (After 5–6 Weeks)

	Pb		Cu		Zn	
	G	A	G	A	G	A
Adsorbed	15.9	22.9	13.4	17.1	20.3	21.0
Exoskeleton	21.4	29.4	15.4	26.4	27.4	26.0
Soft tissue	62.7	47.6	71.1	56.5	52.2	53.0

Note: G = *Gammarus pulex*, A = *Asellus aquaticus*.

uptake. The significant role of soft tissue as a target area is also demonstrated by the approximate doubling of the Zn levels over the full post-storm exposure period to a value which exceeds the exoskeleton level by a factor of over two. A similar accumulation of Zn in *G. pulex* soft tissue is exhibited at the downstream site 5, where Pb and Cu also show a distinct preference for this tissue fraction (Figure 3b).

The adsorbed metal fraction patterns for *A. aquaticus* exposed to overflows are similar to those noted in *G. pulex* although the immediate post-storm increases are less pronounced. The transfer of metal from the adsorbed fraction to the soft

Table 3. Equilibrium Metal Accumulation Levels, Cumulative Percentage Mortalities, and Mortality Rates for *Gammarus pulex* and *Asellus aquaticus* Exposed to Combined Sewer Overflows (Site 6) and Downstream River Waters (Site 5)

		Copper				Lead				Zinc			Cumulative percentage mortality	Mortality rate (% week^{-1})
		A	S	E	T	A	S	E	T	A	S	E	T		
Gammarus pulex	Overflow 1	21	35	27	83	23	37	26	86	42	145	67	254	48	8.0
	Overflows 2/3	20	22	18	60	44	43	28	115	156	129	54	329	88	29.3
	Site 5	14 ± 1.3	72 ± 5.4	17 ± 2.2	103 ± 4.7	10 ± 1.4	117 ± 7.9	11 ± 0.6	138 ± 13.0	43 ± 5.3	111 ± 2.5	58 ± 8.3	213 ± 8.8	53	8.8
Asellus aquaticus	Overflow 1	13	21	25	59	20	27	55	102	82	138	78	298	30	5.0
	Overflows 2/3	19	26	22	67	29	41	42	112	120	141	76	337	16	5.3
	Site 5	16 ± 1.5	55 ± 7.6	26 ± 1.5	97 ± 5.7	26 ± 1.5	55 ± 2.5	34 ± 1.5	115 ± 10.3	51 ± 4.2	127 ± 22.0	62 ± 5.3	240 ± 19.0	22	3.7

Equilibrium metal accumulation levels (μg g^{-1})

Note: A = adsorbed metal
S = soft tissue metal
E = exoskeleton metal
T = total tissue metal

Organism exposure times:
Overflow 1: 40 days
Overflow 2/3: 21 days
Site 5: 42 days

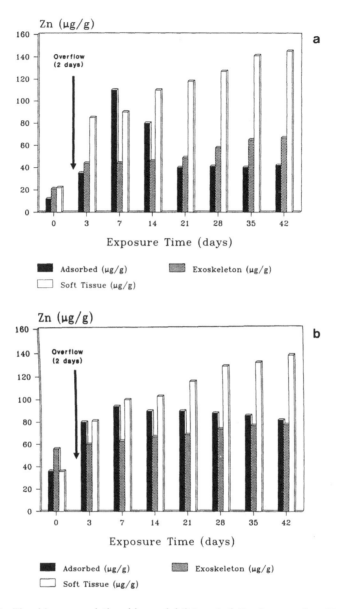

FIGURE 4. Zinc bioaccumulation (dry weight) trends following overflow No. 1 for (a) *Gammarus pulex* and (b) *Asellus aquaticus*.

tissue and exoskeleton of *A. aquaticus* is clearly evident for Zn, with the soft tissue again acting as the major metal sink at equilibrium (Figure 4b). This ultimate metal distribution is replicated at the downstream site (Figure 3d) but in the overflows, Cu and Pb tend to show selective bioaccumulation in the exoskeleton of *A. aquaticus* (Table 3).

D. Metal Uptake Mechanisms

The variation in metal accumulation at the combined sewer overflow and immediately downstream sites can be explained by differences in the duration of exposure to sources of severe contamination. At site 6 the organisms are exposed to acute episodes of metal pollution during and immediately following a storm event, whereas immediately downstream (site 5) there is also chronic exposure to the contaminated sediment and dry weather flow phases which is exacerbated by the intermittent storm and overflow discharge events. The cages located at the overflow site were suspended in the water column but at the downstream river site they were attached to the substrate, exposing the captive organisms to resuspended metal-contaminated sediment particles and interstitial waters. Although sediment levels at the downstream site are high for Cu (247 μg g^{-1}), Pb (219.5 μg g^{-1}), and Zn (445.3 μg g^{-1}), water levels are only high for Zn (particulate fraction, 226.6 μg l^{-1}, dissolved fraction, 155.4 μg gl^{-1}). The lower Cu and Pb total tissue levels in organisms exposed directly to the overflows are therefore to be expected given the absence of a complementary chronic source of contamination, but they reflect uptake essentially from the aqueous phase, which is clearly a significant source of metals during overflow events. At the river site the soft tissue metal levels are maintained throughout the exposure period by a continuous uptake from the less contaminated environment, whereas after the initial acute influence of the overflows a progressive transfer of Cu and Pb from the soft tissue to the exoskeleton occurs.

E. Cumulative Mortalities and Mortality Rates

The cumulative percentage mortalities and mortality rates for each of the species at site 6 are compared to those noted at the downstream site for similar exposure periods in Table 3. The patterns of increasing mortality during exposure to an overflow are illustrated in Figure 5. *A. aquaticus* mortality rates show an increase between the downstream site (3.7%/week) and the overflows which produced consistent rates (5.0 and 5.3%/week). *G. pulex* mortality rates show the greatest response to toxic metal exposure, with the cumulative influence of the second and third overflow events producing an 88% cumulative mortality which corresponds to a rate of 29.3%/week. The comparable value during the first 3 weeks of overflow 1 is 12.7%/week, which exceeds the average value at the downstream site (8.8%/week). In a similar study, Seager and Abrahams[4] found cumulative *G. pulex* mortality of 15.5%/week at a distance of 10 m below a combined sewer overflow. The higher sensitivity of *G. pulex* to particulate and dissolved metal contaminated water is emphasized by comparison of the estimated exposure times to kill 20% of the organism population. LT$_{20}$ values for *G. pulex* at the overflow and downstream river sites are 24 h and 13 days, respectively, with the corresponding values for the more tolerant *A. aquaticus* being 30 days and 41 days.

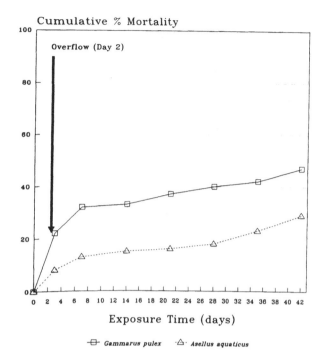

FIGURE 5. Cumulative percentage mortalities for organisms exposed to overflow No. 1.

The mortality rates of *A. aquaticus*, though elevated by exposure to storm sewer overflows, do not exceed 6%/week and indicate the suitability of both species for monitoring chronic toxicity. Although *G. pulex* mortality may be influenced by short exposure periods to depressed dissolved oxygen levels and elevated ammonia concentrations, there is a clear correlation between increases in aqueous metal levels, particularly Zn, and mortality rates. This supports the suitability of caged populations of *G. pulex* for monitoring acute episodes of toxic metal discharges. Free-living organisms have been used to monitor the impact of acute toxic episodes by studying recolonization rates[13] and the pairing behavior of precopula *G. pulex*.[14,15]

The laboratory-based metal accumulation experiments show that the variation with mortality is virtually independent of the solution metal concentration, with only Zn showing a significantly decreased accumulation in the most dilute solution (Figures 6a to 6d).[9] The lack of discrimination between solutions of different metal concentrations is unexpected and suggests, in the case of the laboratory experiments, the existence of a threshold concentration above which a uniform mortality-metal uptake relationship exists. The gradients of the plots for each of the laboratory experiments indicate a greatly enhanced toxicity over the 1-week time period of these tests. The slopes of the lines for *A. aquaticus*

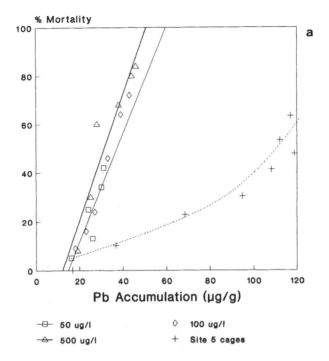

FIGURE 6. Soft tissue metal accumulation (dry weight) and mortality for (a) and (b) *Gammarus pulex* and (c) and (d) *Asellus aquaticus*.

are less steep than for *G. pulex*, indicating the greater tolerance of *A. aquaticus* to elevated levels of these metals, in agreement with previous toxicity studies.[16,17] In this study, the water hardness remained at 230 mg l^{-1} $CaCO_3$ throughout the laboratory experiments and field test concentrations were consistently similar, which would be expected to alleviate the toxic impact of the metals.

The increased toxicity demonstrated by the laboratory experiments may be partly explained by a greater metabolic activity resulting from the higher temperatures to which the test species were subjected. The laboratory organisms were continuously exposed to constant metal concentrations in comparison to the pulsed doses received by in situ animals. In addition, the organisms were presumably being exposed almost entirely to free metal ions, whereas only a small proportion of the soluble metal at site 5 would be present in this most bioavailable and toxic form. Previous studies of Cu, Pb, and Zn speciation in urban river systems[18] have shown that only Zn has a tendency to exist in the free ionic or very weakly complexed forms, with up to 50% of the soluble metal occurring as these species. The metal uptake and toxicity characteristics at site 5 are not, therefore, being controlled entirely by the bioavailable metal concentration, but are being more generally influenced by the physicobiochemical interactions between the organism and its enveloping sediment-water interface.

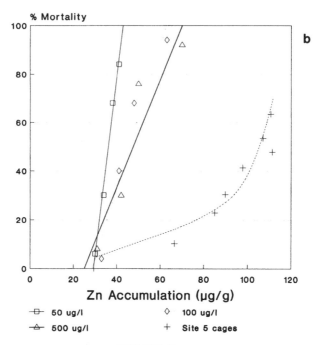

FIGURE 6b.

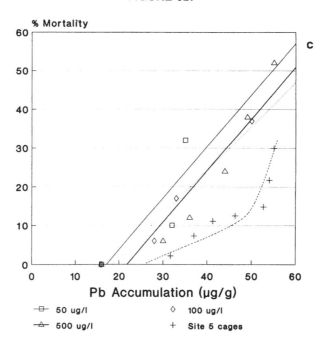

FIGURE 6c.

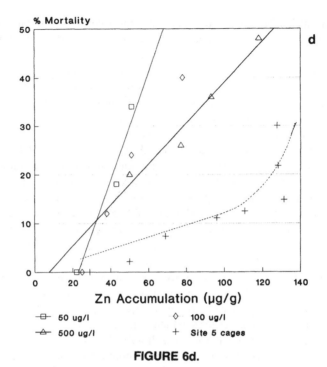

% Mortality

Zn Accumulation (µg/g)

- □ 50 ug/l
- △ 500 ug/l
- ◇ 100 ug/l
- + Site 5 cages

d

FIGURE 6d.

The form of the metal curves at site 5, as shown in Figures 6a to 6d, can be interpreted in terms of an initial linear accumulation to critical levels, which is followed by extremely rapid uptake and associated mortality rates which mirror the gradients obtained in the laboratory toxicity tests. The tissue levels of Pb and Zn associated with an *in situ* mortality rate of 20% are 70 and 80 µg g^{-1} for *Gammarus pulex* and 55 and 130 µg g^{-1} for *Asellus aquaticus*.[9] These trends may well be reflecting chronic exposure effects, whereas the laboratory toxicity tests indicate the influence of acute events.

A similar approach for assessing *G. pulex* behavior to the storm overflows produces a steeper initial response line, with the relationship between percentage mortality and Zn uptake remaining effectively uniform for the entire dose response curve. The more rapid dose response characteristics at the overflow site are consistent with the increased toxicity of intermittent highly polluted discharges.

IV. CONCLUSIONS

The results described in this paper demonstrate the inherent and complex problems associated with the prediction of toxicity or mortality rates from tissue pollutant accumulation rates. Water quality standards have traditionally been derived by consideration of LC_{50} values for a chosen test species over a 96-h period, but such laboratory experiments cannot accurately simulate all aspects of the natural sediment-water environment, particularly for short duration periods of between 1 to 12 h. Relationships between metal tissue levels and mortality can, however, provide a realistic indication of how the organism interacts with variable metal concentrations over varying time periods. Laboratory-based experiments, on the other hand, are inadequate in reproducing characteristic metal concentrations for the critical water, suspended particulate, and sediment phases. The in-stream metal accumulation experiments therefore provide a much more realistic indication of metal-induced mortality. A knowledge of the metal accumulation, detoxification, and excretion rates and soft tissue metal levels which correspond to the transition in mortality between the chronic and acute phases in an appropriate monitoring organism may therefore provide an advanced warning of a potentially severe toxic pollution problem in the receiving water system. The importance of sediment quality criteria as distinct from ambient water phase quality is also of significance in determining chronic exposure rates for benthic organisms in polluted urban receiving waters.

The described biomonitoring procedures can also be used to calibrate and verify the output models of sewer overflow and treatment plant effluents to ensure they will not prejudice environmental quality standards. The development of ecological performance criteria which can act as system design targets for hydraulic and management purposes is highly desirable. In addition, greater attention to the kinetics of toxic processes at both acute and chronic levels will help to reduce existing dependence on the use of generalized application factors of 96/168 h LC_{50} data and provide realistic acute criteria which possess adequate safety margins, such as LC_{20} values for very short-term exposures and short event return periods.

V. ABSTRACT

The levels of copper, lead, and zinc in the dissolved and particulate phases of a combined sewer overflow and in the receiving waters immediately downstream of this and other urban discharges have been determined together with the metal accumulation rates and associated mortalities in caged macroinvertebrates exposed to these aquatic environments. Total body metal burdens as well as selected tissue concentrations are discussed for both *Gammarus pulex* (L.) and *Asellus aquaticus* (L.). Zinc is consistently the most rapidly bioaccumulated metal to the highest levels, with metal equilibrium concentrations being established in the soft tissue fraction at four to five times the background control values after 5 to 6 weeks. Controlled toxicity tests show higher metal accumulation rates in comparison to the field bioassays, with the latter indicating a possible discrimination between chronic and acute impacts.

REFERENCES

1. **Williams, K., Green, D., and Pascoe, D.,** Toxicity testing with freshwater macroinvertebrates: methods and application in environmental management, in *Freshwater Biological Monitoring,* Pascoe, D. and Edwards, R. W., Eds., Pergamon Press, Oxford, 1984, 81.
2. **Ellis, J. B., Ed.,** *Urban Discharge and Receiving Water Quality Impacts,* Pergamon Press, Oxford, 1989, 2.
3. **Bascombe, A. D., Ellis, J. B., Revitt, D. M., and Shutes, R. B. E.,** The role of invertebrate biomonitoring for water quality management within urban catchments, in *Hydrological Processes and Water Management in Urban Areas,* Hooghart, J. C., Ed., IHP/UNESCO, The Netherlands, 1989, 403.
4. **Seager, J. and Abrahams, R.,** The impact of storm sewer discharges on the ecology of a small urban river, in *Urban Stormwater Quality and Ecological Effects upon Receiving Waters, Water Sci. Technol.,* 22(10/11), 22, 163, 1990.
5. **Martin, H. H. and Coughtrey, P. J.,** *Biological Monitoring of Heavy Metal Pollution: Land and Air,* Elsevier, Amsterdam, 1982.
6. **Mance, G.,** *Pollution Threat of Heavy Metals in Aquatic Environments,* Elsevier, Amsterdam, 1987.
7. **Maltby, L., Naylor, C., and Calow, P.,** Effect of stress on a freshwater benthic detritivore: scope for growth in *Gammarus pulex, Ecotoxicol. Environ. Saf.,* 19, 285, 1990.
8. **Naylor, C., Pindar, L., and Calow, P.,** Inter- and intraspecific variation in sensitivity to toxins; the effects of acidity and zinc on the freshwater crustaceans *Asellus aquaticus* (L) and *Gammarus pulex* (L), *Water Res.,* 24(6), 757, 1990.
9. **Bascombe, A. D., Ellis, J. B., Revitt, D. M., and Shutes, R. B. E.,** The development of ecotoxicological criteria in urban catchments, urban stormwater quality and ecological effects upon receiving waters, *Water Sci. Technol.,* 22(10/11), 173, 1990.
10. **Standing Committee of Analysts,** Acute Toxicity Testing with Aquatic Organisms, London, HMSO, 1983.
11. **Green, D. W. J. and Williams, K. A.,** A continuous flow toxicity testing apparatus for macroinvertebrates, *Lab. Pract.,* 32, 74, 1983.
12. **Abel, T. and Barlocher, F.,** Uptake of cadmium by *Gammarus fossarum* (Amphipoda) from food and water, *J. Appl. Ecol.,* 25, 223, 1988.
13. **Edwards, R. W., Ormerod, S. J., and Turner, C.,** Field experiments to assess the biological effects of pollution episodes in streams, *Verh. Int. Verein. Limnol.,* in press.
14. **Poulton, M. and Pascoe, D.,** Disruption of precopula in *Gammarus pulex* (L) — development of a behavioural bioassay for evaluating pollutant and parasite-induced stress, *Chemosphere,* 20, 403, 1990.
15. **McCahon, C. P. and Pascoe, D.,** Episodic pollution: causes, toxicological effects and ecological significance, *Functional Ecol.,* 4, 375, 1990.
16. **Cowley, C.,** The Influence of Road Runoff on the Benthic Macroinvertebrates of an Unpolluted Chalk Stream, Ph.D. thesis, Polytechnic of Central London, U.K., 1985.
17. **Martin, T. R. and Holditch, D. M.,** The acute lethal toxicity of heavy metals to peracarid crustaceans (with particular reference to freshwater Asellids and Gammarids), *Water Res.,* 20, 1137, 1986.

18. **Morrison, G. M. P., Revitt, D. M., Ellis, J. B., Balmer, P., and Svensson, G.,** The physicochemical speciation of Zn, Cd, Pb and Cu in urban stormwater, in *Planning and Control of Urban Storm Drainage,* Balmer, P., et al., Eds., Chalmers University, Gotheburg, Sweden, 1984.

CHAPTER 13

Evolution of Resistance and Changes in Community Composition in Metal-Polluted Environments: A Case Study on Foundry Cove

Paul L. Klerks and Jeffery S. Levinton

TABLE OF CONTENTS

0-87371-734-1/93/$0.00 + $.50
© 1993 by Lewis Publishers

I. INTRODUCTION

Environmental pollutants may have a variety of effects on communities. An alteration of the taxonomic composition is a commonly observed impact. For example, a reduction in the number of species has been reported for waters receiving metal-rich waste,[1-5] as well as for sites subject to both organic and inorganic pollutants.[6-8] Such a change in taxonomic composition of a community may be due to any of a wide variety of effects. These effects will be different for the multitude of species in a polluted site (due to differences in sensitivity among species or by virtue of the unique position that each species occupies in a foodweb), and may be as obvious as an increased mortality or reduced fecundity. For example, laboratory life-cycle exposures of crustaceans have shown reductions of the intrinsic rate of population growth by affecting survival, time till sexual maturity, number of broods, and brood development time.[9,10] But effects may be more subtle. For example, a toxicant may elicit an avoidance response by nonsessile organisms. This was demonstrated for the addition of hydrogen and aluminum to natural streams, which resulted in an increased emigration by several insect species.[11] A toxicant may also have an indirect effect on an organism. For example, an increase of predation in metal-dosed systems was shown for the predation of caddisflies by a stonefly,[12] and the predation of grass shrimp by killifish.[13] On the other hand, a reduction in growth of rainbow trout in insecticide-treated ponds was shown to be due to a reduction in prey abundance,[14] while an algal bloom in an insecticide-treated microcosm was due to the removal of grazers.[15]

The continued presence of a species in a polluted environment does not necessarily mean that pollutants are not having an effect (be it direct or indirect) on this species. For example, immigration of conspecifics from nearby unpolluted sites may be the basis of a population's survival in the polluted site. Another possibility is the occurrence of acclimation, where individuals become more resistant during their lifetime. Acclimation has been found for several populations and can be due, for instance, to the induction of detoxifying enzymes.[16,17] Finally, the continued presence of a species in a polluted environment despite an initial deleterious effect may be due to the evolution of resistance to the pollutant(s). This factor has obvious implications for long-term effects of both local pollution and global environmental changes, and will receive an in-depth treatment in this chapter.

The evolution of resistance is best known for resistance to pesticides in pest species. At least 504 species of insects or mites are known to be resistant to pesticides,[18] which demonstrates that resistance evolves commonly and rapidly. Factors that determine the occurrence and speed with which resistance to environmental pollutants evolves are expected to be similar to those identified for the development of resistance to pesticides.[19] The presence of heritable variation for resistance, selection pressures, immigration rates, etc. will all influence the evolution of resistance in a natural population that is exposed to an environmental pollutant. Resistance to environmental pollutants has been reported for several toxicants and species; for example, resistance to sulfur dioxide in *Geranium carolinianum*,[20] resistance to metals in several species of plants on abandoned mines,[21,22] resistance to metals in bacterial strains isolated from tanks used in the metal-processing industry,[23] and resistance to metals in an aquatic crustacean.[24] Yet insufficient data are available, especially for populations of animals, to determine how likely it is that resistance will evolve in polluted environments.[25,26] In addition, insight is lacking on the interplay between the evolution of resistance and changes at the community level brought about by an environmental pollutant. This study looked at both aspects of the effects of metals in invertebrates exposed to the very high levels of a few metals in Foundry Cove.

Foundry Cove is a tidal freshwater marsh on the Hudson River, located near Cold Spring (NY) approximately 87 km upriver from the mouth of the Hudson River. Waste from a nickel-cadmium battery factory was discharged into Foundry Cove from 1953 to 1971. Cobalt was released into the cove during some periods when it was being used as an additive. In spite of dredging in 1972 and 1973 to remove the most polluted sediment, cadmium levels as high as 50,000 μg Cd per g dry sediment were found in 1975.[27] Generally lower Cd concentrations were found in surface sediment in 1983, though surface concentrations approaching 40,000 μg/g Cd and subsurface Cd concentrations of 225,000 μg/g Cd were encountered.[28] Cadmium, nickel, and cobalt in Foundry Cove sediment generally occur in constant molar ratios of approximately 18:20:1, respectively.[28] The cadmium in Foundry Cove is bioavailable to plants, fish, and invertebrates.[29-31] Klerks[31] also reported accumulation of nickel by the benthic macrofauna, but did not find evidence that the cobalt was bioavailable. A nearby cove, South Cove, was used as a control site. Cadmium levels in this cove averaged about 19 μg/g, somewhat above baseline Cd levels but well below levels found in Foundry Cove.[28]

This study investigated effects of the metals in Foundry Cove on both the development of resistance and effects on the taxonomic composition of the benthic macrofauna. Effects on the composition of the macrobenthos were assessed by two taxonomic surveys. The development of resistance was determined by quantifying resistance in the oligochaete *Limnodrilus hoffmeisteri* and the chironomid *Tanypus neopunctipennis* collected from Foundry Cove, their offspring (for the oligochaete species), and animals collected from the control site. In addition, the presence of genetic variation for resistance was determined in *L. hoffmeisteri*, by quantifying the amount of heritable variation for metal resistance in the control population and by performing a laboratory selection experiment.

II. EFFECTS OF METALS IN FOUNDRY COVE ON THE COMMUNITY COMPOSITION

In an initial survey, benthic macrofauna samples were collected by taking sediment cores (surface area 35 cm^2) over a depth of 0 to 5 cm. Replicate samples were taken in the control area (19 μg/g Cd) and at three sites in Foundry Cove (approximately 500, 7000, and 52,000 μg/g Cd). At 4 of the sites, 5 cores were taken at 3 subsites that were approximately 10 m apart, while at the site with the highest Cd level (F.C.3, the original outfall site) only 1 set of 5 cores was taken as this site covered only a small area. The macrofauna was separated from the sediment using a 500 μm sieve, preserved, and later sorted under a dissecting microscope. The most abundant taxa, oligochaetes and chironomids, were usually identified to the species or genus level. Other taxonomic groups were generally identified at the family level. Results are reported here for the last two (October 11 and December 12, 1984) out of five sampling dates, since site F.C.3 was not included in the first three sampling dates. The existence of extremely high Cd levels (at F.C.3) was not discovered till the Fall of 1984. Analyses used subsite averages (means of five cores) and data for the two sampling dates were combined in the absence of significant differences among sampling dates and significant interactions between sampling date effects and site effects.

Densities of macrobenthic organisms did not differ significantly among the various sites (Figure 1a). This result held for the combination of all taxa, as well as the combination of oligochaetes and chironomids ($p > 0.05$ in ANOVA on log-transformed data). There is therefore no evidence that the high metal levels in Foundry Cove reduced the total number of organisms present. The situation was different however when the number of taxa were compared among the sites (Figure 1b). Both for all taxa combined as well as for oligochaetes and chironomids, the site with the highest Cd level showed a significantly reduced taxonomic diversity relative to all other sites ($p < 0.05$ in *a posteriori* pairwise comparisons on square root transformed data). The combination of an unchanged abundance and a reduced number of species means that some taxa are more

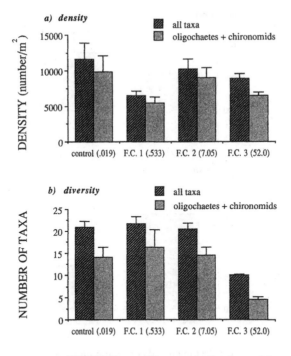

FIGURE 1. Densities of benthic macrofauna (a) and number of taxa (b) for all benthic organisms and the combination of oligochaetes + chironomids, in a site in the control area and three sites in Foundry Cove. Values are mean + S.E. for two sampling dates in 1984. Note that sediment Cd concentrations at the collection sites are reported in mg/g.

abundant at the outfall site (F.C.3). This was the case, for example, for the nematodes, the chironomid *T. neopunctipennis,* and the oligochaete *L. hoffmeisteri.* Absent from the outfall site were, for example, the chloromid tribe Tanytarsini and genera *Polypedilum, Einfeldia,* and *Cryptochironomus.* The chironomid genus *Procladius* also showed a statistically lower abundance at the outfall site, yet was present at this site. A possible complication of the conclusions on the taxonomic analyses could have been formed by differences in habitat type; the outfall site was located in a small cul-de-sac in Foundry Cove, while the other sites were located in the main parts of the cove. This left open the possibility that this difference in habitat-type was responsible for the reduced diversity in the outfall site rather than differences in metal levels.

To determine if the observed difference in diversity held up among sites with different metal levels yet similar in habitat-type, samples were collected on September 19, 1986 in both the outfall site and a nearby site also located in a cul-de-sac. In each site, 15 cores were taken. Sediment Cd concentrations in

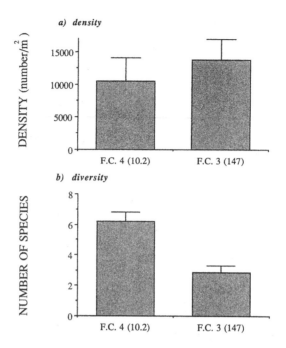

FIGURE 2. Densities of oligochaetes and chironomids combined (a) and number of chironomid and oligochaete species (b) in two sites in Foundry Cove with different metal levels. Values are mean + S.E. for samples collected in 1986. Sediment Cd concentrations were determined in the <500 μm fraction.

the <500 μm fraction of the sediment collected together with the macrobenthos, were 147,000 μg/g for the outfall site and 10,200 μg/g for the other site (F.C.4). Remember that the earlier survey had shown that the taxonomic diversity at a site with a Cd concentration of 7050 μg/g could not be distinguished from that at a control site. Results, based on averages per core, are similar to the results reported for the earlier survey. Densities of the combination of oligochaetes and chironomids (Figure 2a) did not differ among the sites ($p > 0.05$ in ANOVA on log-transformed data). Taxonomic diversities as expressed by the number of species (Figure 2b), were again lower at the site with the highest metal levels ($p < 0.001$ in ANOVA on $\sqrt{n}$-transformed data). It thus appears that the reduced diversity in the outfall site is indeed a consequence of the high metal levels at this site.

III. RESISTANCE TO METALS IN TWO SPECIES OF MACROBENTHOS OF FOUNDRY COVE

To determine if some of the macrobenthos species present in Foundry Cove exhibited an elevated resistance, laboratory bioassays were performed with two species: the chironomid *Tanypus neopunctipennis* and the oligochaete *Limnodrilus hoffmeisteri*. Animals collected in the control area and in the Foundry Cove site, with sediment Cd levels of about 7000 µg/g, were exposed in the laboratory to sediments from each site. *L. hoffmeisteri* was exposed for 28 days, using three replicates of 10 worms for each treatment. The chironomids were exposed for 14 days (to allow for their short life-cycle), starting the experiment with three replicates of 50 first-instar larvae (0 to 1 day old) per treatment. These larvae came from laboratory cultures that used sediment from the same site as the chironomids with which the cultures were started. All organisms were fed during the experiments. Survival was determined at the end of the exposure periods.

Results for *L. hoffmeisteri* (Figure 3a) show that these worms from Foundry Cove have an elevated resistance to the Cd-rich sediment; the difference in survival between worms from the two sites was highly significant ($p < 0.001$ in ANOVA on square root of number of survivors). None of the worms from the control area survived the exposure to Foundry Cove sediment, whereas 90% of the worms from Foundry Cove survived in this sediment. Experiments exposing worms from Foundry Cove and the control area to 8.9 μM Cd, 10.2 μM Ni, and 0.5 μM Co in water (the same ratio as observed for these metals in Foundry Cove sediment) confirmed that they are more resistant to the combination of Cd, Ni, and Co.[32]

The results were different for *T. neopunctipennis* (Figure 3b); no difference in survival could be detected ($p > 0.05$ in ANOVA on square root of number of survivors), though the Foundry Cove sediment resulted in a drastic reduction of the number of surviving chironomid larvae. Additional experiments with this species, exposing them for a longer part of their life-cycle (from larvae to adults or egg to adult stage), using more replicates per group and using sediment with Cd levels as high as 45,000 µg/g, did not result in the detection of differences in resistance between *T. neopunctipennis* from Foundry Cove and their conspecifics from the control area.[31]

As noted in the introduction, an increased resistance in organisms collected from a polluted site can be due to either acclimation or adaptation. To determine if the increased resistance to Cd-rich sediment in *L. hoffmeisteri* collected from Foundry Cove had a genetic basis, resistance in laboratory-reared offspring was investigated.[32] Worms from Foundry Cove were kept in clean sediment for two generations. Resistance to Cd-rich sediment was then compared among these offspring and worms collected from Foundry Cove and the control site. This was done in the same 28-day exposure procedure described earlier, except that sediments with a wider range in Cd concentrations were used in this experiment.

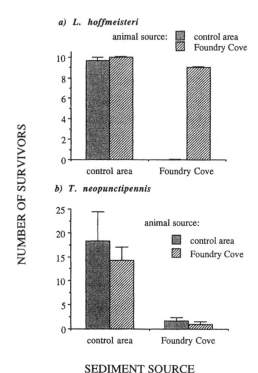

a) *L. hoffmeisteri*

b) *T. neopunctipennis*

NUMBER OF SURVIVORS

SEDIMENT SOURCE

FIGURE 3. Resistance comparison for the oligochaete *L. hoffmeisteri* (a) and the larval stage of the chironomid *T. neopunctipennis* (b) between animals from the control site and individuals from Foundry Cove. Organisms from both sites were exposed to control sediment and metal-rich Foundry Cove sediment. Values are mean + S.E. (n = 3). Oligochaetes were exposed for 28 days (with 10 worms per replicate on day 0), while chironomids were exposed for 14 days (starting with 50 first-instar larvae). (Figure 3a is modified from Klerks, P. L. and Levinton, J. S., *Biol. Bull.,* 176, 135, 1989. With permission.)

Sediments with Cd concentrations ranging from about 20,000 to 90,000 μg/g were obtained by collecting sediment from various sites in Foundry Cove, sieving this to a less than 250 μm particle size, combining aliquots of sediment from different sites in various ratios, and stirring the combined sediment manually. Results of this experiment (Figure 4) show that very few of the worms survived the exposure to the sediment with Cd levels in excess of 50,000 μg/g. However, worms collected from Foundry Cove and their second generation offspring exhibited overall a significantly higher survival than the worms from the control area (in a multiple *t*-test at α = 0.05 on square root of number of survivors). The results of this experiment thus show that the resistance in *L. hoffmeisteri* from Foundry Cove is largely genetically determined. Overall,

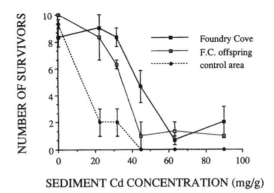

FIGURE 4. Resistance comparisons among *L. hoffmeisteri* from the control site, Foundry Cove, and second generation offspring of the Foundry Cove worms reared in the laboratory in control sediment. Worms were exposed for 28 days to sediment with various metal levels. Values are mean ± S.E. (n = 3, with 10 worms per replicate on day 0). (From Klerks, P. L. and Levinton, J. S., *Biol. Bull.*, 176, 135, 1989. With permission.)

no statistically significant difference in survival was detected between the worms collected from Foundry Cove and their offspring. However, the offspring tended to show a somewhat lower survival rate than the field-collected Foundry Cove worms. This reduction was statistically significant at one of the intermediate Cd levels, indicating a small environmental component in the resistance or a reduction in resistance upon relaxation of the selection pressure.

IV. GENETIC VARIATION FOR RESISTANCE IN *LIMNODRILUS HOFFMEISTERI* FROM THE CONTROL AREA AS DETERMINED BY HERITABILITY ESTIMATES AND A SELECTION EXPERIMENT

The genetic adaptation of *L. hoffmeisteri* to the high metal levels in Foundry Cove demonstrated that genetic variation for this resistance was initially present in this population. Frequency distributions of resistance among worms collected from specific sites in Foundry Cove (with resistance being determined as survival time upon exposure to a combination of Cd, Ni, and Co in reconstituted fresh water), indicated that the resistance in *L. hoffmeisteri* was a polygenic trait.[31] However, we do not have any more direct evidence on the number of genes involved in this resistance. Research addressing this issue is currently underway. For assessing the speed at which the evolution of resistance can occur, it is important to quantify the amount of genetic variation for resistance. This variation in a population not subjected to an intense selection was assessed by determining

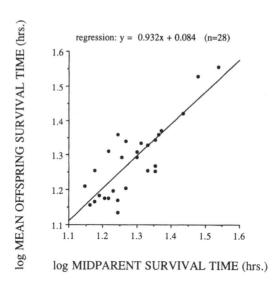

FIGURE 5. Relation between resistance in pairs of *L. hoffmeisteri* from the control site and their offspring. Resistance determined as survival time in water with 8.9 μM Cd, 10.2 μM Ni, and 0.5 μM Co.

the heritability of resistance to the combination of Cd, Ni, and Co in the control population, using standard quantitative genetics procedures.[33] The heritability (also called narrow-sense heritability) of a trait is the proportion of the phenotypic variation that consists of additive genetic variation (i.e., the variation that selection acts upon).

Animals were reared in pairs, and offspring were regularly separated from the parents. Metal resistance was then quantified by determining survival times of parents and offspring, in water spiked with 8.9 μM Cd, 10.2 μM Ni, and 0.5 μM Co. The regression of mean offspring survival times on midparent survival times, shown in Figure 5, translates into a heritability of 0.93 $\pm$ 0.12 (S.E.). This is very close to the theoretical maximum of 1. Applying a weighting factor (based on the number of offspring in each family) to offspring mean resistance values resulted in a virtually unchanged heritability estimate of 0.92 $\pm$ 0.12. Another estimate, obtained from the regression of individual offspring survival times on individual parent values, was very high as well: 1.07 $\pm$ 0.10 (n = 455). Following the selection experiment (see below), a heritability estimate was obtained from the regression of the selection response on the cumulative selection differential (a cumulative measure of the selection applied). This realized heritability of 0.59 $\pm$ 0.14 was somewhat lower than the estimates obtained from the resemblances between relatives, but again indicated that a large part of the variation for resistance to the combination of Cd, Ni, and Co in the control population consisted of (additive) genetic variation. This means that the resistance in Foundry Cove could have evolved very rapidly, assuming similar heritabilities in this population prior to the metal waste discharge. For example, at a heritability of 0.6 and the top 40% in resistance of the population contributing to the next generation, the resistance observed in Foundry Cove would have evolved in about four generations.

The potential for the evolution of resistance in a control population was also addressed by performing a laboratory selection experiment. Three selection lines and one control line were started, with 441 juveniles each. Juveniles of the four lines came from the same source, pooled laboratory populations of worms from the control area. The three selection lines were exposed as adults to either metal-rich Foundry Cove sediment (in generation 0) or to Cd, Ni, and Co in water (in subsequent generations, this change being implemented to speed up the selection process). Surviving adults were allowed to reproduce in control sediment. Subsequently, the next generation was then started with 500 of their offspring, in each line. For the control line, the number of adults was reduced in each generation to the same (average) number that survived the selection exposures in the other lines. This was done to obtain the same effective population size (and thus inbreeding coefficient) in the control line as in the selected lines. The selection was continued for three generations and, on average, the individuals belonging to the top 14% in resistance were selected to contribute to the next generation. In each generation, resistance was quantified by determining survival times of adults exposed to 8.9 μM, 10.2 μM Ni, and 0.5 μM Co in reconstituted fresh water. The response to the selection was very rapid; a strong increase in resistance was obtained by the second generation (Figure 6).

The final difference in resistance between the selected lines and the control lines amounted to about 66% of the difference in resistance that was found between *L. hoffmeisteri* collected from Foundry Cove and their conspecifics from the control site. At this rate, the field-observed differences could have been obtained in four to five generations of selection at the conditions used in our laboratory selection procedure.

V. DISCUSSION

The demonstration of a reduced number of macrobenthic taxa at the most cadmium-, nickel-, and cobalt-polluted site in Foundry Cove is consistent with other studies in metal-polluted environments.[1-5] Surprisingly, no difference was found among the control area and the sites in the main part of Foundry Cove, despite sediment Cd levels as high as 7000 $\mu g/g$ in the latter area. The fact that no difference in total number of taxa among these sites was detectable could be due to either a low metal-bioavailability, a low sensitivity of this parameter, or migration among sites. Other aspects of our study favor the explanation involving a low sensitivity of changes in the total number of taxa. It has been pointed out that such broad indices as species diversity are not very sensitive in detecting responses to stress, even though specific changes in species composition are among the earliest detectable responses.[34]

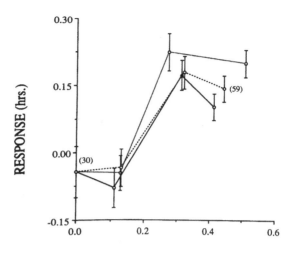

CUMULATIVE SELECTION DIFFERENTIAL

FIGURE 6. Response to selection in *L. hoffmeisteri* from the control site selected
for an increased resistance to Cd, Ni, and Co during three generations.
Response expressed as the difference in resistance (determined as
survival time in 8.9 μM Cd, 10.2 μM Ni, and 0.5 μM Co in water)
between a selected line and the control line. Values are mean (with
95% confidence interval) of the difference with the control line, with
the three different line symbols each referring to a different selection
line; n = 60 for the determinations of survival times, except where
noted otherwise. (From Klerks, P. L. and Levinton, J. S., *Biol. Bull.*,
176, 135, 1989. With permission.)

Our study did not detect any differences in total density of the macrobenthos
among the differently polluted sites. This contrasts with studies conducted in
Foundry Cove during 1973 and 1974.[35] The difference might reflect the reduction
of Cd levels in Foundry Cove between 1974 and 1984, or could be related to
the development of resistance in *L. hoffmeisteri*. If this species in Foundry Cove
has become more resistant since 1974, a further increase in its abundance would
have offset the lower abundance of other species. As mentioned earlier, our
survey in 1984 showed that *L. hoffmeisteri* is much more abundant at the most
polluted site than at the other sites.

The oligochaete *L. hoffmeisteri* in Foundry Cove was shown to have evolved
a resistance to the high metal levels in this marsh. This adaptation has taken
place in less than 30 years, demonstrating that response to selection in natural
populations may be very rapid. This finding agrees with observations on the
development of resistance in natural populations of pest species treated with
pesticides. A review of the development of resistance to pesticides reported a
rather narrow range of typically 5 to 50 generations of selection till resistance
evolved, among numerous pesticide resistance cases.[36] Both the result of our

selection experiment and the estimates of the heritability of resistance (of 0.59 to 1.07) demonstrate that resistance can evolve very rapidly. Similarly high heritabilities of resistance to stress have been reported by others: a heritability of 0.50 for resistance to sulfur dioxide in *Geranium carolinianum*,[37] and a heritability of 0.64 of resistance to desiccation in *Drosophila melanogaster*.[38]

The resistance in *L. hoffmeisteri* from Foundry Cove is not due to a reduced metal accumulation,[39] though this mechanism has been shown to be responsible for some cases of an increased metal resistance. For example, a copper-resistant strain of the alga *Ectocarpus siliculosus* was shown to accumulate less copper than their copper-sensitive conspecifics.[40] It appears that in worms from Foundry Cove, more Cd is detoxified by binding to a metallothionein-like protein than in *L. hoffmeisteri* from the control area.[39] In addition, Cd occurred in granules in worms from the metal-resistant population. Small (1-μm scale) granules and larger (up to 30 μm) granular aggregates were found in which Cd was possibly present as CdS. In addition, Cd was present in other small granules, possibly as a mixed calcium-cadmium-iron phosphate.[39]

Our resistance comparisons showed that *L. hoffmeisteri* in Foundry Cove has adapted to the high metal levels, whereas no resistance differences could be detected for the chironomid *T. neopunctipennis*. This difference between the two species could be due to the much lower metal accumulation in the chironomid than in the oligochaete.[31] Moreover, it is also expected that the amount of gene flow among sites will be higher for the chironomid (with a flying adult stage) than for the worm species. The choice of species to be included in the resistance comparisons was based on the availability of sufficient individuals for conducting experiments, as well as the success of maintaining laboratory cultures. For example, of the several species of chironomids brought into the laboratory, only *T. neopunctipennis* reproduced under the conditions that were used. The use of *L. hoffmeisteri* and *T. neopunctipennis* in the resistance comparisons meant that these comparisons were performed only with the two species that were most abundant at the site with the highest metal levels. Though the organisms used in the resistance comparisons were collected in a site in Foundry Cove where metal levels were lower (and where the two species in question did not exhibit an elevated abundance), it is unlikely that the resistance comparisons were representative of all the species present in the control area. Consequently, our study showed that some species may evolve a resistance to environmental pollutants, rather than demonstrating that resistance will evolve in many of the species in a polluted environment. Moreover, it has been pointed out that in order to assess the consequences of environmental change, one should worry about the species that do not evolve.[41] Now it has been shown by several investigations that some species can adapt to a changed environment, more insight might be obtained by investigating species that have disappeared from a changed environment.

VI. FUTURE DIRECTIONS

It has now been well established that evolutionary responses to toxic substances are common, but uneven in effects. Current research, however, has not
been complete enough usually to understand the genetic architecture of adaptation
to metals, nor have studies led to an understanding of potential interactions with
neighboring populations. Our current understanding of the rapid evolution of
resistance in *L. hoffmeisteri* permits exploration of some of these problems.

A. Cost of Adaptation

There clearly is much allelic variation for resistance in populations in relatively clean environments. This allows a rapid response to selection, as has
been observed in Foundry Cove and in our selection experiments. We do not
know, however, whether there is a **cost** to the adaptation to metals. We define
cost as the reduction in fitness of a metal-adapted genotype. Owing to genotype-
environment interactions we expect that the costs can be of two types: (I) costs
associated with exposure to the metal; and (II) costs associated with the evolution
of the adaptation itself. In the first case, the ability to sequester large amounts
of cadmium may have extensive metabolic costs. Production of large amounts
of the protein metallothionein may be costly, as may be the suite of transport
processes used to sequester and eventually excrete the metal. Even when the
animal is not exposed to metal, the possession of metal-resistance genes may
have type II costs, since they may automatically induce a set of stereotyped
physiological reactions which cost energy. It is possible, for example, that resistant strains automatically produce more sequestering organelles, which re-
moves energy that might be allocated to other functions. Type II costs may also
include changes in genes not related to metal resistance, but related to the
selection process itself. For example, strong natural selection for metal resistance
might involve the breakup of a strongly coadapted genome. In addition, genes
linked to the genes involved in resistance might hitchhike to increased frequency
in the population. This might be tested by an allozyme survey of the selected
population, as compared to a metal-vulnerable population.

An understanding of these costs is important in assessing the potential for
gene flow with other populations. If type I costs matter, resistant genotypes may
do poorly in polluted habitats, and will contribute fewer offspring and hence
fewer colonists to adjacent habitats. If type II costs are present, the resistant
genotypes may disperse to other habitats, but they will not persist in very great
abundance, since they are at an energetic disadvantage relative to local genotypes.

Type I costs would be encumbered only when the animal is exposed to
metals. Type II costs may not be present, since the adaptation may possibly be
explained by gene amplification. There is only limited evidence for amplification
in insecticide resistance.[42] In *Drosophila melanogaster*, cases of gene duplication
for the metallothionein gene are known, and these might influence adaptation

to metals.[43] If gene multiples are selected in resistant populations, then the type II costs of adaptation may just involve the possession of more copies of a gene, which only involves synthesizing a little bit more DNA. However, type II costs may be substantial, even if resistance is due to gene amplification (for example, where the amplification results in elevated basal levels of metallothioneins).

B. Number of Genes Explaining the Trait

It is useful to understand the quantitative genetics of resistance, as we have demonstrated above for *L. hoffmeisteri*. The presence of resistant and susceptible populations also allows a study of the number of segregating genetic control elements, which we are now performing (Levinton and Martinez, in preparation). Such an analysis of resistance has been performed for the resistance to arsenic in the grass *Agrostis capillaris*.[44]

An algorithm to determine the number of segregating factors was first developed by Wright[45] and has been discussed succinctly with several applications.[46] Briefly, one crosses populations with different values of a trait to produce an F1 generation, and then calculates the increase of variance in the trait found in the F2. Lande[46] found that many morphological traits are underpinned by five to ten factors, which is a minimum estimate for the number of controlling genes. In cases where rapid evolution occurs, there is reason to believe that an allele at one locus might be recruited and would be the major source of evolutionary change. Major genes with modifiers, as observed for the As resistance in *A. capillaris*,[44] may be a common genetic architecture for traits under strong selection.

If only one factor underlies the resistance, then the trait is segregating as a Mendelian gene with two alleles. If we find this for *L. hoffmeisteri*, it is reasonable to conclude that the metallothionein gene is completely responsible for the genetic aspects of the evolution of resistance.

C. Molecular Studies of the Metallothionein Gene

Because a metallothionein-like protein is found in very high activity in the resistant strain of *L. hoffmeisteri*, we plan to investigate the molecular biology of adaptation by examining directly possible changes in the metallothionein gene(s), which have not yet been investigated in oligochaetes. Such genes have been identified and sequenced in several organisms, including the nematode *Caenorhabditis elegans*,[47] the fungus *Candida glabrata*,[48] the sea urchin *Strongylocentrotus purpuratus*,[49] and the fruit fly *Drosophila melanogaster*.[50] Using either protein purification and cloning or PCR, it should be possible to sequence the gene, and to search for cases of duplicates or sequence differences in the metallothionein genes of metal-adapted individuals.

ACKNOWLEDGMENTS

This research was financed, in part, through a research grant from the Hudson River Foundation for Science and Environmental Research, Inc. (New York, NY). This is contribution No. 816 from the Graduate Program in Ecology and Evolution of the State University of New York at Stony Brook.

REFERENCES

1. **Carpenter, K. E.**, A study of the fauna of rivers polluted by lead mining in the Aberystwyth district of Cardiganshire, *Ann. Appl. Biol.*, 11, 1, 1924.

2. **Winner, R. W., Boesel, M. W., and Farrell, M. P.**, Insect community structure as an index of heavy-metal pollution in lotic ecosystems, *Can. J. Fish. Aquat. Sci.*, 37, 647, 1980.

3. **Bryan, G. W. and Gibbs, P. E.**, Heavy metals in the Fal Estuary, Cornwall: A study of long-term contamination by mining waste and its effects on estuarine organisms, Mar. Biol. Assn. U.K., Plymouth, 1983.

4. **Malueg, K. W., Schuytema, G. S., Krawczyk, D. F., and Gakstatter, J. H.**, Laboratory sediment toxicity tests, sediment chemistry and distribution of benthic macroinvertebrates in sediments from the Keweenaw Waterway, Michigan, *Environ. Toxicol. Chem.*, 3, 233, 1984.

5. **Newell, R. C., Maughan, D. W., Trett, M. W., Newell, P. F., and Seiderer, L. J.**, Modification of benthic community structure in response to acid-iron wastes discharge, *Mar. Pollut. Bull.*, 22, 112, 1991.

6. **Butcher, R. W.**, Relation between the biology and the polluted condition of the Trent, *Int. Ver. Theor. Angew. Limnol. Verh.*, 12, 823, 1955.

7. **Reish, D. J.**, An ecological study of pollution in Los Angeles — Long Beach Harbors, California, *Occas. Pap. Allan Hancock Found.*, 22, 1, 1959.

8. **Becker, D. S., Bilyard, G. R., and Ginn, T. C.**, Comparisons between sediment bioassays and alterations of benthic macroinvertebrate assemblages at a marine superfund site: Commencement Bay, Washington, *Environ. Toxicol. Chem.*, 9, 669, 1990.

9. **Gentile, J. H., Gentile, S. M., Hoffman, G., Heltshe, J. F., and Hairston, N., Jr.**, The effects of a chronic mercury exposure on survival, reproduction and population dynamics of *Mysidopsis bahia*, *Environ. Toxicol. Chem.*, 2, 61, 1983.

10. **Van Leeuwen, C. J., Niebeck, G., and Rijkeboer, M.**, Effects of chemical stress on the population dynamics of *Daphnia magna:* A comparison of two test procedures, *Ecotoxicol. Environ. Saf.*, 14, 1, 1987.

11. **Hall, R. J., Driscoll, C. T., and Likens, G. E.**, Importance of hydrogen ions and aluminum in regulating the structure and function of stream ecosystems: An experimental test, *Freshwater Biol.*, 18, 17, 1987.

12. **Clements, W. H., Cherry, D. S., and Cairns, J., Jr.**, The influence of copper exposure on predator-prey interactions in aquatic insect communities, *Freshwater Biol.*, 21, 483, 1989.

13. **Kraus, M. L. and Kraus, D. B.,** Differences in the effects of mercury on predator avoidance in two populations of the Grass Shrimp *Palaemonetes pugio, Mar. Environ. Res.,* 18, 277, 1986.

14. **Crossland, N. O.,** A method for evaluating effects of toxic chemicals on the productivity of freshwater ecosystems, *Ecotoxicol. Environ. Saf.,* 16, 279, 1988.

15. **Taub, F. B.,** Demonstration of pollution effects in aquatic microcosms, *Int. J. Environ. Stud.,* 10, 23, 1976.

16. **Benson, W. H. and Birge, W. J.,** Heavy metal tolerance and metallothionein induction in fathead minnows: results from field and laboratory investigations, *Environ. Toxicol. Chem.,* 4, 209, 1985.

17. **Roesijadi, G., Drum, A. S., Thomas, J. M., and Fellingham, G. W.,** Enhanced mercury tolerance in marine mussels and relationship to low molecular weight, mercury-binding proteins, *Mar. Pollut. Bull.,* 13, 250, 1982.

18. **Georghiou, G. P.,** Overview of insecticide resistance, in *Managing Resistance to Agrochemicals,* Green, M. B., LeBaron, H. M., and Moberg, W. K., Eds., American Chemical Society, Washington, D.C., 1990, 18.

19. **Georghiou, G. P. and Taylor, C. E.,** Factors influencing the evolution of resistance, in *Pesticide Resistance: Strategies and Tactics for Management,* N.R.C. Committee on Strategies for the Management of Pesticide Resistant Pest Populations, Ed., National Academy Press, Washington, D.C., 1986, 157.

20. **Taylor, G. E., Jr. and Murdy, W. H.,** Population differentiation of an annual plant species, *Geranium carolinianum* L., in response to sulfur dioxide, *Bot. Gaz.,* 136, 212, 1975.

21. **Antonovics, J., Bradshaw, A. D., and Turner, R. G.,** Heavy metal tolerance in plants, *Adv. Ecol. Res.,* 7, 1, 1971.

22. **Verkleij, J. A. C. and Prast, J. E.,** Cadmium tolerance and co-tolerance in *Silene vulgaris* (Moench.) Garcke [= *S. cucubalus* (L.) Wib.], *New Phytol.,* 111, 637, 1989.

23. **Schmidt, T. and Schlegel, H. G.,** Nickel and cobalt resistance of various bacteria isolated from soil and highly polluted domestic and industrial wastes, *Microbiol. Ecol.,* 62, 315, 1989.

24. **Brown, B. E.,** Observations on the tolerance of the isopod *Asellus meridianus* Rac. to copper and lead, *Water Res.,* 10, 555, 1976.

25. **Klerks, P. L. and Weis, J. S.,** Genetic adaptation to heavy metals in aquatic organisms: a review, *Environ. Pollut.,* 45, 173, 1987.

26. **Klerks, P. L.,** Adaptation to metals in animals, in *Heavy Metal Tolerance in Plants: Evolutionary Aspects,* Shaw, A. J., Ed., CRC Press, Boca Raton, FL, 1989, 313.

27. **Kneip, T. J. and Hazen, R. E.,** Deposit and mobility of cadmium in a marsh-cove ecosystem and the relation to cadmium concentration in biota, *Environ. Health Perspect.,* 28, 67, 1979.

28. **Knutson, A. B., Klerks, P. L., and Levinton, J. S.,** The fate of metal contaminated sediments in Foundry Cove, New York, *Environ. Pollut.,* 45, 291, 1987.

29. **Hazen, R. E. and Kneip, T. J.,** Biogeochemical cycling of cadmium in a marsh ecosystem, in *Cadmium in the Environment. Part I: Ecological Cycling,* Nriagu, J. O., Ed., John Wiley & Sons, New York, 1980, 399.

30. **Hazen, R. E.,** *Cadmium in an Aquatic Ecosystem,* Ph.D. thesis, New York University, N.Y., 1981.

31. **Klerks, P. L. M.**, *Adaptation to Metals in Benthic Macrofauna*, Ph.D. thesis, State University of New York at Stony Brook, Stony Brook, 1987.

32. **Klerks, P. L. and Levinton, J. S.**, Rapid evolution of metal resistance in a benthic oligochaete inhabiting a metal-polluted site, *Biol. Bull.*, 176, 135, 1989.

33. **Falconer, D. S.**, *Introduction to Quantitative Genetics*, Longman, London, 1981.

34. **Schindler, D. W.**, Detecting ecosystem responses to anthropogenic stress, *Can. J. Fish. Aquat. Sci.*, 44 (Suppl. 1), 6, 1987.

35. **Occhiogrosso, T. J., Waller, W. T., and Lauer, G. J.**, Effects of heavy metals on benthic macroinvertebrate densities in Foundry Cove on the Hudson River, *Bull. Environ. Contam. Toxicol.*, 22, 230, 1979.

36. **May, R. M. and Dobson, A. P.**, Population dynamics and the rate of evolution of pesticide resistance, in *Pesticide Resistance: Strategies and Tactics for Management*, N.R.C. Committee on Strategies for the Management of Pesticide Resistant Pest Populations, Ed., National Academy Press, Washington, D.C., 1986, 170.

37. **Taylor, G. E., Jr.**, Genetic analysis of ecotypic differentiation within an annual plant species, *Geranium carolinianum* L., in response to sulfur dioxide, *Bot. Gaz.*, 139, 362, 1978.

38. **Parsons, P. A.**, Environmental stresses and conservation of natural populations, *Annu. Rev. Ecol. Syst.*, 20, 29, 1989.

39. **Klerks, P. L. and Bartholomew, P. R.**, Cadmium accumulation and detoxification in a Cd-resistant population of the oligochaete *Limnodrilus hoffmeisteri*, *Aquat. Toxicol.*, 19, 97, 1991.

40. **Hall, A., Fielding, A. H., and Butler, M.**, Mechanisms of copper tolerance in the marine fouling alga *Ectocarpus siliculosis*. Evidence for an exclusion mechanism, *Mar. Biol.*, 54, 195, 1979.

41. **Holt, R. D.**, The microevolutionary consequences of climatic change, *Trends Ecol. Evol.*, 5, 311, 1990.

42. **Devonshire, A. L. and Field, L. M.**, Gene amplification and insecticide resistance, *Annu. Rev. Entomol.*, 36, 1, 1991.

43. **Maroni, G., Wise, J., Young, J. E., and Otto, R.**, Metallothionein gene duplication and metal tolerance in natural populations of *Drosophila melanogaster*, *Genetics*, 117, 739, 1987.

44. **Watkins, A. J. and Macnair, M. R.**, Genetics of arsenic tolerance in *Agrostis capillaris* L., *Heredity*, 66, 47, 1991.

45. **Wright, S.**, *Evolution and the Genetics of Populations. Vol. 1, Genetic and Biometric Foundations*, University of Chicago Press, Chicago, 1968.

46. **Lande, R.**, The minimum number of genes contributing to quantitative variation between and within populations, *Genetics*, 99, 541, 1981.

47. **Imagawa, M., Onozawa, T., Okumura, K., Osada, S., Nishihara, T., and Kondo, M.**, Characterization of metallothionein cDNAs induced by cadmium in the nematode *Caenorhabditis elegans*, *Biochem. J.*, 268, 237, 1990.

48. **Mehra, R. K., Garey, J. R., Butt, T. R., Gray, W. R., and Winge, D. R.**, *Candida glabrata* metallothioneins, *J. Biol. Chem.*, 264, 19749, 1989.

49. **Harlow, P., Watkins, E., Thornton, R. D., and Nemer, M.,** Structure of an ectodermally expressed sea urchin metallothionein gene and characterization of its metal-responsive region, *Mol. Cell. Biol.,* 9, 5445, 1989.

50. **Mokdad, R., Debec, A., and Wegnez, M.,** Metallothionein genes in *Drosophila melanogaster* constitute a dual system, *Proc. Natl. Acad. Sci. U.S.A.,* 84, 2658, 1987.

SECTION III

Terrestrial Environments

General Considerations
14. Strategies of Metal Detoxification in Terrestrial Invertebrates

Terrestrial Gastropods
15. Budgeting the Flow of Cadmium and Zinc Through the Terrestrial Gastropod, *Helix pomatia L.*

16. Quantitative Aspects of Zinc and Cadmium Binding in *Helix pomatia:* Differences Between an Essential and a Nonessential Trace Element

Terrestrial Oligochaetes
17. Metal Relationships of Earthworms

Terrestrial Isopods
18. Deficiency and Excess of Copper in Terrestrial Isopods

19. Metal Contamination Affects Size-Structure and Life-History Dynamics in Isopod Field Populations

Terrestrial Insects
20. Metal Bioaccumulation in a Host Insect (*Lymantria dispar* L., Lepidoptera) During Development -- Ecotoxicological Implications

Terrestrial Invertebrate Communities
21. Soil and Sediment Quality Criteria Derived from Invertebrate Toxicity Data

CHAPTER 14

Strategies of Metal Detoxification in Terrestrial Invertebrates

Reinhard Dallinger

TABLE OF CONTENTS

0-87371-734-1/93/$0.00 + $.50
© 1993 by Lewis Publishers

I. INTRODUCTION

Different species of terrestrial invertebrates clearly exhibit different dispositions to accumulate and eliminate essential and nonessential trace metals.[1,2] This applies even to closely related species, as shown, for instance, in two terrestrial isopods from the same polluted environment.[3] In spite of such species-specific differences, however, the extraordinary capacity for accumulating certain metals within distinct tissues seems to be a consistent feature of some invertebrate phyla. This has been demonstrated for metal-exposed species in polluted environments or in the laboratory — in midgut glands of metal-contaminated terrestrial isopods, extremely high concentrations of copper, zinc, cadmium, and lead have been recorded.[4] Elevated body burdens of metals have been observed in earthworms from mining areas.[5,6] High concentrations of lead, copper, cadmium, and zinc have been detected in tissues of metal-exposed snails in the field[7] or in the laboratory[8-10] as well as in slugs from polluted woodlands and mining areas.[11,12] It is significant, however, that these species also exhibit their excessive metal-accumulating capacity in relatively clean environments. It has been shown by Knutti and colleagues,[13] for example, that earthworms, isopods, terrestrial gastropods, and spiders from an unpolluted forest accumulated larger amounts of cadmium in comparison to other soil invertebrate groups. Similar observations concerning terrestrial arthropods have been made by Boháč and Pospíŝil[14] with additional metals. On the basis of different concentration factors for different metals, these authors have proposed to distinguish, among terrestrial invertebrates, between macroconcentrators, microconcentrators, and deconcentrators.[14] According to this definition, terrestrial gastropods, isopods, earthworms, and possibly some arachnids, belong to the group of macroconcentrators, at least for some of the most common metals such as copper, zinc, and cadmium.

Typically, metals accumulated by these organisms are not evenly distributed in the body which means that some organs and tissues are involved in metal accumulation, but others are not.[8-10] In most cases, digestive tissues such as gut epithelia or digestive glands, which play a crucial role in the nutrition physiology of these organisms (see Chapter 18)[15] are the predominant sites of metal accumulation.[16] It has also been shown that in some species such as gastropods, the metals accumulated can be stored over long periods of time.[8]

At the cellular level, certain metals are directed into specific cells in which they are sequestered by vesicular compartments of sometimes variable composition and shape. These vesicles have been described, for example, as "cuprosomes" or "copper granules" in terrestrial isopods,[17,18] as "cadmosomes" in earthworms,[6] and as "concretions" or "mineral concretions" in intestinal cells of insects.[19,20] Hopkin[16] has classified metal-containing granules in terrestrial invertebrates following the classification scheme of Nieboer and Richardson.[21] In isopods and insects these granules have been identified as lysosomes or as residual bodies,[22-24] representing a cellular storage-detoxification system.[16]

At the molecular level some metals, such as copper, zinc, or cadmium, are bound to cytosolic metallothioneins or to other metal-binding proteins. Metallothioneins are low molecular weight proteins with a distinct amino acid composition and a high content of sulfur by means of which they are able to bind specifically class B or borderline metals,[21] predominantly copper, cadmium, and zinc.[25] In contrast to the vesicular storage-detoxification system which is restricted to some invertebrate phyla, metallothioneins occur throughout the animal kingdom[26] and were supposed to be responsible for the high persistence of cadmium in certain animal tissues.[27] Among terrestrial invertebrates metallothioneins have been reported from earthworms,[28,29] snails,[30] slugs,[31] and insects.[32] The striking accumulation and persistence of cadmium in the midgut gland of terrestrial snails, for example, is due to complexation of this metal by such proteins.[8,33] It has been argued that sequestration and thus inactivation of toxic metals in the cytoplasm of invertebrate cells may be an important biological function of metallothioneins.[30,31] In consequence, these molecules have often been regarded as true detoxification proteins, a view which has also been criticized.[34] Metallothioneins rather may be multifunctional proteins,[26,35] involved in the homeostasis and in the metabolism of essential trace elements such as zinc and copper.[36] Detoxification of nonessential metals such as cadmium may only be one of several functions of these proteins. Bearing in mind, however, that in cadmium-resistant strains of *Drosophila melanogaster* increased cadmium tolerance is achieved by the duplication of the gene encoding this protein,[37] the role of metallothioneins in the evolution of metal resistance (see Chapter 13)[38] is very likely. Many other questions remain when considering the significance of metallothioneins in terrestrial invertebrates. For example, it is not sufficiently clear how the cellular storage-detoxification system interferes with metals bound to metallothioneins.

In the light of such reflections this article will deal with some specific connects of the strategies of metal detoxification in terrestrial invertebrates. For the sake of brevity, we shall restrict ourselves to four metals with high biological and ecotoxicological relevance: copper, zinc, cadmium, and lead. In particular, the following questions will be addressed:

1.   What are the common links of metal accumulation shared by all groups of terrestrial invertebrates?
2.   What is the significance of organ- and cell-specific metal distribution?
3.   What are the specific roles played by the cellular storage-detoxification system on the one hand, and the binding of metals to metallothioneins on the other hand, and how are these two immobilization systems linked?
4.   What are the metal-specific peculiarities of accumulation in different species?
5.   What are the ecological and evolutionary implications of these strategies?

II. METAL ACCUMULATION IN TERRESTRIAL INVERTEBRATES: STORAGE-DETOXIFICATION VERSUS EXCRETION

A. Macroconcentration: a One-Compartment Model

Storage-detoxification of metals is supposed to be a characteristic feature of those terrestrial invertebrates which have been defined as macroconcentrators.[14] However, macroconcentration is only an extreme case of metal accumulation determined by two inverse processes: metal uptake and metal elimination (see Chapter 7).[39] Emphasizing the dynamic nature of this system, one may refer to a model in which the invertebrate organism is regarded as a black box, the concentration of a metal within this black box depending on the rates at which the metal enters (uptake) and leaves (elimination) the system (Figure 1). Although being oversimplified,[2] this model may be helpful in order to explain the basic processes behind accumulation.

In contrast to their freshwater and marine relatives, most terrestrial invertebrates have to gain their nutrients or mineral and trace elements from the food and thus by absorption via the intestinal epithelia.[40] This pathway is also important for the uptake of unwelcome trace metals. In insects, for example, unwanted and toxic ions can readily cross the midgut epithelium towards the hemolymph.[41] Only some gastropod species have been shown to absorb certain metals, such as copper or zinc, via the skin.[42,43] Few studies have focused on the mechanisms by which metals cross the membranes of epithelia in terrestrial invertebrates.[44] It can be assumed that most trace elements enter the cells passively through channels, following a concentration gradient by diffusion or facilitated diffusion. In some cases, endocytotic processes may also be of importance. Whatever the mechanisms involved may be, it is generally accepted that the uptake of trace metals by a terrestrial invertebrate is not under the control of the animal,[15] but depends predominantly on the concentration in the food (Figure 1). Thus in cases of excessive metal supplies in the diet, terrestrial invertebrates are not likely to be able to limit the uptake of a trace metal by preventing it from entering the body.

Under such circumstances the capacity for eliminating unwanted metals turns out to be a decisive factor (Figure 1). One possibility to eliminate nonessential trace elements or essential metals taken up in excess of needs would be to excrete them by ultrafiltration or by active transport mechanisms and to discharge them as solutes together with excretory fluids. However, adaptation to terrestrial life has confronted invertebrates with the risk of dehydration.[16,45] The release of excretory fluids must be restricted to a minimum in order to avoid excessive loss of water. Thus in many terrestrial invertebrates selective pressure has acted towards the formation of solid excretion products, such as uric acid or purines.[45] These substances are only slightly toxic and can be discharged without substantial loss of water. Terrestrial gastropods produce urea and uric acid, but they have developed a storage system by which the excretion products can be retained within the body during temporary periods of dehydration, excretion occurring rapidly following rehydration.[46,47]

In most terrestrial invertebrates the elimination of unwanted metals with excretory fluids would present problems, and this is probably the main reason why these animals have developed a strategy which allows them to inactivate and retain toxic metals or excessive amounts of essential trace elements by intracellular compartmentalization. Such mechanisms have been discovered in all important phyla of terrestrial invertebrates, including nematodes,[48] gastropods,[49,50] oligochaetes,[6] crustaceans,[18,22,23,51,52] arachnids,[1,53] myriapods,[54] and insects.[20,24] In his book on ecophysiology of metals in terrestrial invertebrates, Hopkin[16] has critically reviewed the literature dealing with intracellular compartmentalization of metals in these animals. By sequestration within vesicles, nonessential trace elements can be efficiently inactivated and detoxified. At the same time, excessive amounts of essential trace metals can also be detoxified or stored and possibly remobilized during periods of increased demand.[55,56] These are probably the main advantages of such a strategy.

However, not all species of terrestrial invertebrates retain and store assimilated metals in their bodies. It has been shown, for instance, that different species of soil invertebrates eliminate cadmium to very different degrees.[2] In fact, many terrestrial invertebrates are able to discharge toxic metals and essential trace elements predominantly by cellular processes such as the removal of whole degenerated cells, exocytosis, or extrusion of metal-containing vesicles into the

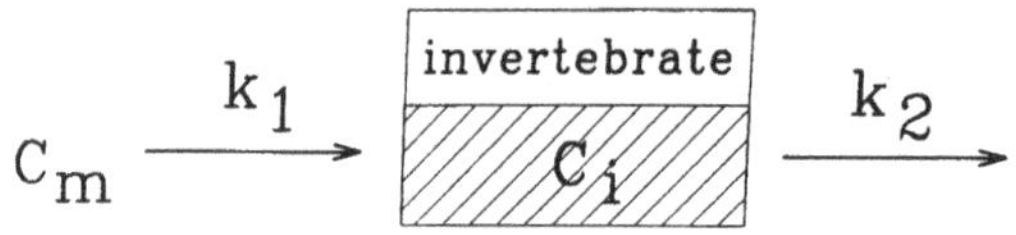

$$C_m \xrightarrow{\quad k_1 \quad} \boxed{\text{invertebrate } /\!/\!/C_i/\!/\!/} \xrightarrow{\quad k_2 \quad}$$

(1) $\quad \dfrac{dC_i}{dt} = k_1 * C_m - k_2 * C_i$

(2) $\quad C_{i,t} = C_{i,\infty} * (1 - e^{-k_2 * t})$

(2a) $\quad \lim\limits_{t \to \infty} (1 - e^{-k_2 * t}) = 1$

(2b) $\quad C_{i,\infty} = a * C_m$

with $\quad a = \dfrac{k_1}{k_2} \quad$ (biol. concentration factor)

$a > 2 \qquad$ macroconcentration

$2 > a > 1 \qquad$ microconcentration

$a < 1 \qquad$ deconcentration

FIGURE 1. Description of metal accumulation in a terrestrial invertebrate by a one-compartment model (inset, above).
Explanation:
(1) The change of metal concentration in an invertebrate is given by metal uptake minus metal elimination, metal uptake depending on the metal concentration in the medium (e.g., the gut fluid) and the assimilation constant, metal elimination depending on metal concentration in the invertebrate and the elimination constant.
(2) The metal concentration in an invertebrate at time t is proportional to the ratio of assimilation constant (k_1) and elimination constant (k_2), to metal concentration in the medium and to a term denoting the metal loss from the one-compartment model. With increasing time ($t \to \infty$) the metal concentration in the invertebrate attains a steady state level which is proportional to the ratio of assimilation constant (k_1) and elimination constant (k_2), and to the metal concentration in the medium. The ratio of assimilation constant (k_1) and elimination constant (k_2) represents the biological concentration factor (a). Figure symbols are as follows:

lumen of the digestive tract.[57] In particular, such elimination mechanisms occur in spiders,[1] insects,[24] gastropods,[58,59] and isopods.[60] In earthworms, metal-containing vesicles derived from the chloragogenous cells can be released into the coelomic fluid and eliminated via the nephridia.[16] Thus it seems that a further advantage of cellular compartmentalization is the possibility for an invertebrate to discharge metal-filled vesicles into the lumen of the gut or the excretory system. In this way, unwelcome metals can be eliminated in a concentrated and nearly solid form without excessive loss of water. However, the ability to excrete metals by these means varies considerably between phyla and species. In some cases, cyclic excretion of metal-containing cells or vesicles has been observed. Such a cyclic discharge can be related to molting activities, as shown for the elimination of lead and cadmium in the collembolan species *Orchesella cincta*.[61] In the slug *Arion ater*, copper is excreted cyclically together with calcium-containing vesicles of the hepatopancreas.[59] In a recent article, Hames and Hopkin[60] have described a daily cycle of apocrine secretion of zinc-containing vesicles in the B cells of terrestrial isopods. The same authors have stated, however, that the S cells of isopods which are also important sites of intracellular metal compartmentalization, do not exhibit cellular excretion activities at all.[60]

In the light of these considerations, storage-detoxification can be regarded as the result of a situation in which an invertebrate is unable to discharge metal-filled vesicles, which thus become stored over long periods of time. Since the metals are compartmentalized and chemically inactivated, they can be considered as a detoxified storage form. In some cases the involvement of metallothioneins strongly binding cadmium, copper, or zinc may complicate the situation. In this way some terrestrial invertebrates are able to concentrate elevated amounts of cadmium, zinc, copper, and lead within their bodies and store these metals over long periods of time. Terrestrial isopods, gastropods, earthworms, and possibly some arachnids, provide examples for this kind of strategy characteristic of "macroconcentrators"[14] (Table 1). On the other hand, there are those terrestrial invertebrates which possess the capacity to readily discharge assimilated metals by excretion of metal-containing vesicles or cells. Some insect species, such as carabid beetles,[2] may represent this kind of strategy, and have been termed "deconcentrators"[14] (Table 1). Many other species may be regarded as being intermediate between the two extreme possibilities (Table 1).

C_m: metal concentration in the environment (e.g., the gut fluid);
C_i: metal concentration in the invertebrate;
$C_{i,t}$: metal concentration in the invertebrate at time t;
$C_{i,\infty}$: metal concentration in the invertebrate under steady state conditions;
t: time;
k_1: assimilation constant;
k_2: elimination constant; and
a: biological concentration factor.

Table 1. List of Selected Terrestrial Invertebrates (Class, Order, Species) Known as Macroconcentrators, Microconcentrators, or Deconcentrators of Cadmium, Lead, Copper, and Zinc

	Class (Order)	Species	Metal	Concentration factor	Ref.
Macroconcentration:	Gastropoda (Pulmonata)	Helix pomatia	Cadmium	6	13
Concentration factor					
>2			Zinc	10	139
			Copper	1–9	139
		Helix aspersa	Cadmium	10	139
			Cadmium	3–8	140
			Copper	1.5–10	140
		Cepaea nemoralis	Cadmium	18–33	141
		Arianta arbustorum	Cadmium	20	142
			Cadmium	6–8	143
			Zinc	7.3	142
			Copper	17.2	142
			Copper	7	143
		Arion lusitanicus	Cadmium	3.5	31
	Clitellata (Oligochaeta)	Lumbricus rubellus	Cadmium	0.6–93	71
			Cadmium	5	13
			Lead	0.01–4	71
			Zinc	0.15–2.8	71
		Lumbricus terrestris	Copper	11.4	144
		Allolobophora calliginosa	Cadmium	12	13
	Crustacea (Isopoda)	Porcellio scaber	Cadmium	6	7
			Cadmium	2–3	3
			Zinc	0.6–7	63
			Zinc	0.04–7	64
			Copper	12	7
			Copper	12–14	3

Group	Species	Metal	Value	Ref
Crustacea (Isopoda)	*Oniscus asellus*	Cadmium	7	137
			5–23	70
		Copper	3	137
			2.6–16	70
Arachnida (Araneae)	*Trachelipus ratzeburgii*	Cadmium	2.7	13
	Ligidium hypnourm	Cadmium	8	13
	Trochosa terricola	Cadmium	3	141
	Pardosa lugubris	Cadmium	6	13
	Coelotes terrestris	Cadmium	9	13
Insecta (Hymenoptera)	*Formica sanguinea*	Cadmium	3	13
	Formica rufa	Cadmium	2.6	13
(Coleoptera)	*Melighetes* sp.	Zinc	8	143
			5.5	139
		Copper	5	143
			7.7	139
	Amara familiaris	Zinc	2.4	139
Microconcentration: Gastropoda (Pulmonata)	*Helix pomatia*	Lead	1.3	139
			1.1	119
Concentration factor 1–2	*Helix aspersa*	Zinc	0.3–2	140
	Arianta arbustorum	Lead	1.6	142
			1	139
	Arion lusitanicus	Zinc	1.5–2	31
Clitellata (Oligochaeta)	*Lumbricus rubellus*	Lead	1.3	145
Crustacea (Isopoda)	*Porcellio scaber*	Cadmium	1.4	65
		Zinc	0.6–2	144
			1.5	7
Arachnida (Acari)	*Plathynothrus peltifer*	Cadmium	1	146

Table 1 (continued). List of Selected Terrestrial Invertebrates (Class, Order, Species) Known as Macroconcentrators, Microconcentrators, or Deconcentrators of Cadmium, Lead, Copper, and Zinc

Class (Order)	Species	Metal	Concentration factor	Ref.
Myriapoda (Chilopoda)	Lithobius forficulatus	Cadmium	2	141
(Diplopoda)	Glomeridae	Cadmium	1.7	13
Insecta (Coleoptera)	Carabus coriaceus	Cadmium	1	13
	Agelastica alni	Lead	1	143
	Amara familiaris	Copper	1.7	139
Deconcentration: Concentration factor <1				
Clitellata (Oligochaeta)	Lumbricus rubellus	Copper	0.01–0.6	71
Crustacea (Isopoda)	Porcellio scaber	Lead	0.43	7
			0.06–0.12	3
	Oniscus asellus	Lead	0.07	65
			0.23	137
			0.2–0.5	70
			0.5	3
		Zinc	0.2	137
			0.75	7
			0.4–0.6	3
Myriapoda (Chilopoda)	Lithobius forficulatus	Cadmium	0.5	13
	Eupolyborthrus tridentinus	Cadmium	0.7	13
(Diplopoda)	Cylindoiulus nitidus	Cadmium	0.8	13
	Tachypodoiulus niger	Cadmium	0.15	13
Insecta (Collembola)	Orchesella cincta	Cadmium	<1	146
(Coleoptera)	Psylliodes chrysocephala	Cadmium	0.6	13

Agelastica alni	Cadmium	0.3	143
Amara familiaris	Cadmium	0.5	139
	Lead	0.6	139

Note: Concentration factors were calculated according to or reported from the references listed in the last column. Concentration factors usually refer to metal concentrations of the soil or of the substrate (leaves, litter) on which invertebrates live or feed; in a few cases, the factors relate to metal concentrations of soil extracts. Macroconcentration, microconcentration, and deconcentration are defined according to Boháč and Pospíšil[14] as follows:

Concentration factor >2: macroconcentration
Concentration factor 1–2: microconcentration
Concentration factor <1: deconcentration

Thus the steady-state concentration of a metal measured within the body of a terrestrial invertebrate is the result of the animal's ability to eliminate the metal by the processes described above. The ratio between the assimilation constant and the elimination constant is defined as the concentration factor (Figure 1) which indicates by how much a metal is concentrated in an animal in relation to environmental concentrations.

B. Tissue-Specific Adaptations

When considering a terrestrial invertebrate as a black box, metal accumulation can only be assessed on a whole-body level. Such an approach may be helpful, for instance, for the use of terrestrial invertebrates as biological indicators of metal pollution.[7] In this case the concentration of a metal in the whole body of an animal is believed to reflect metal concentrations in the environment. Terrestrial isopods, for example, have been adopted successfully as biological indicators for the contamination of soils with copper, zinc, cadmium, and lead.[62-65] In a similar way, earthworms (see Chapter 17)[66] and terrestrial gastropods[67,68] have also been used as biological indicators.

Detailed studies have revealed, however, that in most terrestrial invertebrates, whether they are considered as macroconcentrators or not, trace metals are selectively concentrated in only one or a few organs or in specific sections of a tissue, whereas most of the body is not at all involved in the accumulation process. Typically, these target organs are parts of the digestive tract of an animal, or alternatively, are excretory tissues displaying an open connection with the alimentary canal, such as the Malphighian tubules of insects.

Metal-exposed terrestrial snails and slugs, for instance, store cadmium, zinc, and lead predominantly in the midgut gland,[8,9,69] while the metal concentrations in other organs of these animals remain unaffected (Figure 2). Terrestrial isopods sequester cadmium, copper, zinc, and lead in their hepatopancreas tubules.[23,70] Earthworms have been shown to concentrate assimilated metals such as cadmium, lead, and zinc predominantly in the hepatopancreas.[1,72] In insects, metals are absorbed in specific sections of the alimentary canal. In Collembola, lead is concentrated in the midgut epithelium.[20] In cockroaches the highest concentrations of zinc and other metals are found in the ileum.[73] In Lepidoptera the midgut and the Malpighian tubules are the sites of highest cadmium concentrations,[74] whereas in *Drosophila melanogaster* assimilated cadmium is concentrated in both the anterior and posterior segments of the midgut.[24] In metal-contaminated millipedes the concentrations of cadmium and lead increase mainly in the midgut, whereas zinc is predominantly incorporated and stored in the subcuticular layer.[16]

The prevailing role of digestive epithelia, midgut glands, and Malpighian tubules in concentrating and storing trace metals is probably a further adaptive response of invertebrates to terrestrial life. It may also be the consequence of the fact that in terrestrial animals the predominant pathway of nutrient uptake is via the alimentary epithelia. It has been stressed by Hopkin[16] that the internal

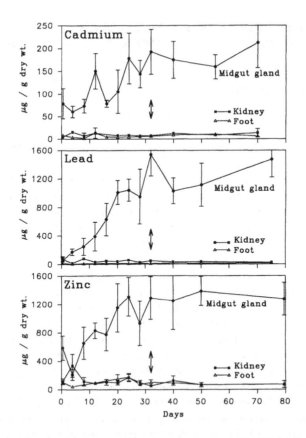

FIGURE 2. Patterns of metal accumulation (μg/g dry weight) in different tissues (hepatopancreas, kidney, and foot) of metal-exposed individuals of *Helix pomatia*. Snails were fed on either cadmium-, lead-, or zinc-containing lettuce as described by Dallinger and Wieser[8] over a period of 32 days, after which time (arrows) metal-exposure was stopped and animals fed on uncontaminated lettuce to the end of the experiment. (Reproduced after Dallinger, R. and Wieser, W., *Comp. Biochem. Physiol.*, 79C, 117, 1984. With permission.)

anatomy of digestive organs is crucial for determining metal accumulation in terrestrial invertebrates. Thus potentially harmful metals can be readily absorbed, concentrated, and inactivated by those specific tissues which are in direct contact with the metals in the gut content, and this fact may prevent other organs and tissues from becoming contaminated. In contrast, toxic metals stored within vesicles of digestive cells can be released directly into the lumen of the alimentary canal or the midgut gland and be discharged with the feces. Thus it seems that elimination via the feces becomes an essential route of metal loss in terrestrial invertebrates.

III. DIFFERENT ROLES OF INTRACELLULAR COMPARTMENTALIZATION AND METALLOTHIONEINS

A. General Function of the Lysosomal System in Accumulating Cellular Waste Products

Sequestration of trace metals within intracellular vesicles does already occur in marine relatives of terrestrial isopods and gastropods.[15] The inheritance of these mechanisms from their marine ancestors may have better preadapted isopods and snails for life on land.

Hopkin[16] distinguished between three types of intracellular metal inclusions in terrestrial invertebrates, calling them "granules". From a scientific point of view, it is not so important whether these inclusions are called "granules" or "vesicles", but it would be of interest if the basic cellular processes leading to their formation could be defined. Compartmentalization being a universal mechanism, the basis of intracellular metal compartmentalization may be provided by the lysosomal system. In fact, lysosomes are membrane-limited vesicles which occur in nearly all animal cells. They function as the digestive system of the cell and typically go through different stages of maturity in which they fuse with endocytotic vesicles or autophagosomes until they form, as a last stage, the residual bodies.[76] A characteristic feature of lysosomes is their varying size and shape. It has been stressed by de Duve and Wattiaux[76] that the polymorphism of lysosomes is due to their digestive function which may cause them to be filled with waste products and a variety of substances and objects in an advanced state of disintegration. Consequently, the shape, size, density, and other features of lysosomes are essentially determined by their contents. They can be recognized by their performance rather than by their morphological appearance.[76] Since it is an intrinsic feature of lysosomes to be involved in the storage of cellular waste products, it is not surprising that nonessential trace elements which have entered a cell are eventually stored in residual bodies. This may also be true for excessive amounts of essential trace elements which cannot actually be utilized by the cell. There are many examples of metal accumulation within lysosomes in animal tissues. In vertebrates, for instance, different species and chemical forms of trace metals have been found to accumulate in lysosomes of various tissues from rats,[77,78] hamsters,[79] and humans.[80] Similar observations have been made in different species and tissues of fish.[81-83] In invertebrates the involvement of the lysosomal system in the formation of metal-binding granules has been suggested by Brown.[84] Metal-accumulating lysosomes have been observed in tissues of marine invertebrates such as mussels[85-87] and crustaceans.[88] In the freshwater crayfish *Orconectes propinquus,* iron and lead are stored in lysosomes of the hepatopancreas and the antennal gland.[89]

Metal-containing lysosomes have also been reported in terrestrial invertebrates. Lauverjat et al.[24] found a large number of cadmium-sequestering lysosomes in midgut cells of *Drosophila melanogaster.* Similarly, lysosomes storing cadmium, copper, lead, and zinc have been described and characterized in the hepatopancreas of the woodlouse *Porcellio scaber*[22,23] (Figures 3 and 4). It is generally accepted that these lysosomes are identical with the type B granules reported by Hopkin.

In the classification scheme of Hopkin[16] there are two further granule types of uncertain origin: the type A granules and the type C granules. Are they also lysosomes? Type C granules have been discovered in the hepatopancreatic B cells of isopods and in digestive cells of spider hepatopancreas.[1] In both cases, they contain predominantly the degradation products of ferritin and are thus sites of iron accumulation. In contaminated isopods they also contain zinc and lead.[18] In the spider *Dysdera crocata*, these vesicles are deposited in excretory vacuoles which are cyclically voided into the lumen of the alimentary canal.[1] All these facts strongly suggest that the type C granules also belong to the lysosomal system.

Type A granules occur in the basophil cells of the hepatopancreas of terrestrial gastropods. As a rule they are concentrically structured[16] and contain large proportions of calcium phosphate,[58] but may also sequester trace elements such as zinc,[90] manganese,[49] or copper.[59] In the marine gastropod *Littorina littorea*, the same granules have been found to contain cadmium.[91] Typically, the shape of these granules is quite different from that of the lysosomes described above. However, it is not the appearance by which a cellular vesicle is characterized as a lysosome, but by its origin and function.[76] Lysosomal vesicles are formed by the cellular Golgi apparatus[92] sometimes in close association with the endoplasmatic reticulum.[93] This seems to be true for the calcium phosphate granules of gastropods.[57] Since excessive amounts of free calcium ions in the cytoplasm are toxic to the cells,[94] calcium sequestration within these granules can also be regarded as a detoxification process. Apart from their origin, lysosomes are also characterized by their enzymatic activities, among which acid phosphatase and hydrolase activities are predominant.[76] Interestingly, Janssen[95] has found the calcium phosphate granules in the terrestrial slug *Arion ater* to exhibit significant hydrolase activities. In consequence, he considered these granules to be part of the lysosomal system.

If this is true then it would appear that all the metal-sequestering granules mentioned above have a common link: they are part of the lysosomal system which functions as an intracellular lytic compartment through which refuse material becomes disintegrated and nondigestible wastes are trapped. In the light of these suggestions, storage-detoxification of metals would appear to be part of a strategy which is based on universal cellular processes common to all animals. Terrestrial invertebrates have successfully adapted these mechanisms to the needs of trace element detoxification under terrestrial conditions.

The variable appearance of metal-containing granules in different species might be explained by different developmental stages of lysosomes and by their varying contents of metallic and nonmetallic substances. In some species the lysosomes may have gone through a process of differentiation towards a more efficient mechanism of sequestering specific cations. In any case it has to be

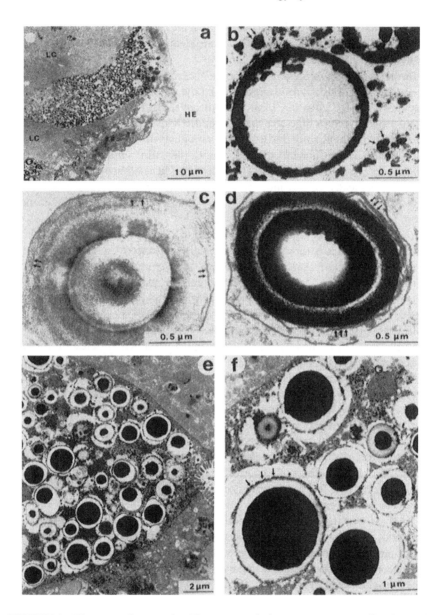

FIGURE 3. Electron micrograph of lysosomes in hepatopancreas cells of metal-exposed *Porcellio scaber* from a contaminated mining area (Braubach, Germany). For methods see Reference 22.

a: Cross section through a hepatopancreas (HE) tubule: nonspecific precipitation of metals with hydrogen sulfide and silver staining in a hepatopancreas cell. Large cells (LC) are free of metals.

b: Magnification of metal precipitations within a lysosome from a;

assumed that it is the cell and not the lysosome which decides, by means of specific or nonspecific ligands, which metals are taken up and eventually stored in residual bodies.

B. Occurrence and Function of Metallothioneins

In some species of terrestrial invertebrates the presence of metallothioneins may complicate the process of metal accumulation. For instance, the extraordinary capacity of some terrestrial gastropods to accumulate and to retain cadmium in their midgut gland is not the result of intracellular compartmentalization, but a consequence of metal complexation by cytosolic metallothioneins.[30,31,96] The high capacity of earthworms for accumulating and storing cadmium is probably also due to metal sequestration by metallothioneins.[28] Such proteins have been isolated, for instance, from the posterior alimentary tract of *Dendrodrilus rubidus* and *Lumbricus rubellus*.[29]

Metallothioneins are low molecular weight proteins with a high content of cysteine and small amounts, if any, of aromatic amino acids.[25] They have a strong affinity to some borderline and class B metal ions[21] such as cadmium, zinc, and copper, which are bound by the sulfur atoms of the cysteine residues.[97] A classification of these proteins has been proposed by Fowler et al.[98]

So far metallothioneins have been found to occur in a large number of eukaryotic species including fungi, plants, and animals.[26] As far as terrestrial invertebrates are concerned, metallothioneins or metallothionein genes have been reported from nematodes,[99] snails,[30,33] slugs,[31] earthworms,[28,29] and several insect species including fruit flies,[100] flesh flies,[101] and cockroaches.[102] The primary structure of metallothioneins has so far been resolved in only a few species of terrestrial invertebrates. The first amino acid sequence of a terrestrial invertebrate metallothionein has been deduced from the gene nucleotide sequence by Lastowski-Perry et al.[100] in the fruit fly *Drosophila melanogaster*. A second, but similar, metallothionein gene has been discovered in cadmium-resistant individuals of this species.[37,103] The primary structure of a metallothionein in the nematode *Caenorhabditis elegans* has also been deduced from the nucleotide sequence of cDNAs.[99]

Recently, the amino acid sequence of a metallothionein from the terrestrial gastropod *Helix pomatia* has been analyzed by Dallinger and colleagues.[104] A

c: Normal fixation of a lysosome (reduced osmium tetroxide), showing membrane fragments and associated endoplasmatic reticulum (arrows);
e: Evidence of acid phosphatase activity in lysosomes (inner dense matrix) of a cell; and
f: Magnification of e, showing acid phosphatase activity also along the edge of lysosomes (arrows).
(Reproduced after Prosi, F. and Dallinger, R., *Cell Biol. Toxicol.*, 4, 81, 1988. With permission.)

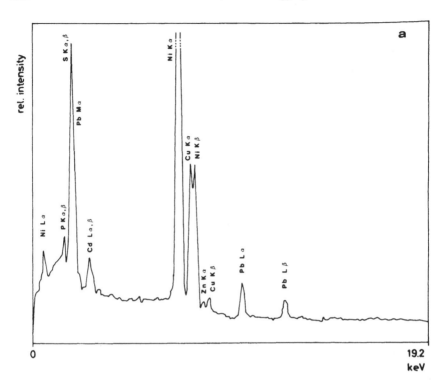

FIGURE 4. EDX-spectrograms of elements within cells of metal-exposed *Porcellio scaber*. For methods see Reference 22.
a: EDX-spectrogram of granular metal precipitations inside a lysosome of a cell, as shown in Figure 3b.
b: EDX-spectrogram of small metal concretions within the cytoplasm of a cell, as shown in Figure 3b (arrows).
(Reproduced after Prosi, F. and Dallinger, R., *Cell Biol. Toxicol.,* 4, 81, 1988. With permission.)

preliminary sequence of this protein isolated from cadmium-exposed snails is reported in Figure 5. The amino-terminal part of the protein is blocked, probably by acetylation of the first amino acid. We know, however, that the blocked fragment contains seven additional amino acids, none of which is a cysteine. The protein shows considerable sequence relationships with the primary structure of metallothioneins recently decoded in the marine bivalves *Mytilus edulis*[105] and *Crassastrea virginica*.[106] It is interesting that these molluscan metallothioneins, including that of *Helix pomatia*, exhibit a higher degree of homology with mammalian metallothioneins than with those reported from other terrestrial invertebrates, such as *Drosophila melanogaster*[100] or *Caenorhabditis elegans*.[99]

The most obvious structural links shared by all metallothioneins are their arrangements of cysteine residues, among which Cys-X-Cys, Cys-X-X-Cys and

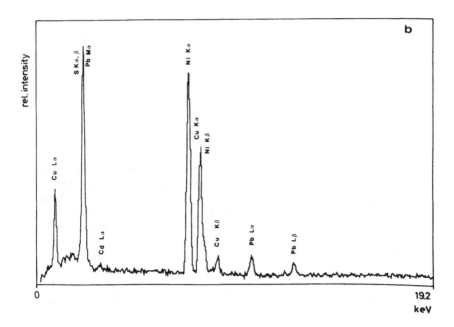

FIGURE 4b.

Cys-X-X-X-Cys segments are common, X denoting any amino acid except cysteine (Figure 5). This indicates that in the metallothionein of *Helix pomatia*, metals such as cadmium, copper, and zinc are bound by thiolate clusters in a similar manner as in mammalian metallothioneins. It has been stressed by Kägi and Kojima[107] that the stability of the thiolate-metal complexes in these proteins is exceptionally high, cadmium being bound more strongly than zinc by a factor of 10,000. The binding situation is probably similar in metallothioneins of terrestrial gastropods, since it has been shown that in metal-exposed snails, cadmium is always a predominant constituent of the protein, while only low amounts of zinc are present.[30,33] Since the protein of *Helix pomatia* possesses 18 cysteine residues instead of 20 as in mammalian metallothioneins, it is assumed to bind 6 metal atoms per molecule. Preliminary data[108] support this assumption.

As the synthesis of metallothioneins can be induced by cadmium exposure, it has often been argued that detoxification of this and other toxic metals may be one of the predominant functions of these proteins. The involvement of metallothioneins in metal detoxification has been suggested, for instance, in earthworms[29] and in terrestrial gastropods.[30,31,109] Maroni et al.[37] have shown that cadmium- and copper-resistant larvae of *Drosophila melanogaster* possess functional duplications of the metallothionein genes. Individuals which carry these gene duplications are able to produce more metallothionein encoding mRNA[37] and higher amounts of protein.[24] In both cadmium-resistant and cadmium-susceptible strains of *Drosophila melanogaster*, the metal, apart from being bound to metallothioneins, is sequestered within lysosomes of midgut cells.[24] Thus it

			11			15				20					25			
BI*-	Lys	Cys	Thr	Ser	Ala	Cys	Arg	Ser	Glu	Pro	Cys	Gln	Cys	Gly	Ser	Lys	Cys	Gln

		26			30				35				40
Cys	Gly	Glu	Gly	Cys	Thr	Cys	Ala	Ala	Cys	Lys	Thr	Cys	Asn	Cys

41				45					50				55
Thr	Ser	Asp	Gly	Cys	Lys	Cys	Gly	Lys	Gly	Cys	Thr	Gly	Pro	Asp

56			60				65	
Ser	Cys	Lys	Cys	Gly	Ser	Ser	Cys	Ser	Cys	Lys

FIGURE 5. Preliminary amino acid sequence of blocked *Helix pomatia* metallothionein. The protein has been purified by chromatographic techniques, including reverse-phase HPLC as described in Reference 96. The purified protein was S-methylated according to Hunziker[147] and cleaved by means of different enzymes including trypsin, endoproteinase Glu-C, endoproteinase Asp-N, and endoproteinase Arg-C (all enzymes: Boehringer, Germany). The cleavage products were fractionated by HPLC (RP-300 column, Aquapore, U.S.) and sequenced in an Applied Biosystems (U.S.) sequencer (model 470A) using the manufacturer's supplied standard chemistry.[104] The blocked peptide fragment (BI*-) containing seven amino acids (but no cysteine) was stored for further analysis.

may not be the complexation of cadmium to cytosolic metallothioneins, but the sequestration of this metal within lysosomes which accounts for the resistance of *Drosophila* individuals against cadmium intoxication.[24] If this is true, then it must be assumed that cadmium-loaded metallothioneins are picked up by lysosomal vesicles where the metal could be retained after the degradation of the protein. In the cockroach *Blatella germanica* mercury is bound by polymerized metallothioneins residing within the lysosomes of ileum cells.[24,110] In the midgut gland of metal-exposed mussels, polymerized metallothioneins within lysosomes are also involved in the binding of copper.[87] In contrast, some terrestrial gastropods, such as *Helix pomatia*, do not at all accumulate cadmium within lysosomes, the metal remaining bound to cytosolic metallothioneins over long periods of time (Figure 6). Thus it seems that the route of cadmium and possibly that of other trace metals from metallothioneins into lysosomal vesicles or granules is not a pathway common to all terrestrial invertebrates. In consequence, the connections between metal fractions bound to cytosolic metallothioneins and those sequestered by lysosomes are largely unknown.

It has also been argued that the biological function of metallothioneins is not restricted to the detoxification of potentially harmful metals, be they essential or nonessential. Some authors have postulated that metallothioneins play an important role in the homeostatic regulation of intracellular copper activity and in the storage of zinc.[26] Engel and Brouwer[35] have suggested that in the decapod crustacean *Callinectes sapidus*, metallothioneins are responsible for the regu-

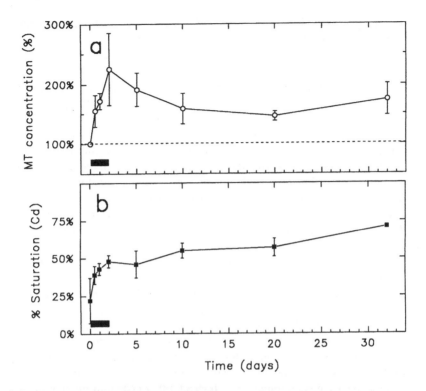

FIGURE 6. Quantification of metallothionein synthesis in hepatopancreas of cadmium-exposed *Helix pomatia* and percentage saturation of the protein with cadmium. Relative concentrations of metallothionein and percentage cadmium saturation were assessed by treating protein extracts with acetonitrile and binding of excess cadmium to Chelex-100, as described by Bartsch et al.[148] and modified by Berger et al.[108]

a: Relative metallothionein concentration in hepatopancreas of *Helix pomatia*, expressed as percentage of controls (unexposed animals), over a period of 34 days. During the first 2 days (solid bar) snails were exposed to cadmium via metal-contaminated agar plates (275.5 ± 5.7 μg Cd/g dry wt., n = 10); subsequently the animals were fed on uncontaminated lettuce till the end of the experiment.

b: Percentage saturation of metallothionein with cadmium. For metal exposure and duration see above.

lation of the essential metals copper and zinc, but that in cadmium-exposed individuals the inducible proportion of surplus metallothionein may also account for the detoxification of this metal. More recent findings suggest that metallothioneins are involved in the regulation of genes whose transcription depends on the interaction of metallothionein with zinc finger proteins.[111,112] Thus the most convincing hypothesis is that the metallothioneins are multifunctional proteins involved in the storage, transport, and homeostasis of essential trace metals, but also in gene regulation and in the detoxification of cadmium and other nonessential trace metals.

In principle this may also apply to terrestrial invertebrates. However, there are indications that in certain species or phyla of terrestrial invertebrates the function of metallothioneins may be somewhat different from that in most vertebrates. One indication for this is that in some terrestrial invertebrates metallothioneins predominantly bind cadmium and only small amounts of zinc. In the snail *Helix pomatia*, for example, no interactions of cadmium and zinc could be found after exposure of animals to a combination of both metals (see Chapter 15)[10]. In fact, a detailed study of cadmium and zinc binding in the midgut gland of this species revealed that the predominant proportion of the cadmium is always bound to metallothionein, while only a small proportion of zinc is sequestered by this protein, most of the metal being associated with a low molecular weight ligand and with particulate fractions.[33] Even in supernatants of zinc-exposed animals, a large proportion of the metal was always associated with low molecular weight fractions not corresponding to metallothioneins[33] (also see Chapter 16, this volume). This leads us to suggest that in contrast to the vertebrate situation, the function of metallothioneins in terrestrial gastropods may have shifted, perhaps at the expense of regulation of essential trace metals, towards a more efficient detoxification of cadmium. There is another, more general, argument supporting such a hypothesis: whereas the ubiquitous presence of metallothioneins in vertebrate tissues[26] indicates an essential regulatory function of this protein, there are several examples of invertebrate species in which no traces of metallothioneius have been found.[113-115] Instead, other cadmium-binding ligands have been reported. In the marine gastropod *Buccinum tenuissimum*, for example, two cadmium-binding glycoproteins have been characterized from the midgut gland.[116] A cadmium-binding glycoprotein has also been isolated from the stonefly *Pteronarcys californica*,[117] whereas Martoja and colleagues[113] discovered an inducible, cadmium-binding glycoprotein in the grasshopper *Locusta migratoria*. Recently, Dallinger and colleagues[118] have isolated a cadmium-binding protein which probably is a glycoprotein from the terrestrial isopod *Porcellio scaber* (Table 2). Thus, in terrestrial invertebrates, metallothioneins may or may not be present. In some species, however, they account for high accumulations of cadmium which is strongly bound by these proteins.

IV. METAL- AND SPECIES-SPECIFIC PECULIARITIES IN MACROCONCENTRATION: TWO EXAMPLES

A. Macroconcentration in *Helix pomatia*

1. Cadmium

The terrestrial gastropod *Helix pomatia* is a strong concentrator of cadmium (Table 1). This applies also to individuals from relatively clean environments:

Table 2. Amino Acid Composition of a Cadmium-Binding Protein, Probably a Glycoprotein, With an Apparent Mol Wt of 12,000 Da, Isolated From Cadmium-Exposed Individuals of *Porcellio scaber*.[118]

Amino acid	*Porcellio scaber* protein (Res.%)[118]	*Agaricus macrospours* protein (Res.%)[154]
Lys	4.9 ± 1.4	4.6
Leu	3.9 ± 2.8	2.5
Ile	3.6 ± 2.9	1.6
Val	3.7 ± 1.5	3.6
Ala	8.7 ± 5.3	8.2
Gly	15.1 ± 1.7	16.5
Pro	6.2 ± 1.7	7.1
Glx	19.8 ± 2.7	20.3
Ser	7.6 ± 1.8	7.8
Thr	6.2 ± 2.3	4.3
Asx	17.7 ± 3.0	15.0
Cys	2.6 ± 0.4[a]	0
Arg	0	1.9
Met	0	0
Phosphoserine	n.d.	4.2
His	0	1.7
Phe	0	1.0

Note: Woodlice were collected in the vicinity of Innsbruck, Austria, and reared on cadmium-loaded litter (96.3 ± 28.5 μg Cd mg^{-1} dry weight, n = 11) for a period of 20 weeks. For amino acid analysis, 380 animals (approx. 10 g fresh weight) were homogenized in 50 mM Tris-HCl buffer (pH 7.5) containing 10 mM mercaptoethanol and protease inhibitors (anitipain, pepstatin, and phenylmethylsulfonylfluoride). Homogenates were centrifuged twice (15 min × 30,000 g and 1 h × 100,000 g) and separated by three subsequent fractionation steps using gel chromatography (Sephacryl S-200 and Sephacryl S-100) and ion exchange chromatography (DEAE-cellulose). Cadmium-containing fractions derived from the last purification step were pooled and oxidized as described by Dallinger et al.[30] Amino acid analysis was carried out on a Liquimat III apparatus (Labotron, Germany) according to the method described by Dallinger et al.[30] For *Porcellio scaber* protein, average percentile residues (Res.%) with standard deviations of 4 repeated analyses are given. For comparison, the amino acid composition of a glycoprotein isolated from the fungus *Agaricus macrosporus*, is also reported.[154]

[a] Determined as cysteic acid; n.d.: not determined.

Knutti et al.[13] have shown that *Helix pomatia* and other pulmonates from an unpolluted woodland area exhibited cadmium concentrations which were higher than those of most other soil invertebrate species by a factor of about ten. The capacity for accumulating cadmium is most pronounced in individuals of *Helix pomatia* collected from metal-contaminated sites in the field,[119] as well as in snails fed on cadmium-containing diets in the laboratory.[8,10] The target organ of cadmium accumulation in this species clearly is the midgut gland (Figure 2). In control as well as in cadmium-exposed individuals, 55 to 75% of the assimilated metal are concentrated in this organ.[8,10] As shown in Figure 2, the high concentration of cadmium in the midgut gland persisted even in animals which after a short period of metal exposure were reared on an uncontaminated diet for several weeks.

After centrifugal fractionation of midgut gland homogenates, most of the cadmium (75 to 90%) was always associated with the supernatant,[33,96] whereas only minor proportions of the metal could be found in the sediment. Apart from other cellular components the sediment contains calcium-rich pyrophosphate granules derived from the calcium cells and large lipofuscin vesicles from the digestive cells. Interestingly, the lysosomal activity and the production of lipofuscin granules in digestive cells was stimulated by cadmium feeding.[120] Preliminary efforts to detect cadmium in these granules by means of EELS techniques failed.[120] However, since it has been shown that in other gastropods some cadmium was sequestered by the pyrophosphate granules,[91] it can be assumed that this also applies to the sediment-bound cadmium in *Helix pomatia*.

The predominant part of cadmium in the midgut gland was bound to a cytosolic metallothionein which we characterized by its amino acid sequence (Figure 5). In the terrestrial slug *Arion ater* cadmium was not associated with any specific organelles, but was exclusively bound to a cytosolic protein showing typical features of metallothioneins.[69] Similar results have been reported for the slug *Arion lusitanicus*.[31]

As shown by means of Chelex experiments,[108] the synthesis of metallothionein in *Helix pomatia* can be induced within a few hours after cadmium exposure, the amount of the protein increasing rapidly with continued cadmium feeding (Figure 6a). Even after one single cadmium pulse, the metal remained bound to the metallothionein over several weeks, the percentage cadmium saturation of the protein increasing steadily until the end of the experiment (Figure 6b). The reason for this is that the cadmium released from degraded metallothionein was picked up, in the absence of vesicular sequestration, by the remaining portion of the cytosolic protein.

Thus the accumulation and long-term storage of cadmium in *Helix pomatia* is predominantly accounted for not by intracellular compartmentalization, but by association of the metal with a cytosolic metallothionein in the midgut gland. This may be regarded as part of a detoxification strategy, indicated by the fact that in terms of lethality *Helix pomatia* is exceptionally tolerant against high cadmium concentrations in the food (see Chapter 15)[10]. However, the complexation of cadmium by metallothioneins in terrestrial invertebrates does not inevitably lead to high accumulations of this metal. Some insect species, for example, are able to eliminate the metal by cellular excretion processes.[24] Thus the macroconcentration of cadmium in *Helix pomatia* can also be interpreted as the animal's inability to eliminate this metal.

2. Lead

Helix pomatia is also a microconcentrator of lead (Table 1). In individuals from polluted areas in the vicinity of a metal smelter industry, highly elevated concentrations of this metal were detected.[119] As shown in Figure 2, the most important organ involved in lead accumulation is again the midgut gland, in which the metal can be retained over extended periods of time. In lead-contaminated snails, the midgut gland contains up to 90% of the assimilated metal.[8]

In midgut glands of control animals after centrifugal fractionation a high amount of lead was associated with the supernatant, whereas in the metal-loaded individuals the predominant proportion of the metal was detected in the sediment.[33] This can be explained in the light of findings that in several species of gastropods metal intoxication stimulates lysosomal activity.[59,120,121] In fact, in lead-contaminated individuals of *Helix pomatia* an increasing number of lipofuscin granules was observed in the digestive cells of the midgut gland. Microscopical and ultrastructural examinations have revealed that lead is accumulated within these vesicles (Figure 7). These granules grow by fusion with other vesicles of lysosomal origin[122] and are voided into the lumen of hepatopancreatic tubules by disruption of the digestive cells.[120] In lead-contaminated individuals, however, this kind of cellular excretion occurs at very low rates.[122]

In exposed snails considerable amounts of lead were also detected in the posterior section of the alimentary tract[8] where the metal was first accumulated in the mucous cells and then voided into the lumen of the alimentary canal together with mucous extrusions (Figure 8). The occurrence of lead in mucous cells may be due to these cells containing an elevated proportion of SH groups, important constituents of the mucous.[123] Since lead is a borderline metal with considerable class B properties,[21] the metal is assumed to bind to the sulfur atoms of the mucous substances.

Thus it seems that in contaminated *Helix pomatia*, cellular excretion processes are involved in the elimination of lead in the midgut gland and in the digestive tract. However, these excretory activities do not match the high accumulation capacity of the midgut gland which thus must be considered as a long-term storage organ for this metal as well.

3. Zinc

Zinc is an essential trace element involved in the formation and function of several key enzymes.[124] It is not surprising, therefore, that even uncontaminated individuals of *Helix pomatia* contain relatively high amounts of zinc in their soft

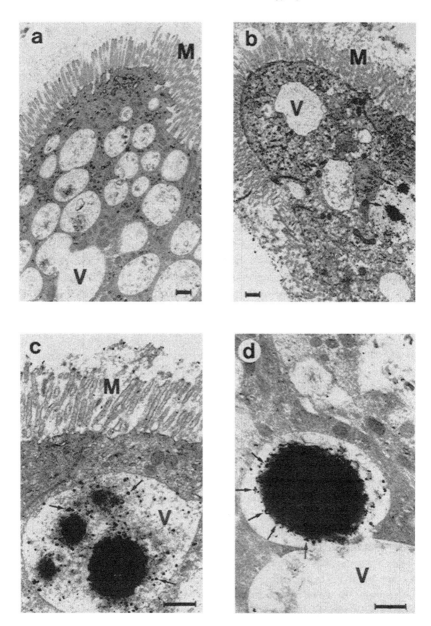

GURE 7. Electron micrographs showing electron-dense metal precipitations in lysosomes located within vacuoles of hepatopancreas digestive cells from lead-exposed *Helix pomatia* and from uncontaminated control snails. Exposed animals had been fed on lead-containing lettuce (870 μg/g dry weight) for a period of 40 days. Small pieces of hepatopancreas were fixed for 2 h in 2% glutaraldehyde, dissolved in 0.01 *M*

tissues, among which the midgut gland is the organ with the highest contents, accounting for more than 50% of the zinc in this species.[8,10] In zinc-contaminated animals, up to 70% of the body metal is concentrated in the midgut gland, reaching concentrations of more than 1000 μg/g dry weight.[8,10] Since zinc levels in the midgut gland of *Helix pomatia* remain elevated over long periods of time (Figure 2), effective mechanisms are supposed to be involved in the sequestration of this metal. *Helix pomatia* can therefore also be regarded as a macroconcentrator of zinc (Table 1).

In contaminated snails, 50% of the zinc in the midgut gland was associated with supernatant fractions, while in metal-fed animals this proportion was only 20 to 30%.[33,96] In contrast to cadmium, however, most of the supernatant zinc in the midgut gland of *Helix pomatia* was not associated with metallothionein, but with a homogenous, nonmetallothionein component showing a molecular weight between 1000 and 4000 Da (see Chapter 16).[96] It is significant that this zinc-binding ligand was also predominant in uncontaminated snails.[33] The nature of this component remains to be elucidated.

cacodylate buffer (pH 7.4) and saturated with hydrogen sulfide; metal precipitates were developed with 10% silver nitrate; after repeated washings in 0.1 M Tris-maleate buffer (pH 7.4) and in 0.01 M cacodylate buffer (pH 7.4) samples were postfixed for 2 h in 1% OsO_4 (pH 7.4) and again washed repeatedly in 0.01 M cacodylate buffer; after dehydration in methanol and propylene oxide samples were embedded in araldite. Ultrathin sections were obtained by cutting with a diamond knife and observed on a Zeiss EM 9 or EM 10 electron microscope (previously unpublished data).[122] In order to prove specifically the presence of lead in lysosomes, hepatopancreas aliquots of lead-exposed snails were also fixed for light microscopy in formaldehyde and 3% potassium dichromate for several days, and lead specifically precipitated as lead chromate by the method described by Prosi et al.[155] Paraffine cross-sections were observed by polarized light microscopy on a dark background (results not shown).

a: Electron micrograph of digestive cells from hepatopancreas of a control snail, showing large vacuoles (V) within the cells and the microvilli border (M) towards the lumen. Bar: 1 μm.

b: Electron micrograph of digestive cells from hepatopancreas of a lead-contaminated snail, showing the microvilli border (M) on the side of the lumen, vacuoles (V) within the cells with electron-dense metal precipitations aggregated around lysosomes (arrows). Bar: 1 μm.

c: Electron micrograph showing a detail of a digestive cell with the microvilli border (M), including a large vacuole (V) containing several lysosomes with dense metal precipitations (arrows). Bar: 1 μm.

d: Electron micrograph showing a detail of a digestive cell with vacuols (V) and a large lysosome with metal precipitations on its surface (arrows). Bar: 1 μm.

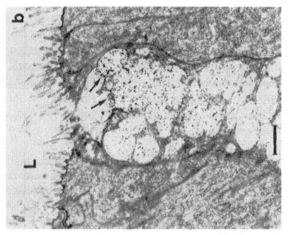

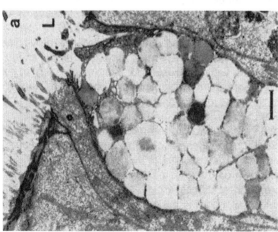

FIGURE 8. Electron micrographs of mucous cells from the digestive tract (midgut) of a control and a lead-fed individual of *Helix pomatia*. For experimental conditions and sample preparation see Figure 7 caption.
a: Electron micrograph of a mucous cell in a control snail, opening towards the lumen (L) of the digestive tract (midgut). Bar: 1 μm.
b: Electron micrograph of a mucous cell in a lead-exposed animal, containing granular metal precipitations (arrows), and opening towards the lumen (L) of the digestive tract (midgut). Bar: 1 μm.

A significant or even a predominant proportion of hepatopancreas zinc, however, was always bound to the sediment fractions which in zinc-exposed animals contained 70 to 80% of the metal.[33,96] If it is assumed that sediment-bound zinc is associated with lysosomal fractions, two possible candidates must be considered: the lipofuscin granules in the digestive cells on the one hand, and the pyrophosphate granules in the calcium cells on the other hand. Our own histochemical and ultrastructural observations have shown that in midgut glands of zinc-exposed *Helix pomatia*, metal precipitations were normally found within the pyrophosphate granules (Figure 9), but not in the lipfuscin vesicles of the digestive cells. This is corroborated by the fact that in the closely related species *Helix aspersa* zinc has also been found to be enriched within the pyrophosphate granules of the calcium cells.[49,50]

It is concluded that *Helix pomatia* is a macroconcentrator of zinc, the highest amounts of this metal being found in the midgut gland. In contrast to cadmium and lead, however, a significant proportion of hepatopancreas zinc is bound to a low molecular weight ligand, while only a small amount of the metal is chelated by metallothioneins. In zinc-contaminated snails, a predominant proportion of the metal is sequestered by calcium-rich pyrophosphate granules within the calcium cells.

4. Copper

Helix pomatia is also a macroconcentrator of copper (Table 1). This trace element is an essential constituent of the snail's oxygen-carrying hemocyanin. Thus copper concentrations in tissues of uncontaminated snails may at least in part be attributed to this protein.[33,125] In both uncontaminated and copper-exposed *Helix pomatia*, the metal does not show a clear preference for any organ, but is rather evenly distributed within the body.[8,126] The foot and mantle tissues contain the highest amounts of copper, while the midgut gland seems not to play an important role in concentrating this metal. The metal is not stored in any of these organs but is quickly released. At the same time copper concentrations in the alimentary tract increase slightly,[8] suggesting that the intestine is involved in the elimination of copper.

It is interesting that a large proportion of copper in the midgut gland was always associated with supernatant fractions. Among these, a low molecular weight component which probably is a metallothionein (see Chapter 16) was predominant in control snails. A considerable proportion of copper was always associated with high molecular weight fractions which contained the copper-binding hemocyanin.[33] Apart from its other functions this protein may thus serve as a transitory storage site of surplus copper. About 20 to 30% of the hepatopancreas copper was associated with sediment fractions.[33] These fractions contain, among other cellular components, the calcium-rich pyrophosphate granules which are the sites of cellular copper sequestration, as shown by means of histochemical rubeanate precipitation.[127] This is corroborated by the fact that in

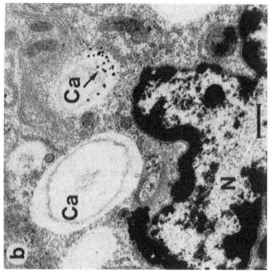

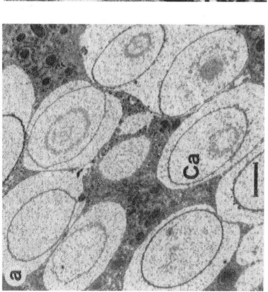

FIGURE 9. Electron micrographs of calcium cells from the midgut gland of a control and a zinc-fed individual of *Helix pomatia*. Exposed animals had been fed on zinc-containing lettuce (1067 µg/g dry weight) for a period of 50 days. For sample preparation see Figure 7 caption.

a: Electron micrograph of a calcium cell in the midgut gland of a control snail, showing calcium granules (Ca) with concentric layers. Bar: 1 µm.

b: Electron micrograph of a calcium cell in a zinc-exposed animal, showing the large nucleus (N) besides smaller calcium granules (Ca), one of which containing electron-dense metal precipitations (arrow). Bar: 1 µm.

the closely related species, *Helix aspersa*, copper was also found in these granules.[128] Similar results were reported for the slug *Arion ater*.[59] In this species cyclic elimination of copper was observed due to the secretion of copper-containing pyrophosphate granules.[59]

It is concluded that copper-exposed *Helix pomatia* is capable of eliminating this metal efficiently by means of excretion processes in the intestine and in the midgut gland. On the whole-body level, however, *Helix pomatia* is a macroconcentrator of copper.

B. Storage-Detoxification in *Porcellio scaber*

1. Cadmium

Elevated concentrations of cadmium were found in individuals of *Porcellio scaber* living in metal-polluted areas,[3,22] or fed on cadmium-containing litter in the laboratory.[23,118,129] Cadmium concentrations of *Porcellio scaber* in the field have been shown to depend, among other factors, on metal concentrations in the substrate on which isopods live, concentration factors always exceeding the value of 1 (Table 1).[63,65] In consequence, *Porcellio scaber* has been used successfully as a biological indicator of environmental cadmium pollution in metal-polluted regions.[63,65,68]

In both uncontaminated and metal-loaded *Porcellio scaber*, a large fraction of assimilated cadmium (55 to 80%) was concentrated in the hepatopancreas.[23] Some cadmium was also shown to accumulate in the posterior part of the alimentary tract, and it has been suggested that this section of the intestine may be involved in metal uptake.[130]

After subcellular fractionation of hepatopancreas homogenates by means of density gradient centrifugation, most of the cadmium in this organ was found in the supernatant fractions, but a significant amount of this proportion may have originated from broken lysosomes which normally also contain cadmium.[23] After gentle homogenization and repeated centrifugation of hepatopancreas homogenates, Donker and colleagues[115] also found 70 to 80% of the hepatopancreas cadmium in *Porcellio scaber* to be associated with supernatant fractions. Thus in spite of some cadmium originating from broken lysosomes, a considerable amount of metal is bound to cytosolic components. This is also indicated by the fact that by means of X-ray microanalysis, cadmium was found scattered throughout the cytoplasm of S cells[22] (Figure 4). Among cytosolic components, possible candidates accounting for cadmium binding may be metallothioneins. However, efforts to isolate metallothioneins from the hepatopancreas of *Porcellio scaber* have failed so far.[114,115] Recently a non-metallothionein cadmium binding ligand was isolated and characterized from metal-contaminated *Porcellio scaber* in our laboratory.[118] This ligand has an apparent molecular weight of 12,000 Da and is probably a glycoprotein, as indicated by characteristic features such as low cysteine content, low cadmium binding affinity, and extremely high susceptibility to oxidizing conditions.[118] The amino acid composition of the novel protein is shown in Table 2. This ligand is assumed to be involved in the detoxification of cadmium. A considerable fraction of cytosolic cadmium was associated with low molecular weight components ($\approx$ 1500 Da) of unknown structure.[118] Low molecular weight cadmium binding ligands from the same species have also been reported by Donker and colleagues.[115]

A minor but significant proportion of cadmium in the hepatopancreas of *Porcellio scaber* was shown to be associated with lysosomal fractions.[23] As indicated by X-ray microanalysis, this fraction of the metal accumulated within lysosomes of S cells, and was associated with other trace elements such as lead, zinc, and copper[22] (Figure 4). Phosphorus and sulfur were also detected within these vesicles, but X-ray mapping showed cadmium to be associated with sulfur rather than with phosphorus.[131]

Metal-loaded individuals of *Porcellio scaber* fed on uncontaminated leaves over a period of 20 weeks lost cadmium from their hepatopancreas at low, but constant rates.[3] Thus it seems that *Porcellio scaber* is a macroconcentrator of cadmium (Table 1) but is also able to eliminate a certain fraction of the metal slowly. This might be explained by assuming two different cadmium pools in the hepatopancreas of *Porcellio scaber*: one pool of the metal being immobilized by storage-detoxification within lysosomes, which cannot be excreted;[1] and a second, more mobile, cadmium pool which can be excreted at low rates, and which may be represented by the cytosolic metal fractions described above.

2. Lead

In *Porcellio scaber* from an ancient mining region in Germany lead concentrations of more than 1000 μg/g dry weight were detected.[22] Elevated concentrations of lead were also recorded in isopods around metal smelting works,[1,63,119] and in woodlice from contaminated urban environments.[65] High concentrations of this metal were also reported from *Porcellio scaber* fed on lead-contaminated litter in the laboratory.[23] However, the concentration factors for lead in *Porcellio scaber* are usually below 1,[3,65] hence this species must be considered as a deconcentrator of lead (Table 1).

In both uncontaminated and metal-loaded *Porcellio scaber* a predominant proportion of lead (70 to 80%) was shown to be concentrated in the hepatopancreas.[23] In contaminated individuals considerable amounts of this metal were also detected in the gut,[3] particularly within the posterior sections of this tissue.[130]

Subcellular density gradient fractionation of hepatopancreas homogenates from *Porcellio scaber* revealed that in uncontaminated isopods up to 30% of lead was associated with lysosomal fractions, whereas in metal-exposed individuals, up to 60% of the metal accumulated in these organelles.[23] This is likely to be a consequence of the fact that in hepatopancreas cells of *Porcellio scaber* lysosomal activity is activated by metal exposure.[132] Thus it is not surprising that the activity of acid phosphatase, as shown by histochemical techniques, was clearly more elevated in hepatopancreas S cells of metal-exposed animals compared to control individuals.[22,132] As shown in Figure 3, the sites of acid phos-phatase activity within these cells are the metal-accumulating granules. X-ray microanalysis revealed that, apart from lead, phosphorus was also present in these vesicles (Figure 4), and it has been suggested that lead may be associated with phosphate groups.[22] In hepatopancreas granules of spiders, for instance, zinc coprecipitated with phosphorus and calcium.[131] This may also be true for lead in the hepatopancreas lysosomes of *Porcellio scaber*, the metal being in-corporated into calcium phosphate compounds. One indication for this is provided by the fact that in metal-exposed individuals of *P. scaber*, a clear correlation was observed between lead and calcium uptake.[133] Such interactions have been explained on the basis of similarities in the outer electron shell structure between the two elements.[133] However, lead is a borderline metal with considerable class B properties[21] and could therefore bind to sulfur-carrying ligands. Since X-ray signals for lead interfere with those for sulfur,[131] this question cannot be decided at the moment.

In both uncontaminated and metal-loaded woodlice, a considerable amount of hepatopancreas lead was found in small aggregates within the cytoplasm of S cells (Figure 4), and this may account for those proportions of the metal which after subcellular fractionation have been found to be associated with cytosolic fractions.[23] It is not known, however, which compounds are responsible for the binding of lead in the cytoplasm.

In hepatopancreas of lead-contaminated isopods fed on a metal-free diet over a period of 20 weeks, more than 50% of the metal was lost during the experiment.[3] This indicates that in the hepatopancreas of this species two lead pools may occur, as suggested for cadmium. While the more mobile form of the metal is eliminated, the inactivated fraction of lead is sequestered and stored within lysosomes.

It is concluded that *Porcellio scaber* is not a concentrator of lead, since concentration factors for this metal are normally below 1 (Table 1). Nevertheless, considerable amounts of lead are inactivated by storage-detoxification within the hepatopancreas, which organ can retain these metal fractions over long periods of time.

3. Zinc

Porcellio scaber is a facultative concentrator of zinc: the concentration factor for this metal in isopods varies over a large range between 0.04 and 7 (Table 1). It seems that concentration factors decrease with increasing metal contami-nation of the environment.[3,64] This means that in unpolluted areas, *Porcellio scaber* is likely to be a macroconcentrator, and in metal-polluted sites, a de-concentrator of zinc.

In both uncontaminated and metal-exposed woodlice zinc was predominantly concentrated in the hepatopancreas, whereas a minor proportion of the metal also accumulated in the gut.[3] Chromatographic fractionation studies have shown that cytosolic hepatopancreas zinc is predominantly bound to low molecular weight compounds of unknown structure.[114,115,118]

A significant proportion of hepatopancreas zinc, however, may be associated with lysosomal granules. In metal-loaded *Porcellio scaber* from a mining area, zinc was detected within the lysosomes of B cells by X-ray microanalysis.[22] Hopkin and colleagues[131] have shown by X-ray mapping that within these granules the metal is normally associated with phosphorus. Thus lysosomal zinc is likely to be a constituent of calcium phosphate compounds. This is corroborated by the fact that in other invertebrate species zinc also showed a strong affinity to calcium phosphate granules[131] (Figure 9).

When zinc-loaded individuals of *Porcellio scaber* were fed uncontaminated leaves for a period of 20 weeks, an insignificant amount of hepatopancreas zinc was lost,[3] which means that *Porcellio scaber* retains most of the hepatopancreas zinc by storage-detoxification. In contrast, *Oniscus asellus* was able to excrete, by means of apocrine secretion, considerable proportions of the metal which in this species also accumulated within the vesicles of S cells.[60] Thus the long-term storage of zinc in the hepatopancreas of *Porcellio scaber* may also be regarded as reflecting the animal's inability to excrete the metal.

4. Copper

Porcellio scaber is a strong concentrator of copper, concentration factors for this metal normally reaching values higher than 2 (Table 1). The extraordinary capacity of terrestrial isopods in concentrating copper has been first described by Wieser,[134] who recognized the crucial role of this essential trace element in the metabolism of woodlice.[55,135] Wieser and colleagues[40,136] have also hypothesized that terrestrial isopods might meet their copper requirements by coprophagy, a view which has been criticized.[16,137] Recently, Hopkin has shown that in uncontaminated sites terrestrial isopods suffer from copper deficiency, unless they ingest copper-concentrating fungal hyphae growing on the surface of leaves on which the animals feed.[15] (See this volume, Chapter 18.)

In laboratory experiments it has been shown that copper assimilation by *Porcellio scaber* increased with increasing metal concentration in the food,[136] a predominant proportion of the metal being concentrated in the hepatopancreas.[3,23,138] After subcellular fractionation of hepatopancreas homogenates from uncontaminated and metal-exposed *Porcellio scaber*, 12 to 30% of copper was found in lysosomal fractions, while the rest of the metal was associated with the cytosol.[23] Similar distribution patterns have been reported by Donker and colleagues.[115] However, a considerable proportion of cytosolic copper may be accounted for by broken lysosomes which thus may contain more than 30% of this metal.[23] Nevertheless, copper was also detected by X-ray microanalysis within small concretions scattered throughout the cytoplasm of S cells (Figure 4).

Chromatographic fractionation studies revealed that a minor amount of cytosolic copper in the hepatopancreas of *Porcellio scaber* was associated with high molecular weight fractions which probably represent hemocyanin.[118] The remaining proportion of cytosolic copper was bound to unidentified, low molecular weight compounds.[118]

Apparently the proportion of copper within lysosomal granules increases with increasing metal contamination of the animals,[23] indicating that the lysosomal system is stimulated by copper exposure. This lysosomal copper fraction may be identified as the lipid-soluble compartment postulated by Wieser and Klima in the related species *Oniscus asellus*.[17]

Thus it appears that *Porcellio scaber* is an avid concentrator of copper, a proportion of which is sequestered by storage-detoxification in the hepatopancreatic S cells. This copper pool may not be available to the animal, since copper-containing granules are largely insoluble in water and alcohol.[16] The remaining copper in the hepatopancreas is bound to hemocyanin and to low molecular weight compounds, and this proportion of the metal is suggested to represent a more mobile pool which is not excreted but retained as a transitory reserve for periods of increased requirements during molting[55] or other stressful events.

V. SPECULATIVE CONCLUSIONS: ECOTOXICOLOGICAL AND EVOLUTIONARY CONSIDERATIONS

Since micro- and macroconcentrators represent biological sinks for potentially toxic metals in terrestrial foodwebs, the question arises as to whether storage-detoxification in these organisms influences the availability of detoxified metals to predators and detritus feeders. In this context, the term "biomagnification" has been introduced to characterize the accumulation of toxic material along food chains. Some authors state that biomagnification of metals does not normally occur in soil-living animals.[149] The question of biomagnification as a possible hazard to predators is, however, a somewhat artificial problem, since even low transfer factors for metals (<1) may be sufficient to exert an adverse effect: for example, cadmium and lead have been reported to reach critical concentrations in the kidney of soricine shrews feeding on metal-loaded earthworms.[150]

An aspect that requires closer consideration is that there may be a difference between the extents to which the two main cellular pathways involved in storage-detoxification (sequestration within vesicles on the one hand, and binding to cytosolic metallothioneins on the other hand) render available the detoxified metals. Carnivorous marine invertebrates, for instance, do not accumulate metals associated with pyrophosphate granules in the digestive glands of their prey. It has therefore been argued that the detoxification of metals results in the reduction of metal bioavailability in marine food chains.[151] This could also hold for terrestrial food chains in which predators feed on contaminated invertebrates able to sequester metals within granules. The metallothionein-bound fraction of cadmium, however, may be more easily available to potential predators, since this metal is released from the protein under acidic conditions[25] likely to occur within the digestive tracts of consumers.

An additional problem related to metal accumulation in terrestrial inverte-brates concerns the question of whether storage-detoxification might provide a basis for metal resistance in populations living on contaminated sites. In certain strains of *Drosophila melanogaster* the functional duplication of the metallothi-onein-encoding gene appears to confer increased cadmium and copper resis-tance.[24] In vertebrates, on the other hand, the function of metallothionein may primarily be related to the metabolism of essential trace elements;[36,152] the high affinity of this protein to the nonessential element cadmium may be a fortuitous effect.[34] It has also been argued that in most habitats metal contamination is too low to exert a significant selective pressure towards the evolution of a special detoxification system based on metal complexation by metallothioneins.[153]

These arguments may be true for vertebrates which are able to prevent toxic metals from entering their body by several avoidance strategies. However, the situation may be different in invertebrates, which in highly contaminated areas are likely to be exposed to toxic metal concentrations sufficient to exert a real selective pressure. Aquatic oligochaetes have been reported to acquire genetically based cadmium resistance in rivers highly contaminated with this metal, and it has been speculated that this is due to an increased production of metallothionein (see Chapter 13).[38] Individuals of *Helix pomatia* exposed to cadmium at con-centrations which are likely to occur at highly contaminated field sites,[7] are shown in the present article (Figure 6) to respond rapidly by an increased pro-duction of metallothionein. Indeed, the high resistance of *Helix pomatia* to this metal (see Chapter 15)[10] even at high body concentrations, indicates that me-tallothionein exerts a protective function against cadmium toxicity in this species. Moreover, cadmium-loaded individuals of *Helix pomatia* possess additional ca-pacities for the handling of the essential trace element zinc by binding of this metal to specific, low molecular weight ligands (see Chapter 16).[36]

Thus the generally accepted theory that the function of metallothioneins is primarily related to the metabolism of essential trace elements may only be true, if at all, for vertebrates. In sediment-inhabiting aquatic and in terrestrial inver-tebrates exposure to toxic metals may be high enough in certain habitats to exert a real selective pressure. Under such circumstances, metallothioneins may de-velop a detoxifying function which otherwise would not be of importance.

ACKNOWLEDGMENTS

The author thanks Professor W. Wieser, University of Innsbruck, for re-viewing the manuscript and for helpful discussion. The study was supported by the Austrian "Fonds zur Förderung der wissenschaftlichen Forschung in Öster-reich", project Nr. P7815.

REFERENCES

1. **Hopkin, S. P.**, Critical concentrations, pathways of detoxification and cellular ecotoxicology of metals in terrestrial arthropods, *Functional Ecol.*, 4, 321, 1990.
2. **Janssen, M. P. M., Bruins, A., De Vries, T. H., and Van Straalen, N. M.**, Comparison of cadmium kinetics in four soil arthropod species, *Arch. Environ. Contam. Toxicol.*, 20, 305, 1991.
3. **Hopkin, S. P.**, Species-specific differences in the net assimilation of zinc, cadmium, lead, copper and iron by the terrestrial isopods *Oniscus asellus* and *Porcellio scaber*, *J. Appl. Ecol.*, 27, 460, 1990.
4. **Hopkin, S. P. and Martin, M. H.**, Heavy metals in woodlice, in *The Biology of Terrestrial Isopods*, Sutton, S. L. and Holdich, D. M., Eds., Symp. Zool. Soc. London, No. 53, 1984, 143.
5. **Ireland, M. P. and Richards, K. S.**, The occurrence and localisation of heavy metals and glycogen in the earthworms *Lumbricus rubellus* and *Dendrobaena rubida* from a heavy metal site, *Histochemistry*, 51, 153, 1977.
6. **Morgan, A. J. and Morris, B.**, The accumulation and intracellular compartmentation of cadmium, lead, zinc and calcium in two earthworm species (*Dendrobaena rubida* and *Lumbricus rubellus*) living in highly contaminated soil, *Histochemistry*, 75, 269, 1982.
7. **Martin, M. H. and Coughtrey, P. J.**, *Biological Monitoring of Heavy Metal Pollution: Land and Air*, Applied Science Publishers, London, 1982.
8. **Dallinger, R. and Wieser, W.**, Patterns of accumulation, distribution and liberation of Zn, Cu, Cd, and Pb in different organs of the land snail *Helix pomatia* L., *Comp. Biochem. Physiol.*, 79C, 117, 1984.
9. **Berger, B. and Dallinger, R.**, Accumulation of cadmium and copper by the terrestrial snail *Arianta arbustorum* L.: kinetics and budgets, *Oecologia*, 79, 60, 1989.
10. **Berger, B., Dallinger, R., Felder, E., and Moser, J.**, Budgeting the flow of cadmium and zinc through the terrestrial gastropod, *Helix pomatia* L., this volume, chap. 15.
11. **Greville, R. W. and Morgan, A. J.**, The influence of size on the accumulated amounts of metals (Cu, Pb, Cd, Zn, and Ca) in six species of slugs sampled from a contaminated woodland site, *J. Mol. Stud.*, 56, 355, 1990.
12. **Greville, R. W. and Morgan, A. J.**, Concentrations of metals (Cu, Pb, Cd, Zn, Ca) in six species of British terrestrial gastropods near a disused lead and zinc mine, *J. Mol. Stud.*, 55, 31, 1989.
13. **Knutti, R., Bucher, P., Stengl, M., Stolz, M., Tremp, J., Ulrich, M., and Schlatter, C.**, Cadmium in the invertebrate fauna of an unpolluted forest in Switzerland, *Environ. Toxicol. Ser.*, 2, 171, 1988.
14. **Boháč, J. and Pospíšil, J.**, Accumulation of heavy metals in invertebrates and its ecological aspects, in *Heavy Metals in the Environments, Vol. 1*, Vernet, J.-P., Ed., CEP Consultants Ltd, Edinburgh, 1989, 354.
15. **Hopkin, S. P.**, Deficiency and excess of copper in terrestrial isopods, this volume, chap. 18.
16. **Hopkin, S. P.**, *Ecophysiology of Metals in Terrestrial Invertebrates*, Elsevier Applied Science, Barking, U.K., 1989.
17. **Wieser, W. and Klima, J.**, Compartmentalization of copper in the hepatopancreas of isopods, *Mikroskopie*, 24, 1, 1969.

18. **Hopkin, S. P. and Martin, M. H.**, The distribution of zinc, cadmium, lead and copper within the hepatopancreas of a woodlouse, *Tissue & Cell*, 14, 703, 1982.

19. **Sohal, R. S., Peters, P. D., and Hall, T. A.**, Origin, structure, composition and age-dependence of mineralized dense bodies (concretions) in the midgut epithelium of the adult housefly *Musca domestica, Tissue & Cell*, 9, 87, 1977.

20. **Humbert, W.**, Cytochemistry and X-ray microprobe analysis of the midgut of *Tomocerus minor* Lubbock (Insecta: Collembola) with special reference to the physiological significance of the mineral concretions, *Cell Tissue Res.*, 187, 397, 1978.

21. **Nieboer, E. and Richardson, D. H. S.**, The replacement of the nondescript term 'heavy metals' by a biologically and chemically significant classification of metal ions, *Environ. Pollut. Ser. B*, 1, 3, 1980.

22. **Prosi, F. and Dallinger, R.**, Heavy metals in the terrestrial isopod *Porcellio scaber* Latreille. I. Histochemical and ultrastructural characterization of metal containing lysosomes, *Cell Biol. Toxicol.*, 4, 81, 1988.

23. **Dallinger, R. and Prosi, F.**, Heavy metals in the terrestrial isopod *Porcellio scaber* Latreille. II. Subcellular fractionation of metal-accumulating lysosomes from hepatopancreas, *Cell. Biol. Toxicol.*, 4, 97, 1988.

24. **Lauverjat, S., Ballan-Dufrancais, C., and Wegnez, M.**, Detoxification of cadmium. Ultrastructural study and electron-probe microanalysis of the midgut in a cadmium-resistant strain of *Drosophila melanogaster, Biol. Metals*, 2, 97, 1989.

25. **Kägi, J. H. R. and Schäffer, A.**, Biochemistry of metallothionein, *Biochemistry*, 27, 8509, 1988.

26. **Hamer, D. H.**, Metallothionein, *Annu. Rev. Biochem.*, 55, 913, 1986.

27. **Margoshes, M. and Vallee, B. L.**, A cadmium protein from equine kidney-cortex, *J. Am. Chem. Soc.*, 79, 4813, 1957.

28. **Suzuki, K. T., Yamamura, M., and Mori, T.**, Cadmium-binding proteins induced in the earthworm, *Arch. Environ. Contam. Toxicol.*, 9, 415, 1980.

29. **Morgan, J. E., Norey, C. G., Morgan, A. J., and Kay, J.**, A comparison of the cadmium-binding proteins isolated from the posterior alimentary canal of the earthworms *Dendrodrilus rubidus* and *Lumbricus rubellus, Comp. Biochem. Physiol.*, 92C, 15, 1989.

30. **Dallinger, R., Berger, B., and Bauer-Hilty, A.**, Purification of cadmium-binding proteins from related species of terrestrial Helicidae (Gastropoda, Mollusca): A comparative study, *Mol. Cell. Biochem.*, 85, 135, 1989.

31. **Dallinger, R., Janssen, H. H., Bauer-Hilty, A., and Berger, B.**, Characterization of an inducible cadmium-binding protein from hepatopancreas of metal-exposed slugs (Arionidae, Mollusca), *Comp. Biochem. Physiol.*, 92C, 355, 1989.

32. **Maroni, G. and Watson, D.**, Uptake and binding of cadmium, copper and zinc by *Drosophila melanogaster* larvae, *Insect Biochem.*, 15, 55, 1985.

33. **Dallinger, R. and Wieser, W.**, Molecular fractionation of Zn, Cu, Cd, and Pb in the midgut gland of *Helix pomatia* L., *Comp. Biochem. Physiol.*, 79C, 125, 1984.

34. **Cosson, R. P., Amiard-Triquet, C., and Amiard, J.-C.**, Metallothioneins and detoxification. Is the use of detoxification protein for MTs a language abuse?, *Water, Air, Soil Pollut.*, 57/58, 555, 1991.

35. **Engel, D. W. and Brouwer, M.**, Metallothionein and metallothionein-like proteins: physiological importance, *Adv. Comp. Environ. Phys.*, 5, 53, 1989.

36. **Brady, F. O.**, The physiological function of metallothionein, *Trends Biol. Sci.*, 7, 143, 1982.
37. **Maroni, G., Wise, J., Young, J. E., and Otto, E.**, Metallothionein gene duplications and metal tolerance in natural populations of *Drosophila melanogaster*, *Genetics*, 117, 739, 1987.
38. **Klerks, P. L. and Levinton, J. S.**, Evolution of resistance and changes in community-composition in metal-polluted environments: a case study on Foundry Cove, this volume, chap. 13.
39. **Rainbow, P. S. and Dallinger, R.**, Metal uptake, elimination and regulation in freshwater invertebrates, this volume, chap. 7.
40. **Wieser, W.**, Conquering terra firma: the copper problem from the isopdod's point of view, *Helgol. Wiss. Meeresunters.*, 15, 282, 1967.
41. **Maddrell, S. H. P., Whittembury, G., Mooney, R. L., Harrison, J. B., Overton, J. A., and Rodriguez, B.**, The fate of calcium in the diet of *Rhodnius prolixus:* Storage in concretion bodies in the Malpighian tubules, *J. Exp. Biol.*, 157, 483, 1991.
42. **Ryder, T. A. and Bowen, I. D.**, The slug food as a site of uptake of copper molluscicide, *J. Invertebr. Pathol.*, 30, 381, 1977.
43. **Ireland, M. P.**, Sites of water, zinc and calcium uptake and distribution of these metals after cadmium administration in *Arion ater* (Gastropoda: Pulmonata), *Comp. Biochem. Physiol.*, 73A, 217, 1982.
44. **Maddrell, S. H. P., and Gardiner, B. O. C.**, The permeability of the cuticular lining of the insect alimentary canal, *J. Exp. Biol.*, 85, 227, 1980.
45. **Barnes, R. S. K., Calow, P., and Olive, P. J. W.**, *The Invertebrates: a New Synthesis*, Blackwell Scientific, Oxford, 1989.
46. **Potts, W. T. W.**, Excretion in the molluscs, *Biol. Rev.*, 42, 1, 1967.
47. **Riddle, W. A.**, Hemolymph osmoregulation and urea retention in the woodland snail, *Anguispira alternata* (Say) (Endodontidae), *Comp. Biochem. Physiol.*, 69A, 493, 1981.
48. **Jenkins, T.**, Histochemical and fine structure observations of the intestinal epithelium of *Trichuris suis* (Nematoda: Trichinoidea), *Z. Parasitenk.*, 42, 165, 1973.
49. **Howard, B., Mitchell, P. C. H., Ritchie, A., Simkiss, K., and Taylor, M. G.**, The composition of intracellular granules from the metal-accumulating cells of the common garden snail *(Helis aspersa), Biochem. J.*, 194, 507, 1981.
50. **Taylor, M. G., Greaves, G. N., and Simkiss, K.**, Biotransformation of intracellular minerals by zinc ions *in vivo* and *in vitro*, *Eur. J. Biochem.*, 192, 783, 1990.
51. **Witkus, R., Horgan, M. J., Dowling, P., Klein, M., and Faso, L.**, Comparative elemental analysis of the S and B cells of the hepatopancreas of *Armadillidium vulgare*, a terrestrial isopod, *Comp. Biochem. Physiol.*, 87C, 149, 1987.
52. **Tomita, M., Heisey, R., Witkus, R., and Vernon, G. M.**, Sequestration of copper and zinc in the hepatopancreas of *Armadillidium vulgare* Latreille following exposure to lead, *Bull. Environ. Contam. Toxicol.*, 46, 894, 1991.
53. **Martoja, R. and Martoja, M.**, Sur des accumulations naturelles d'aluminium et de silicium chez quelques invertebrés, *C.R. Acad. Sci. Paris*, 276D, 2951, 1973.
54. **Hubert, M.**, Localization and identification of mineral elements and nitrogenous waste in Diplopoda, in *Myriapod Biology*, Camatini, M., Ed., Academic Press, London, 1979, 127.

55. **Wieser, W.,** Uber die Häutung von *Porcellio scaber* Latr., *Verh. Dtsch. Zool. Ges. Kiel,* 178, 1964.

56. **Marshall, A. T.,** X-ray microanalysis of copper and sulphur-containing granules in the fat body cells of homopteran insects, *Tissue & Cell,* 15, 311, 1983.

57. **Simkiss, K. and Taylor, M. G.,** Convergence of cellular systems of metal detoxification, *Mar. Environ. Res.,* 28, 211, 1989.

58. **Mason, A. Z. and Simkiss, K.,** Sites of mineral deposition in metal-accumulating cells, *Exp. Cell Res.,* 139, 383, 1982.

59. **Marigómez, J. A., Angulo, E., and Moya, J.,** Copper treatment of the digestive gland of the slug, *Arion ater* L. 1. Bioassay conduction and histochemical analysis, *Bull. Environ. Contam. Toxicol.,* 36, 600, 1986.

60. **Hames, C. A. C. and Hopkin, S. P.,** A daily cycle of apocrine secretion by the B cells in the hepatopancreas of terrestrial isopods, *Can. J. Zool.,* 69, 1931, 1991.

61. **Joosse, E. N. G. and Buker, J. B.,** Uptake and excretion of lead by litter dwelling Collembola, *Environ. Pollut.,* 18, 235, 1979.

62. **Wieser, W., Busch, G., and Büchel, L.,** Isopods as indicators of the copper content of soil and litter, *Oecologia,* 23, 107, 1976.

63. **Hopkin, S. P., Hardisty, G., and Martin, M. H.,** The woodlouse *Porcellio scaber* as a "biological indicator" of zinc, cadmium, lead and copper pollution, *Environ. Pollut.,* 11B, 271, 1986.

64. **Hopkin, S. P., Hames, C. A. C., and Bragg, S.,** Terrestrial isopods as biological indicators of zinc pollution in the Reading area, South East England, *Monit. Zool. Ital.,* 4, 477, 1989.

65. **Dallinger, R., Berger, B., and Birkel, S.,** Terrestrial isopods: Useful biological indicators of urban metal pollution, *Oecologia,* 89, 32, 1992.

66. **Morgan, A. J., Morgan, J. E., Turner, M., Winters, C., and Yarwood, A.,** Metal relationships of earthworms, this volume, chap. 17.

67. **Berger, B. and Dallinger, R.,** Terrestrial snails as quantitative indicators of environmental metal pollution, *Biol. Monit. Assess.,* in press.

68. **Dallinger, R. and Berger, B.,** Bio-monitoring in the urban environment, in *Biological Indicators for Environmental Monitoring,* Serono Symp. Rev., Nr. 27, Bonotto, S., Nobili, R., and Revoltella, R. P., Eds., 227, 1992.

69. **Ireland, M. P.,** Uptake and distribution of cadmium in the terrestrial slug *Arion ater* (L.), *Comp. Biochem. Physiol.,* 68A, 37, 1981.

70. **Hopkin, S. P. and Martin, M. H.,** The distribution of zinc, cadmium, lead and copper within the woodlouse *Oniscus asellus* (Crustacea, Isopoda), *Oecologia,* 54, 227, 1982.

71. **Morgan, J. E. and Morgan, A. J.,** The distribution of cadmium, copper, lead, zinc and calcium in the tissues of the earthworm *Lumbricus rubellus* sampled from one uncontaminated and four polluted soils, *Oecologia,* 84, 559, 1990.

72. **Hopkin, S. P. and Martin, M. H.,** Assimilation of zinc, cadmium, lead, copper and iron by the spider *Dysdera crocata,* a predator of woodlice, *Bull. Environ. Contam. Toxicol.,* 34, 183, 1985.

73. **Ballan-Dufrancais, C.,** Accumulations minérals et puriques chez trois espèces d'insectes Dictyoptères, *Cellule,* 70, 315, 1974.

74. **Suzuki, K. T., Aoki, Y., Nishikawa, M., Masui, H., and Matsubara, F.,** Effect of cadmium-feeding on tissue concentrations of elements in a germ-free silkworm *(Bombyx mori)* larvae and distribution of cadmium in the alimentary canal, *Comp. Biochem. Physiol.,* 79C, 249, 1984.

75. **Nott, J. A.**, Cytology of pollutant metals in marine invertebrates: a review of microanalytical applications, *Scanning Electron Microsc.*, 5, 191, 1991.

76. **De Duve, C. and Wattiaux, R.**, Functions of lysosomes, *Annu. Rev. Physiol.*, 28, 435, 1966.

77. **Plattner, H. and Henning, R.**, Isolation of hepatocyte-derived rat liver lysosomal fractions, *Exp. Cell Res.*, 91, 333, 1975.

78. **Berry, J.-P., Poupon, M.-F., Galle, S., and Escaig, F.**, Role of lysosomes in gallium concentration by mammalian tissues, *Biol. Cell*, 51, 43, 1984.

79. **Sütterlin, U., Thies, W.-G., Haffner, H., and Seidel, A.**, Comparative studies on the lysosomal association of monomeric ^{239}Pu and ^{241}Am in rat and chinese hamster liver: Analysis with sucrose, metrizamide, and Percoll density gradients of subcellular binding as dependent on time, *Radiat. Res.*, 98, 293, 1984.

80. **Pääkkö, P., Anttila, S., Sutinen, S., and Hakala, M.**, Lysosomal gold accumulations in pulmonary macrophages, *Ultrastruct. Pathol.*, 7, 289, 1984.

81. **Lanno, R. P., Hicks, B., and Hilton, J. W.**, Histological observations on intrahepatocytic copper-containing granules in rainbow trout reared on diets containing elevated levels of copper, *Aquat. Toxicol.*, 10, 251, 1987.

82. **Baatrup, E. and Danscher, G.**, Cytochemical demonstration of mercury deposits in trout liver and kidney following methyl mercury intoxication: differentiation of two mercury pools by selenium, *Ecotoxicol. Environ. Saf.*, 14, 129, 1987.

83. **Sauer, G. R. and Watabe, N.**, Ultrastructural and histochemical aspects of zinc accumulation by fish scales, *Tissue & Cell*, 21, 935, 1989.

84. **Brown, B. E.**, The form and function of metal-containing "granules" in invertebrate tissues, *Biol. Rev.*, 57, 621, 1982.

85. **George, S. G.**, Heavy metal detoxication in *Mytilus* kidney - an *in vitro* study of Cd- and Zn-binding to isolated tertiary lysosomes, *Comp. Biochem. Physiol.*, 76C, 59, 1983.

86. **Jeantet, A.-Y., Ballan-Dufrancais, C., and Martin, J.-L.**, Recherche des mécanismes de détoxication du cadmium par l'huitre *Crassostrea gigas* (Mollusque, Bivalve). II. Sites intracellulaires d'accumulation du métal dans les organes absorbants et excréteurs, *C.R. Acad. Sci. Paris*, 301, 177, 1985.

87. **Viarengo, A., Moore, M. N., Pertica, M., Mancinelli, G., Zanicchi, G., and Pipe, R. K.**, Detoxification of copper in the cells of the digestive gland of mussel: The role of lysosomes and thioneins, *Sci. Total Environ.*, 44, 135, 1985.

88. **Goudard, F., Durand, J.-P., Galey, J., Piéri, J., Masson, M., and George, S.**, Subcellular localisation and identification of ^{95m}Tc- and ^{241}Am-binding ligands in the hepatopancreas of the lobster *Homarus gammarus*, *Mar. Biol.*, 108, 411, 1991.

89. **Roldan, B. M. and Shivers, R. S.**, The uptake and storage of iron and lead in cells of the crayfish *(Orconectes propinquus)* hepatopancreas and antennal gland, *Comp. Biochem. Physiol.*, 86C, 201, 1987.

90. **Schötti, G. and Seiler, H. G.**, Uptake and localisation of radioactive zinc in the visceral complex of the land pulmonate *Arion rufus*, *Experientia*, 26, 1212, 1970.

91. **Nott, J. A. and Langston, W. J.**, Cadmium and the phosphate granules in *Littorina littorea*, *J. Mar. Biol. Assoc. U.K.*, 69, 219, 1989.

92. **Goldfischer, S.**, The internal reticular apparatus of Camillo Golgi: A complex, heterogeneous organelle, enriched in acid, neutral, and alkaline phosphatases, and involved in glycosylation, secretion, membrane flow, lysosome formation, and intracellular digestion, *J. Histochem. Cytochem.*, 30, 717, 1982.

93. **Novikoff, A. B. and Holtzman, E.,** *Cells and Organelles,* 2nd ed., Modern Biology Series, Holt, Rinehart & Winston, New York, 1976.
94. **Taylor, M. G., Simkiss, K., Greaves, G. N., and Harries, J.,** Corrosion of intracellular granules and cell death, *Proc. R. Soc. London,* B234, 463, 1988.
95. **Janssen, H. H.,** Some histophysiological findings on the mid-gut gland of the common garden snail, *Arion rufus* (L.) (Syn. *A. ater rufus* (L.), *A. empiricorum* Férrussac), Gastropoda: Stylommatophora, *Zool. Anz. Jena,* 215, 33, 1985.
96. **Dallinger, R., Berger, B., and Gruber, A.,** Quantitative aspects of zinc and cadmium binding in *Helix pomatia:* Differences between an essential and a non-essential trace element, this volume, chap. 16.
97. **Nordberg, M. and Kojima, Y.,** Metallothionein and other low molecular weight metal binding proteins, in *Metallothionein,* Kägi, J. H. R. and Nordberg, M., Eds., Birkhäuser Verlag, Basel, 1979, 41.
98. **Fowler, B. A., Hildebrand, C. E., Kojima, Y., and Webb, M.,** Nomenclature of metallothionein, *Experientia,* Suppl. 52, 19, 1987.
99. **Imagawa, M., Onozawa, T., Okumura, K., Osada, S., Nishihara, T., and Kondo, M.,** Characterization of metallothionein cDNAs induced by cadmium in the nematode *Caenorhabditis elegans, Biochem. J.,* 268, 237, 1990.
100. **Lastowski-Perry, D., Otto, E., and Maroni, G.,** Nucleotide sequence and expression of a *Drosophila* metallothionein, *J. Biol. Chem.,* 260, 1527, 1985.
101. **Aoki, Y., Suzuki, K. T., and Kubota, K.,** Accumulation of cadmium and induction of its binding protein in the digestive tract of fleshfly *(Sarcophaga peregrina), Comp. Biochem. Physiol.,* 77C, 279, 1984.
102. **Bouquegneau, J. M., Ballan-Dufrancais, C., and Jeantet, A. Y.,** Storage of Hg in the ileum of *Blatella germanica*: Biochemical characterization of metallothionein, *Comp. Biochem. Physiol.,* 80C, 95, 1985.
103. **Mokdad, R., Debec, A., and Wegnez, M.,** Metallothionein gene in *Drosophila melanogaster* constitute a dual system, *Proc. Natl. Acad. Sci. U.S.A.,* 84, 2658, 1987.
104. **Dallinger, R., Berger, B., Hunziker, P., Hauer, Ch., and Kägi, J. H. R.,** in preparation.
105. **Mackay, E. A.,** Polymorphism of Cadmium-Induced Mussel Metallothionein, thesis, University of Aberdeen, 1989.
106. **Unger, M. E., Chen, T. T., Murphy, C. M., Vestling, M. M., Fenselau, C., and Roesijadi, G.,** Primary structure of molluscan metallothioneins deduced from PCR-amplified cDNA and mass spectrometry of purified proteins, *Biochim. Biophys. Acta,* 1074, 371, 1991.
107. **Kägi, J. H. R. and Kojima, Y.,** Chemistry and biochemistry of metallothionein, *Experientia,* Suppl. 52, 25, 1987.
108. **Berger, B., Thomaser, A., and Dallinger, R.,** in preparation.
109. **Janssen, H. H. and Dallinger, R.,** Diversification of cadmium binding proteins due to different levels of contamination in *Arion lusitanicus, Arch. Environ. Contam. Toxicol.,* 20, 132, 1991.
110. **Ballan-Dufrancais, C., Ruste, J., and Jeantet, A. Y.,** Quantitative electron probe microanalysis on insects exposed to mercury. I. Methods. An approach on the molecular form of the stored mercury. Possible occurrence of metallothionein-like proteins, *Biol. Chem.,* 39, 317, 1980.

111. **Zeng, J., Heuchel, R., Schaffner, W., Kägi, J. H. R.,** Thionein (apometallo-thionein) can modulate DNA binding and transcription activation by zinc finger containing factor Sp1, *FEBS Lett.,* 279, 310, 1991.

112. **Zeng, J., Vallee, B. L., and Kägi, J. H. R.,** Zinc transfer from transcription factor IIIA fingers to thionein clusters, *Proc. Natl. Acad. Sci. U.S.A.,* 88, 9984, 1991.

113. **Martoja, R., Bouquegneau, J. M., and Verthe, C.,** Toxicological effects and storage of cadmium and mercury in an insect *Locusta migratoria* (Orthoptera), *J. Invertebr. Pathol.,* 42, 17, 1983.

114. **Dallinger, R. and Prosi, F.,** Fractionation and identification of heavy metals in hepatopancreas of terrestrial isopods: evidence of lysosomal accumulation and absence of cadmium-thionein, in *Proc. 3rd Eur. Cong. Entomology,* Velthuis, H. H., Ed., Nederlaudse Entomologische Vereniging Amsterdam, 1986, 328.

115. **Donker, M. H., Koevoets, P., Verkley, J. A. C., and van Straalen, N. M.,** Comparison of metal binding proteins in hepatopancreas and haemolymph of *Porcellio scaber* (Isopoda) collected from a contaminated and a reference area, *Comp. Biochem. Physiol.,* 97C, 119, 1990.

116. **Dohi, Y., Ohba, K., and Yoneyama, Y.,** Purification and molecular properties of two cadmium-binding glycoproteins from the hepatopancreas of a whelk, *Buccinum tenuissimum, Biochim. Biophys. Acta,* 745, 50, 1983.

117. **Clubb, R. W., Lords, J. L., and Gaufin, A. R.,** Isolation and characterization of a glycoprotein from the stonefly, *Pteronarcys californica,* which binds cadmium, *J. Insect Physiol.,* 21, 53, 1975.

118. **Dallinger, R. and Pittl, M.,** in preparation.

119. **Dallinger, R.,** unpublished data, 1985.

120. **Dallinger, R. and Felder, E.,** in preparation.

121. **Marigómez, J. A., Gil, J. M., and Angulo, E.,** Accumulation of pigment and lipofuscin granules in *Littorina littorea* exposed to sublethal concentrations of cadmium: A histochemical study, *Zool. Jb. Anat.,* 120, 127, 1990.

122. **Dallinger, R.,** unpublished data, 1984.

123. **Triebskorn, R. and Künast, C.,** Ultrastructural changes in the digestive system of *Deroceras reticulatum* (Mollusca: Gastropoda) induced by lethal and sublethal concentrations of the carbamate molluscicide Cloethocarb, *Malacologia,* 32, 89, 1990.

124. **Underwood, E. J.,** *Trace Elements in Human and Animal Nutrition,* Academic Press, New York, 1977.

125. **Sminia, T. and Vlugt van Daalen, J. E.,** Haemocyanin synthesis in pore cells of the terrestrial snail *Helix aspersa, Cell Tissue Res.,* 183, 299, 1977.

126. **Moser, H. and Wieser, W.,** Copper and nutrition in *Helix pomatia* (L.), *Oecologia,* 42, 241, 1979.

127. **Dallinger, R.,** unpublished data, 1984.

128. **Coughtrey, P. J. and Martin, M. H.,** The distribution of Pb, Zn, Cd and Cu within the pulmonate mollusc *Helix aspersa* Müller, *Oecologia,* 23, 315, 1976.

129. **Hames, C. A. C. and Hopkin, S. P.,** Assimilation and loss of ^{109}Cd and ^{85}Zn by the terrestrial isopods *Oniscus asellus* and *Porcellio scaber, Bull. Environ. Contam. Toxicol.,* 47, 440, 1991.

130. **Prosi, F. and Back, H.,** Indicator cells for heavy metal uptake and distribution in organs from selected invertebrate animals, in *Int. Conf. Heavy Metals in the Environment, Proc.,* Lekkas, T. D., Ed., CEP Consultants Ltd., Edinburgh, 1985, 242.

131. **Hopkin, S. P., Hames, C. A. C., and Dray, A.,** X-ray microanalytical mapping of the intracellular distribution of pollutant metals, *Micros. Anal.,* 14, 23, 1989.

132. **Dallinger, R.,** unpublished data, 1989.

133. **Beeby, A.,** Interaction of lead and calcium uptake by the woodlouse, *Porcellio scaber* (Isopoda, Porcellionidae), *Oecologia,* 32, 255, 1978.

134. **Wieser, W.,** Copper in isopods, *Nature,* 191, 1020, 1961.

135. **Wieser, W.,** Copper and the role of isopods in degradation of organic matter, *Science,* 153, 67, 1966.

136. **Dallinger, R. and Wieser, W.,** The flow of copper through a terrestrial food chain. I. Copper and nutrition in isopods, *Oecologia,* 30, 253, 1977.

137. **Coughtrey, P. J., Martin, M. H., Chard, J., and Shales, S. W.,** Micro-organisms and metal retention in the woodlouse *Oniscus asellus, Soil Biol. Biochem.,* 12, 23, 1980.

138. **Wieser, W., Dallinger, R., and Busch, G.,** The flow of copper through a terrestrial food chain. II. Factors influencing the copper content of isopods, *Oecologia,* 30, 265, 1977.

139. **Dallinger, R. and Birkel, S.,** Ökotoxikologisches Teilgutachten zum Standort der Sonderabfalldeponie Enns, in *Abschlussbericht,* Institut für Zoologie, Innsbruck, 1991.

140. **Simkiss, K. and Watkins, B.,** Differences in zinc uptake between snails (*Helix aspersa* (Müller)) from metal- and bacteria-polluted sites, *Functional Ecol.,* 5, 787, 1991.

141. **Kratz, W., Gruttke, H., and Weigmann, G.,** Cadmium accumulation in soil fauna after artificial application of cadmium nitrate in a ruderal ecosystem, *Acta Phytopathol. Entomol. Hung.,* 22, 391, 1987.

142. **Berger, B. and Dallinger, R.,** Factors influencing the contents of heavy metals in terrestrial snails from contaminated urban sites, in *7th Int. Conf. Heavy Metals in the Environment,* Vernet, J. P., Ed., CEP Consultants Ltd., Edinburgh, 1989, 550.

143. **Dallinger, R. and Birkel, S.,** Ökotoxikologisches Teilgutachten zum Standort der Sonderabfalldeponie Aichkirchen/Bachmanning, in *Abschlussbericht,* Institut für Zoologie, Innsbruck, 1991.

144. **Streit, B.,** Effects of high copper concentrations on soil invertebrates (earthworms and oribatid mites): Experimental results and a model, *Oecologia,* 64, 381, 1984.

145. **Janssen, H. H.,** Heavy metal analysis in earthworms from an abandoned mining area, *Zool. Anz. Jena,* 222, 306, 1989.

146. **Van Straalen, N. M., Schobben, J. H. M., and de Goede, R. G. M.,** Population consequences of cadmium toxicity in soil microarthropods, *Ecotoxicol. Environ. Saf.,* 17, 190, 1989.

147. **Hunziker, P. E.,** Cystein modification of metallothionein, in *Methods in Enzymology,* Vol. 205, Riordan, J. S. and Vallee, B. L., Eds., Academic Press, San Diego, 1991, 399.

148. **Bartsch, R., Klein, D., and Summer, K. H.,** The Cd-Chelex assay: a new sensitive method to determine metallothionein containing zinc and cadmium, *Arch. Toxicol.,* 64, 177, 1990.

149. **Laskowski, R.,** Are the top carnivores endangered by heavy metal biomagnification?, *Oikos,* 60, 387, 1991.

150. **Ma, W., Denneman, W., and Faber, J.,** Hazardous exposure of ground-living small mammals to cadmium and lead in contaminated terrestrial ecosystems, *Arch. Environ. Contam. Toxicol.,* 20, 266, 1991.

151. **Nott, J. A. and Nicolaidou, A.,** Transfer of metal detoxification along marine food chains, *J. Mar. Biol. Assoc. U.K.,* 70, 905, 1990.

152. **Bremner, I. and Beattie, J. H.,** Metallothionein and the trace minerals, *Ann. Rev. Nutr.,* 10, 63, 1990.

153. **Karin, M.,** Metallothioneins: Proteins in search of function, *Cell. Vol.,* 41, 9, 1985.

154. **Meisch, H.-U., Bechmann, I., and Schmitt, J. A.,** A new cadmium-binding phosphoglycoprotein, cadmium-mycophosphatin, from the mushroom, *Agaricus macrosporus, Biochim. Biophys. Acta,* 745, 259, 1983.

155. **Prosi, F., Storch, V., and Janssen, H.,** Small cells in the midgut glands of terrestrial isopods: Sites of heavy metal accumulation, *Zoomorphology,* 102, 53, 1983.

CHAPTER 15

Budgeting the Flow of Cadmium and Zinc Through the Terrestrial Gastropod, *Helix pomatia* L.

Burkhard Berger, Reinhard Dallinger, Eduard Felder, and Jürgen Moser

TABLE OF CONTENTS

I. INTRODUCTION

In terrestrial animals the main route of metal uptake is via the food.[1,2] Metal concentrations and metal composition of the food are therefore of basic importance for metal uptake and accumulation. Metals entering cells can cause positive or negative effects on the animal's fitness, depending on the amount of the metal and its biological function (essential or nonessential). Therefore, metal accumulation reflects, on the one hand, the physiological need for the uptake of essential trace elements to meet metabolic requirements, and on the other hand, the inability of an organism to prevent metals from entering the tissues.[3] Toxic effects occur when uptake rates surpass the rates of excretory, metabolic, storage, and detoxification processes.[4] The capacity of these processes may vary between individuals and depend on the life history of the organism and on external factors like temperature, season, chemical speciation, pH, and so on.[5] Boháč and Pospíšil[6] distinguished three classes of accumulation strategies for terrestrial and aquatic invertebrates: macroconcentrators, microconcentrators, and deconcentrators.

Responding to elevated metal levels in the environment[7] or to feeding experiments with metal-loaded food[8-10] terrestrial snails are able to store high amounts of cadmium, zinc, copper, and lead in their bodies. Effective storage and detoxification systems for metals are needed in those animals which follow a macroconcentration strategy. In the case of molluscs, the target organ of metal flux within the body is the midgut gland, where cadmium, zinc, and copper are immobilized by binding to metallothioneins and other metal binding components.[11-14] A detoxifying role of metallothionein has been established in the mussel *Mytilus edulis*[15] and in the marine gastropod *Murex trunculus*.[16,17]

If the binding capacity of metallothionein becomes saturated, two responses may be visualized: (1) toxic effects occur because of excess free metal ions in the cell, interacting with active sites of enzymes, membranes, etc.[18] and (2) the animals switch to other strategies, including avoidance and excretory mechanisms, which should lead to higher metal output rates and lower accumulation efficiencies.

Assuming metallothioneins to be main components of the detoxification/storage system for cadmium in snails, we tested the capacity of this system in specimens of *Helix pomatia* by performing cadmium input/output analyses. It was the aim of this study to establish whether *H. pomatia* is able to respond to very high cadmium influxes by changing its accumulation strategy to a regulatory strategy. Cadmium concentrations in whole bodies and in organs, metal distribution among tissues, and cadmium budgets were considered.

In the first set of feeding experiments, the snails were exposed to increasing concentrations of cadmium in the food; in the second set, cadmium/zinc interactions were tested by exposing the animals to these two elements singly and in combination. Since metallothioneins are proteins with high affinities for both cadmium and zinc, competition between these metals for the limited number of binding sites is to be expected. This is likely to influence the metal accumulation capacity of the tissues involved.

II. MATERIAL AND METHODS

A. Animals and Experimental Design

H. pomatia were obtained from a commercial dealer (Fa. Stein, Lauingen, Germany). The animals were reared in groups of about 50 to 100 individuals in plastic containers with soil as substrate at a temperature of 18°C. Three times a week the snails were moistened with tap water and fed on lettuce *(Lactuca sativa)*.

The feeding experiments were carried out at a temperature of 18°C and a constant photoperiod of 12:12. At the beginning of the experiments a starvation period of three days induced evacuation of the snail's intestine. During the experiments the snails were reared in plastic boxes (18 × 15 × 9 cm), each box containing two or three snails which were fed three times a week with agar plates of known fresh weight and metal concentration (see below). Adult animals belonging to the same weight class were used in the feeding experiments.

B. Calculation of Metal Budgets

Agar (Sigma Powder Type IV) and cadmium or zinc standard solutions (Titrisol, Merck) were mixed in order to obtain 4% agar solutions containing known amounts of metals. The mixtures were boiled and diluted 1:1 with commercial vegetable juice. Then 10 ml of these solutions were filled in one-way plastic dishes (diameter = 9 cm). After a constant cooling period of 15 min the agar plates were weighed (fresh weight) and offered to the snails as food: every second or third day the two snails of one box were fed with one agar plate. At

Table 1. Summary of Relevant Parameters of the Feeding Experiments (FE)

			Food		
	Feeding period	Snails (boxes)	Cadmium conc.	Zinc conc.	Ratio
Experiment	[d]	[n]	[μg/g]	[μg/g]	dw:fw
FE:Cd-0	21	9(3)	0.4 ± 0.1 (9)	n.d.	0.053 ± 0.003 (24)
FE:Cd-10	21	9(3)	10.6 ± 0.4 (9)	n.d.	0.054 ± 0.002 (24)
FE:Cd-100	21	9(3)	92.6 ± 3.9 (9)	n.d.	0.053 ± 0.002 (24)
FE1:Cd	30	12(6)	95.5 ± 4.0 (10)	24 ± 1 (10)	0.053 ± 0.001 (20)
FE2:Zn	30	12(6)	0.5 ± 0.1 (10)	485 ± 16 (10)	0.053 ± 0.001 (20)
FE3:Cd/Zn	30	12(6)	98.8 ± 4.0 (10)	479 ± 21 (10)	0.053 ± 0.001 (20)

Note: Numbers of metal-fed snail (n) refer to individuals at the beginning of the feeding period. The number of boxes in which snails were reared is given in parentheses. Metal concentrations (in μg/g dw) and ratio dry weight (dw) to fresh weight (fw) of the agar plates are expressed as mean ± s.d. and number of determinations (in parentheses); n.d. = not determined.

the same time control plates were prepared in the same manner as described above. After determination of the fresh weight (fw), they were dried at 60°C, weighed again (dry weight; dw) and the ratio of dry weight to fresh weight was calculated (Table 1). From each of the control plates small pieces of agar were digested and prepared for metal analysis as described below. The cadmium and zinc concentrations of the agar plates used in the experiments are listed in Table 1.

Food uptake was calculated on the basis of the agar consumed by weighing the plates before (fw) and after (dw) the feeding interval following the steps listed in Figure 1. Metal input was calculated on the basis of the known concentration of the agar plates. Cadmium or zinc output was determined by collecting and weighing the feces and by measuring their metal contents. Metal assimilation of the snails was calculated from the difference between metal input and metal output.

C. Experimental Protocol

Two sets of experiments are described here, each set consisting of three feeding experiments which were carried out simultaneously (Table 1).

In the first set of experiments the snails were fed on agar plates of different cadmium concentrations. In accordance with the cadmium concentration of the

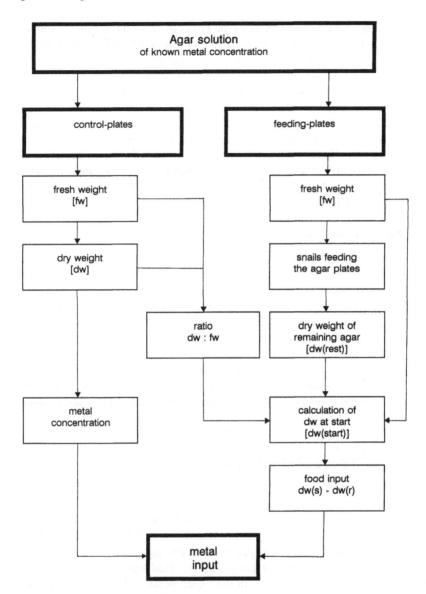

FIGURE 1. Flow diagram of the experimental procedure for the determination of budgets in snails feeding on agar plates.

food (based on dw) these feeding experiments (FE) are abbreviated as FE:Cd-0, FE:Cd-10, and FE:Cd-100 (Table 1). Nine snails were reared in three plastic boxes. Three times a week the three animals of one box were fed with one agar plate. At the end of the 21-day feeding period the snails were dissected and the concentration of the metals measured in the whole body (without shell) or in the hepatopancreas.

In the second set of experiments the effect of zinc on cadmium assimilation and distribution was examined. In each experiment 12 snails were reared in 6 boxes. Over a period of 30 days the two animals of one box were fed with one agar plate three times a week. In feeding experiment 1 (FE1:Cd) cadmium enriched agar plates were used as food, in feeding experiment 2 (FE2:Zn) zinc was added to the agar. The metal composition of the food used in feeding experiment 3 (FE3:Cd/Zn) was made up as in experiment 1 for cadmium and as in experiment 2 for zinc (Table 1). The metal concentration of the whole body (without shell) (WB) was measured in 50% of the snails, in the other 50% concentrations in foot and mantle (F/M), hepatopancreas (H), intestine (I), and remaining organs (R) were analyzed separately. Animals which had been fed uncontaminated lettuce were used as controls.

D. Metal Analyses

Tissues, whole animals (without shell), feces, and agar plates were dried for several days in polypropylene tubes (Greiner, Austria) at 60°C. The samples were digested with a mixture of 1 to 2 ml of nitric acid (Suprapure, Merck) and distilled water (1:1) in a heated aluminum block at 70°C. The concentrations of cadmium and zinc were measured by means of flame atomic absorption spectrophotometry (Perkin Elmer, model 3280) with deuterium background correction. Calibration was carried out using standard solutions (Titrisol, Merck) which were diluted with distilled water and contained 5% nitric acid. Lobster hepatopancreas powder with certified metal contents (TORT 1, NRC) was used as reference. Metal concentrations were found to be within the accepted range for this reference material.

III. RESULTS AND DISCUSSION

A. Mortality

High levels of cadmium, zinc, or copper did not cause increased mortality in terrestrial molluscs under laboratory conditions and only a slight intolerance existed to extremely high dosages of mercury and lead in food.[19,20] In our experiments, the metal concentrations of the food had no obvious effect on mortality of the snails (Table 2). In the experiments with high cadmium levels in the food, mortality varied between 11 and 33%.

Thus mortality is not a sensitive parameter for assessing possible harmful effects of environmental metal pollution in snail populations. More sensitive parameters, such as feeding activity, growth, and reproductive performance have to be selected for such investigations.

B. Feeding Rate

Helicid snails can be fed on uniform food for a long time.[21] In the present experiments the artificial food, consisting of a mixture of agar and vegetable

Table 2. Mortality (in % of the Number of Snails at the Start of the Experiments) of Metal-Fed Snails During the Feeding Experiments

Experiment	Feeding period [days]	Mortality [%]
FE:Cd-0	21	33
FE:Cd-10	21	22
FE:Cd-100	21	11
FE1:Cd	30	33
FE2:Zn	30	25
FE3:Cd/Zn	30	25

Table 3. Feeding Rates and Efficiencies of Food Assimilation of *H. pomatia*

Experiment	f.r.	a.e.
FE:Cd-0	46 ± 24	51
FE:Cd-10	60 ± 9	60
FE:Cd-100	31 ± 7	57
FE1:Cd	52 ± 22	59
FE2:Zn	39 ± 20	61
FE3:Cd/Zn	42 ± 15	61

Note: All values were calculated on the basis of a feeding period of 21 days. Feeding rate (f.r.) is expressed as milligram food (dry weight) per individual and day (mean ± SD; assimilation efficiency (a.e.) is given as % of food input.

juice, was well accepted. In Table 3, mean daily ingestion rates as well as the efficiency of food assimilation are given. For comparative purposes all values were calculated on the basis of a feeding period of 21 days. Mean feeding rates per day and individual ranged between 31 and 60 mg. No effect of metal concentration in the food on feeding rate was observed. For example, the rate of food uptake in the control experiment (FE:Cd-0) was nearly identical with that in FE3:Cd/Zn, in which food containing high dosages of cadmium and zinc was used (Table 3).

Feeding rates in *H. pomatia* fed metal-enriched agar plates were similar to those reported for *Helix lucorum* feeding on *Urtica dioica* (35 mg/day) and *Lactuca sativa* (75 mg/day) in the field.[22]

A cadmium concentration of about 100 µg/g in the food did not reduce the feeding rate of *H. pomatia* in our experiments. In contrast, Russel et al.[19] found marked effects on feeding rate in cadmium-treated specimens of *Helix aspersa*. Food consumption declined with each increase in cadmium concentration and feeding was strongly depressed at 100 µgCd/g and above.

The feeding activity of metal-treated slugs and snails depends on the biological function of the metals added to the food.[19,20] For example, the essential elements copper and zinc did not affect the feeding activity of *Arion ater* at low concentrations. High dosages decreased feeding rates at first, but after a period of acclimation, the feeding rates increased again to rates similar to those observed in control snails. However, nonessential elements caused a lasting decline of feeding rates. Consumption of mercury-treated *A. ater* and of cadmium-treated *H. aspersa* was related to metal dosage, showing decreasing consumption with increasing metal concentration. On the other hand, lead treatment did not affect feeding activity of *A. ater* up to 1000 µg/g in the food.

In order to estimate the effect of metal concentration in the food on feeding rates, additional factors influencing this variable must be considered. For example, feeding rate declines with age. The highest mass-specific values for daily consumption rates were observed in newly hatched and in juvenile snails.[22-24] Season seems to be an additional factor. High ingestion rates and efficiencies of food consumption were observed in periods of maximum activity (spring and early summer) and prior to hibernation (autumn).[22,25] If the effects of a toxicant on food consumption are to be tested, these influencing factors must be taken into account by the experimental design.

C. Efficiency of Food Assimilation

The efficiency of food assimilation of *H. pomatia* was lowest in the control experiment (51% in FE:Cd-0), in all other experiments it was about 60% (Table 3).

Food assimilation efficiencies of *H. pomatia* were lower than those of *Arianta arbustorum* fed an artificial diet similar to that used in the present experiments. In *A. arbustorum* the efficiency of assimilation with cadmium- and copper-enriched agar plates as food varied between 75 and 80%.[10]

In the field the assimilation efficiency of snails was influenced by the composition of the food, time of the year, and physiological state of the animals. *Cepaea hortensis* assimilated about 50% of herbs and about 30% of grasses.[26] In *H. lucorum*, a species closely related to *H. pomatia*,[27] assimilation efficiencies between 59 and 82% were estimated for herbaceous plants.[22] Thus assimilation efficiencies of snails fed artificial food were similar to the efficiencies of snails feeding under natural conditions.

D. Cadmium Concentrations and Cadmium Budgets

In comparison to the control experiment (FE:Cd-0) cadmium contents of the whole body and of the hepatopancreas increased by a factor of approximately

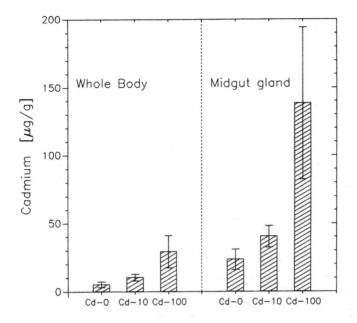

FIGURE 2. Cadmium concentrations in whole body soft tissues (left) and midgut glands (right) of *H. pomatia* after 21-day feeding experiments on agar plates with different cadmium concentrations. Cadmium concentrations are shown as means ± standard deviations, based on dry weight. The number of snails measured was 4 in FE:Cd-0, 7 in FE:Cd-10, and 8 in FE:Cd-100.

two in FE:Cd-10 and by a factor of about six in FE:Cd-100 (Figure 2). In each of the three feeding experiments the cadmium concentration of the midgut gland was four- to fivefold higher than that of the whole body. This ratio seemed to be independent of the total cadmium content of the tissues.

Cadmium contents of the snails after termination of the feeding experiments FE:Cd-0, FE:Cd-10, and FE:Cd-100 were comparable to cadmium contents of snails from different sites in the field.[7]

The cadmium levels of control snails were comparable to those found in *H. pomatia* from an unpolluted forest, or in *H. aspersa* from a reference area.[28,29] The snails of FE:Cd-10 had cadmium concentrations comparable to those occurring in gastropods from moderately contaminated sites, such as cities and roadside areas.[7,25,30] A cadmium concentration of about 30 μg/g in the bodies of snails feeding on food containing 100 μg/g cadmium over a period of three weeks corresponded to cadmium levels of snails living in metal polluted sites near smelters and mines.[31,32] The highest concentrations reported in natural snail populations of about 100 μg/g indicate, however, that the maximum accumulation capacity of *H. pomatia* had not been reached in our experiments.[31,32]

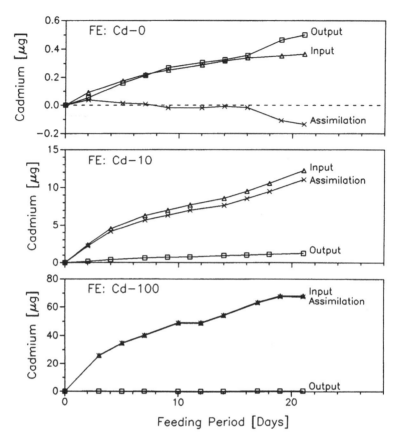

FIGURE 3. Cadmium input, output, and assimilation during a 21-day feeding period, calculated from food uptake and feces production of *H. pomatia*. The values represent cumulative amounts, expressed as μg Cd per snail (means of 10 individuals).

The mean daily uptake of cadmium per snail was below 0.1 mg in FE:Cd-0, 0.6 mg in FE:Cd-10, and 3.2 mg in FE:Cd-100 (Figure 3). In FE:Cd-10 loss of cadmium was slight, 90% of the ingested metal being assimilated. In FE:Cd-100 nearly all of the cadmium taken up (68 μg) was assimilated, the efficiency amounting to 99%.

Our data, as summarized in Figure 3 and 4, allow the following generalizations on the cadmium balance of *H. pomatia*:

1. Cadmium input, assimilation, and assimilation efficiency increased with increasing cadmium contents of the food; and
2. Cadmium output was not related to the amount of ingested cadmium but appeared to be more or less constant in all three feeding experiments.

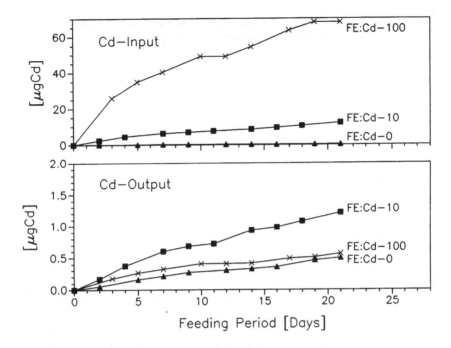

FIGURE 4. Comparison of cadmium input and cadmium output of snails from the three feeding experiments. The values represent cumulative amounts and are expressed as μg cadmium per snail (means of 10 individuals).

Our experiments demonstrate both a high capacity of metal assimilation and a remarkable inability of cadmium elimination in *H. pomatia*. Although molluscs in general are known to accumulate high amounts of metals,[1-8,33-37] *H. pomatia* seems to be the most characteristic representative of such an accumulation strategy.[2] This is illustrated by a comparison with the cadmium budgets of the related species *A. arbustorum* feeding on cadmium enriched agar plates (Table 4).[10] In *H. pomatia* as well as in *A. arbustorum* the common responses to cadmium treatment were high cadmium input rates and increased cadmium concentrations in the tissues, especially in the midgut gland. However, with regard to cadmium output vast differences prevailed. During a feeding period of 20 days one specimen of *A. arbustorum* ingested 1.2 mg cadmium per day. The efficiency of cadmium assimilation decreased continuously from about 90% at the beginning of the experiment to about 55% after 20 days. A marked increase of the cadmium concentration in the feces occurred between the 3rd and the 10th day, reaching a maximum of more than 600 μg/g. Over the whole experimental period, *A. arbustorum* eliminated about 250 μg cadmium per gram (dw) feces, whereas *H. pomatia* eliminated as little as 2 μg cadmium per gram feces. In *H. pomatia* cadmium concentrations in the feces reached a maximum of only 5% compared with the cadmium content in the food, whereas in *A. arbustorum* the same ratio

Table 4. Parameters of Cadmium Input/Output Analyses of
Helix pomatia* and *Arianta arbustorum

Source:	*Helix pomatia* FE:Cd-100	*Arianta arbustorum*
Cd conc. of food [μg/g(dw)]	93 ± 4	166 ± 7
Feeding period [days]	21	20
Feeding rate [mg(dw)/d]	31	8
Cd input rate [μgCd/d]	3.2	1.2
Cd assimilation efficiency [%]	99	55
Cd output per g of feces [μCd/g(dw)]	2	280
Max. Cd conc. of feces [μg/g]	5	600

Data for *Arianta arbustorum* from Berger, B. and Dallinger, R., *Oecologia,* 79, 60, 1989. With permission.

amounted to about 300%. Thus the quantification of metal budgets leads to a discriminating picture of the dynamic responses to metal exposure in different species of snails. Although both species belong to the category of "macroconcentrators"[2] *A. arbustorum* appears to switch to a regulation/elimination strategy after approximately five days of high cadmium input. Such a switch was not observed in *H. pomatia* during an experimental period of 30 days.

E. Cadmium/Zinc Interactions

Interactions between cadmium and zinc were examined with regard to metal uptake, tissue concentrations, and body distribution.

1. Metal Accumulation

Figure 5 shows the amount of assimilated cadmium (top) and zinc (bottom) depending on food input. The linear relationship between food intake and metal accumulation indicates that the efficiency of metal assimilation was invariant in our experiments.

Individual snails assimilated, on average, 104 μg and 79 μg cadmium in FE1:Cd and FE3:Cd/Zn, respectively, this difference being due to the different feeding rates of the snails in the two experiments (Table 3). Calculated on the basis of food input, cadmium assimilation was similar in all experiments (Figure 5). Using the equations of the regression lines listed in Figure 5, a food intake of 1 g resulted in the assimilation of 89 μg cadmium in FE1:Cd and of 93 μg

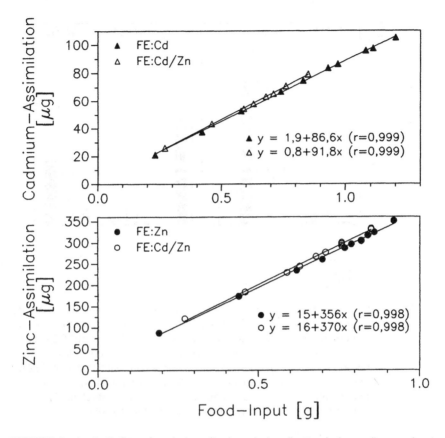

FIGURE 5. Assimilation of cadmium (top) and zinc (bottom) depending on food input in *H. pomatia*. The values represent cumulative amounts and are expressed as μg metal per snail (means of 12 individuals). Results of linear regression analysis are listed for each experiment. Upper graph: comparison of assimilated amounts of cadmium without (FE1:Cd) and with FE3:Cd/Zn) additional zinc in the food. Lower graph: comparison of assimilated amounts of zinc without (FE2:Zn) and with (FE3:Cd/Zn) additional cadmium in the food.

cadmium in FE3:Cd/Zn, both values differing by less than 5%. This indicates that the presence of high amounts of zinc did not influence cadmium assimilation. In the same fashion, the concentration of cadmium in the food had no effect on the rate of zinc assimilation (Figure 5).

2. Metal Concentrations

Compared with the control animals the concentration of cadmium increased significantly in all organs of the snails fed a cadmium-enriched diet. Highest concentrations were detected in the hepatopancreas and in the intestine (Figure 6).

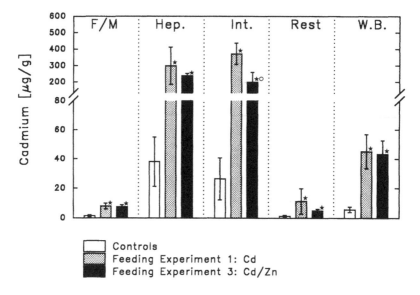

FIGURE 6. Concentrations (means ± SD) of cadmium in tissues of *H. pomatia* after three types of feeding experiments. Levels of control animals are compared to cadmium exposed snails (FE1:Cd and FE3:Cd/Zn). An asterisk indicates a significant change in metal concentration compared to controls, an open circle indicates a difference between FE2:Cd and FE3:Cd/Zn (Student's *t*-test; $p < 0.01$). Abbreviations — F/M: foot and mantle tissues; Hep: hepatopancreas; Int: intestine; Rest: remaining organs; and W.B.: whole body (without shell).

Feeding zinc-enriched food led to a two- to threefold increase of total body concentration. The elevated body burdens were attributed to a marked increase of zinc concentrations in the hepatopancreas. The other organs did not accumulate significant amounts of zinc (Figure 7).

The cadmium concentrations in the organs of the snails were not influenced by the zinc concentration in the food, and cadmium in the diet had no effect on zinc concentrations in the snails.

3. Body Distribution

The relative distribution of cadmium in the body was independent of the amount of cadmium and zinc accumulated. About 75% of cadmium was always stored in the midgut gland (Figure 8).

Zinc exposure affected a shift of percentage zinc distribution. Compared with controls, the portion of zinc stored in the hepatopancreas increased from 58 to about 80% (Figure 9). Cadmium exposure, on the other hand, had no effect on the relative body distribution of zinc.

The patterns of uptake, assimilation, and distribution of cadmium and zinc were similar in *H. pomatia*. Uptake was high in all experiments, and the assim-

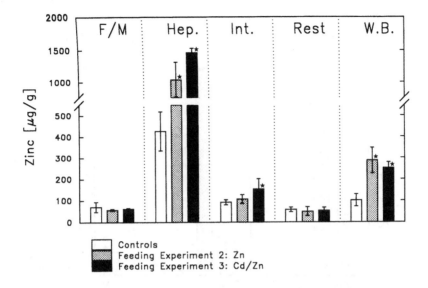

FIGURE 7. Concentrations (means ± SD) of zinc in tissues of *H. pomatia* after three types of feeding experiments. An asterisk indicates a significant change in metal concentration compared to controls (Student's *t*-test; $p < 0.01$). Abbreviations: see Figure 6 caption.

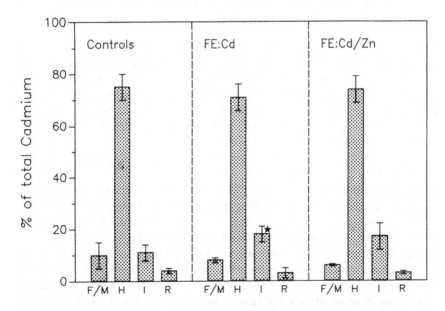

FIGURE 8. Relative distribution of total cadmium in control animals and in cadmium treated snails (FE1:Cd and FE3:Cd/Zn). An asterisk indicates a significant change compared with the controls (Student's *t*-test; $p < 0.01$). Abbreviations: see Figure 6 caption.

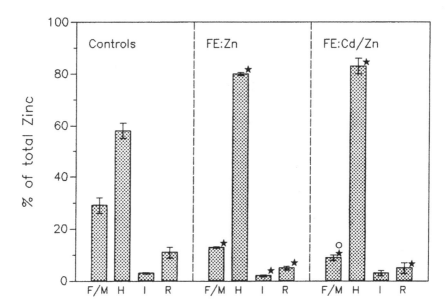

FIGURE 9. Relative distribution of total zinc in control animals and in zinc treated snails (FE2:Zn and FE3:Cd/Zn). An asterisk indicates a significant change compared to the controls, an open circle indicates a difference between FE2:Zn and FE3:Cd/Zn (Student's *t*-test; *p* <0.01). Abbreviations: see Figure 6 caption.

ilation was efficient and constant during the whole feeding period (Figure 5). The predominant target organ for metal storage was the hepatopancreas. The flow of cadmium and zinc through metal-fed *H. pomatia* is shown in Figures 10 and 11, background (BG) levels having been calculated from the controls. In these two experiments, two thirds of the metals taken up by the snails were stored in the midgut gland. The main portion of the remaining cadmium accumulated in the intestine (18%), whereas the rest of the zinc was either excreted or stored in the foot/mantle tissue. This indicates that although body distribution and elimination of cadmium and zinc differed in detail, the main parameters of uptake and assimilation were similar. The lack of interference suggests different pathways and mechanisms of handling of these metals in the tissues and cells.[38]

In vertebrates, Cd/Zn interactions with regard to metal uptake, assimilation, and toxicity are well known.[39] Intestinal cadmium uptake increased after zinc-induced metallothionein production[40] and preexposure to zinc resulted in enhanced cadmium uptake and transfer in the gills of fish.[41] There is strong evidence for a relationship between tissue concentrations of zinc and cadmium, indicating interactions in accumulation and storage processes.[42-45] It has also been shown that toxicity due to cadmium is suppressed by the simultaneous administration of large quantities of zinc.[46] Since both cadmium and zinc are capable of inducing the synthesis of vertebrate metallothioneins and are bound specifically to the

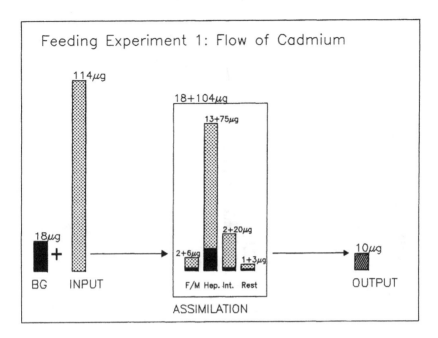

FIGURE 10. Flow of cadmium through the organs of an average specimen of *H. pomatia*, based on relative distributions and on budgets at the end of the FE1:Cd feeding experiment. The amounts of metal are symbolized by bars, at the top of each bar the corresponding values of cadmium (in μg) are given. Assimilated cadmium consists of background contents (BG) plus the amounts due to feeding a cadmium-enriched food. The background contents (black bars) were derived from control animals. Abbreviations: see Figure 6 caption.

sulfhydryl groups of these proteins,[47] metallothioneins offer themselves as the major sites of Cd/Zn interactions. Increased synthesis of zinc-induced metallothionein also increases the number of binding sites available to cadmium, thereby accounting for the positive correlations of the amounts of both metals accumulating in the cells.[44] Zinc-induced metallothionein was shown to bind and detoxify calcium ions entering the cells.[48]

In molluscs, competition of cadmium and zinc for metallothionein binding sites and subcellular distribution of these metals seem to differ from the vertebrate situation. In mussels, high levels of zinc led to a reduction in the accumulation of cadmium.[49] In *Anodonta cygnea* zinc competes with cadmium for metal binding sites of the high molecular weight fraction in gills, whereas metallothionein synthesis was not affected by zinc. After combined exposure to both metals most cadmium was bound to metallothionein and nearly all zinc was associated either with the particulate fraction or was bound to high molecular weight proteins.[50] Different pathways of biochemical and intracellular sequestration of zinc

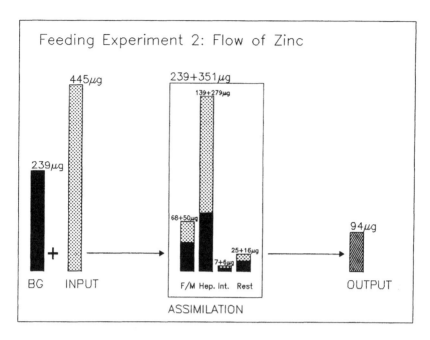

FIGURE 11. Flow of zinc through the organs of one specimen of *H. pomatia*, based on relative distributions and on budgets at the end of the FE2:Zn feeding experiment. The amounts of metal are symbolized by bars, at the top of each bar the corresponding values of zinc (in μg) are given. Assimilated zinc consists of background contents plus the amounts due to feeding on zinc-enriched food. The background contents (black bars) were derived from control animals. Abbreviations: see Figure 6 caption.

and cadmium appear to be present in terrestrial gastropods.[38] In *H. pomatia* fed on cadmium- and zinc-enriched agar plates, 86% of cadmium and 21% of zinc were found in the soluble fraction of the hepatopancreas, while 95% of the soluble cadmium was bound to metallothionein, whereas the soluble zinc was associated with different ligands of low molecular weight. Compared to cadmium, less than 10% of zinc was bound to metallothionein.[38]

Functional aspects of metallothioneins are still a matter of discussion.[51-53] The conserved structure, ubiquity, and inducibility of metallothioneins by a wide range of stimuli, including metals, e.g., cadmium or zinc, and ''stress fractors'', suggests that these proteins do not only play a role in the detoxification of metals, but also in the homeostatic control, metabolism, and transfer of trace metals. They may also act as free radical scavengers and acute phase proteins.[54] We suggest that the accumulation strategy of *H. pomatia* together with the lack of Cd/Zn interactions and the distinct biochemical handling of cadmium and zinc indicate that in terrestrial gastropods and possibly in other molluscs the functions of metallothioneins may differ from those discussed within the framework of the vertebrate model. Therefore, further investigations on trace metal budgets and on functional aspects of metal-binding proteins in snails might enlarge our knowledge on the functions of metallothionein in general.

IV. ABSTRACT

Specimens of the terrestrial gastropod *Helix pomatia* were fed on metal-contaminated agar plates. In the first group of feeding experiments the cadmium concentration of the diet ranged from 0.4 to about 100 μg/g(dw). Cadmium input and cadmium assimilation efficiency increased with increasing cadmium concentrations of the food, whereas cadmium output seemed to be more or less constant in all experiments. The high capacity of metal assimilation and the inability to eliminate cadmium identify *H. pomatia* as one of the most characteristic metal macroconcentrators.[2]

In a second set of experiments Cd/Zn interactions were tested by feeding cadmium- and zinc-loaded food singly or in combination. The concentration of cadmium increased significantly in all organs, the main sites of storage being the midgut gland and the intestine. Zinc accumulated mainly in the midgut gland, reaching concentrations of more than 1000 μgZn/g. A combination of cadmium and zinc in the food did not influence uptake, concentrations, and distribution of the two metals. No synergistic or antagonistic effects between cadmium and zinc were found.

Different cytosolic binding components for cadmium and zinc, as well as the properties of metallothioneins, are discussed as being responsible for this lack of metal interaction.

ACKNOWLEDGMENTS

The authors thank Professor W. Wieser, University of Innsbruck, for reviewing the manuscript and for helpful discussion. The study was supported by the Austrian "Fonds zur Förderung der wissenschaftlichen Forschung in Österreich", project Nr. P7815.

REFERENCES

1. **Hopkin, S. P.,** *Ecophysiology of Metals in Terrestrial Invertebrates,* Elsevier Applied Science, London, 1989.
2. **Dallinger, R.,** Strategies of metal detoxification in terrestrial invertebrates, in *Ecotoxicology of Metals in Invertebrates,* Dallinger, R. and Rainbow, P. S., Eds., Lewis Publishers, ch. 14, Boca Raton, FL, 1993.
3. **Depledge, M. H. and Rainbow, P. S.,** Models of regulation and accumulation of trace metals in marine invertebrates, *Comp. Biochem. Physiol.,* 97C, 1, 1990.
4. **Langston, W. J.,** Toxic effects of metals and the incidence of metal pollution in marine ecosystems, in *Heavy Metals in the Marine Environment,* Furness, R. W. and Rainbow, P. S., Eds., CRC Press, Boca Raton, FL, 1990, chap. 7.
5. **Hopkin, S. P.,** Critical concentrations, pathways of detoxification and cellular ecotoxicology of metals in terrestrial arthropods, *Functional Ecol.,* 4, 321, 1990.
6. **Boháč, J. and Pospíšil, J.,** Accumulation of heavy metals in invertebrates and its ecological aspects, in 7th Int. Conf. Heavy Metals in the Environment, Geneva, September 1989. Conf. Proc., Vol. 1, 1989, 354.
7. **Berger, B. and Dallinger, R.,** Terrestrial snails as quantitative indicators of environmental metal pollution, *Environ. Monit. Assess.,* in press.
8. **Dallinger, R. and Wieser, W.,** Patterns of accumulation, distribution and liberation of zinc, copper, cadmium, and lead in different organs of the land snail *Helix pomatia* L., *Comp. Biochem. Physiol.,* 79C, 117, 1984.
9. **Beeby, A.,** The role of *Helix aspersa* as a major herbivore in the transfer of lead through a polluted ecosystem, *J. Appl. Ecol.,* 22, 267, 1985.
10. **Berger, B. and Dallinger, R.,** Accumulation of cadmium and copper by the terrestrial snail *Arianta arbustorum* L.: kinetics and budgets, *Oecologia,* 79, 60, 1989.
11. **Cooke, M., Jackson, A., Nickless, G., and Roberts, D. J.,** Distribution and speciation of cadmium in the terrestrial snail, *Helix aspersa, Bull. Environ. Contam. Toxicol.,* 23, 445, 1979.
12. **Dallinger, R. and Wieser, W.,** Molecular fractionation of Zn, Cu, Cd, and Pb in the midgut gland of *Helix pomatia* L., *Comp. Biochem. Physiol.,* 79C, 125, 1984.
13. **Dallinger, R., Berger, B., and Bauer-Hilty, A.,** Purification of cadmium binding proteins from related species of terrestrial Helicidae (Gastropoda, Mollusca), *Mol. Cell. Biochem.,* 85, 135, 1989.
14. **Dallinger, R., Janssen, H. H., Bauer-Hilty, A., and Berger, B.,** Characterization of an inducible cadmium-binding protein from hepatopancreas of metal-exposed slugs (Arionidae, Mollusca), *Comp. Biochem. Physiol.,* 92C, 355, 1989.
15. **Roesijadi, G. and Fellingham, G. W.,** Influence of Cu, Cd, and Zn pre-exposure on Hg toxicity in the mussel *Mytilus edulis, Can. J. Fish. Aquat. Sci.,* 44, 680, 1987.
16. **Dallinger, R., Carpené, E., Dalla Via, G. J., and Cortesi, P.,** Effects of cadmium on *Murex trunculus* from the Adriatic sea. I. Accumulation of metal and binding to a metallothionein-like protein, *Arch. Environ. Contam. Toxicol.,* 18, 554, 1989.
17. **Dalla Via, G. J., Dallinger, R., and Carpené, E.,** Effects of cadmium on *Murex trunculus* from the Adriatic sea. II. Oxygen consumption and acclimation effects, *Arch. Environ. Contam. Toxicol.,* 18, 562, 1989.

18. **Janssen, H. H. and Dallinger, R.**, Diversification of cadmium-binding proteins due to different levels of contamination in *Arion lusitanicus, Arch. Environ. Contam. Toxicol.*, 20, 132, 1991.

19. **Russel, L. K., de Haven, J. I., and Botts, R. P.**, Toxic effects of cadmium in the garden snail *(Helix aspersa), Bull. Environ. Contam. Toxicol.*, 26, 634, 1981.

20. **Marigomez, J. A., Angulo, E., and Saez, V.**, Feeding and growth responses to copper, zinc, mercury, and lead in the terrestrial gastropod *Arion ater* (Linne), *J. Mol. Stud.*, 52, 68, 1986.

21. **Frömming, E.**, *Biologie der mitteleuropäischen Landgastropoden*, Duncker & Humbolt, Berlin, 1954.

22. **Staikou, A. and Lazaridou-Dimitriadou, M.**, Feeding experiments on and energy flux in a natural population of the edible snail *Helix lucorum* L. (Gastropoda: Pulmonata: Stylommatophora) in Greece, *Malacologia*, 31, 217, 1989.

23. **Jennings, T. J. and Barkham, J. P.**, Quantitative study of feeding in woodland by the slug *Arion ater, Oikos*, 27, 168, 1976.

24. **Seifert, D. V. and Shutov, S. V.**, The consumption of leaf litter by land molluscs, *Pedobiologia*, 21, 159, 1981.

25. **Williamson, P.**, Variables affecting body burdens of lead, zinc, and cadmium in a roadside population of the snail *Cepaea hortensis* Müller, *Oecologia*, 44, 213, 1980.

26. **Williamson, P. and Cameron, R. A. D.**, Natural diet of the land-snail *Cepaea nemoralis, Oikos*, 27, 493, 1976.

27. **Kerney, M. P., Cameron, R. A. D., and Jungbluth, J. H.**, *Die Landschnecken Nord und Mitteleuropas*, Verlag Paul Parey, Berlin, 1983.

28. **Knutti, R., Bucher, P., Stengl, M., Stoz, M., Tremp, J., Ulrich, M., and Schlatter, C.**, Cadmium in the invertebrate fauna of an unpolluted forest in Switzerland, *Environ. Toxin. Ser.*, 2, 171, 1988.

29. **Coughtrey, P. J. and Martin, M. H.**, The distribution of lead, zinc, cadmium, and copper within the pulmonate mollusc *Helix aspersa* Müller, *Oecologia*, 23, 315, 1976.

30. **Beeby, A. and Richmond, L.**, Adaptation by an urban population of the snail *Helix aspersa* to a diet contaminated with lead, *Environ. Pollut.*, 46, 73, 1987.

31. **Martin, M. H. and Coughtrey, P. J.**, *Biological Monitoring of Heavy Metal Pollution: Land and Air*, Applied Science Publishers, Englewood, NJ, 1982.

32. **Greville, R. W. and Morgan, A. J.**, The influence of size on the accumulated amounts of metals (Cu, Pb, Cd, Zn, and Ca) in six species of slugs sampled from a contaminated woodland site, *J. Mol. Stud.*, 56, 355, 1990.

33. **Martin, M. H. and Flegal, A. R.**, High copper concentrations in squid livers in association with elevated levels of Ag, Cd, and Zn, *Mar. Biol.*, 30, 51, 1975.

34. **Howard, B. and Simkiss, K.**, Metal binding by *Helix aspersa* blood, *Comp. Biochem. Physiol.*, 70A, 559, 1981.

35. **Viarengo, A., Zanicchi, G., Moore, M. N., and Orunesu, M.**, Accumulation and detoxication of copper by the mussel *Mytilus galloprovincialis* Lam.: A study of the subcellular distribution in the digestive gland cells, *Aquat. Toxicol.*, 1, 147, 1981.

36. **Ward, T. J.**, Laboratory study of the accumulation and distribution of cadmium in the Sydney Rock Oyster *Saccostrea commercialis, Aust. J. Mar. Freshwater Res.*, 33, 33, 1982.

37. **Kurihara, Y. and Suzuki, T.,** Removal of heavy metals and sewage sludge using the mud snail, *Cipangopaludina chinensis malleata* Reeve, in paddy fields as artificial wetlands, *Water Sci. Technol.,* 19, 281, 1987.
38. **Dallinger, R., Berger, B., and Gruber, A.,** Quantitative aspects of zinc and cadmium in *Helix pomatia*: Differences between an essential and a nonessential trace element, in *Ecotoxicology of Metals in Invertebrates,* Dallinger, R. and Rainbow, P. S., Eds., Lewis Publishers, ch. 16, Boca Raton, FL, 1993.
39. **Honda, R. and Nogawa, K.,** Cadmium, zinc and copper relationships in kidney and liver of humans exposed to environmental cadmium, *Arch. Toxicol.,* 59, 437, 1987.
40. **Foulkes, E. C. and McMullen, D. M.,** Endogenous metallothionein as determinant of intestinal cadmium absorption: a reevaluation, *Toxicology,* 38, 285, 1986.
41. **Wicklund, A., Norrgren, L., and Runn, P.,** The influence of cadmium and zinc on cadmium turnover in the Zebrafish, *Brachydanio rerio, Arch. Environ. Contam. Toxicol.,* 19, 348, 1990.
42. **Elinder, C.-G. and Piscator, M.,** Cadmium and zinc relationships, *Environ. Health Perspect.,* 25, 129, 1978.
43. **Friel, J. K., Borgman, R. F., and Chandra, R. K.,** Effect of chronic cadmium administration on liver and kidney concentrations of zinc, copper, iron, manganese, and chromium, *Bull. Environ. Contam. Toxicol.,* 38, 588, 1987.
44. **Ewers, E., Turfeld, M., Freier, I., Jermann, E., and Brockhaus, A.,** Interrelationships between cadmium, zinc, and copper in human kidney cortex, *Toxicol. Environ. Chem.,* 27, 31, 1990.
45. **Walsh, P. M.,** The use of seabirds as monitors of heavy metals in the marine environment, in *Heavy Metals in the Marine Environment,* Furness, R. W. and Rainbow, P. S., Eds., CRC Press, Boca Raton, FL, 1990, chap. 10.
46. **Ueda, F., Hiroo, S., Fujiwara, H., Ebara, K., Minomiya, S., and Shimaki, Y.,** Interacting effects of zinc and cadmium on the cadmium distribution in the mouse, *Vet. Hum. Toxicol.,* 29, 367, 1987.
47. **Kägi, J. H. R. and Vallee, B. L.,** Metallothionein: a cadmium- and zinc-containing protein from equine renal cortex, *J. Biol. Chem.,* 235, 3460, 1960.
48. **Din, W. S. and Frazier, J. M.,** Protective effect of metallothionein on cadmium toxicity in isolated rat hepatocytes, *Biochem. J.,* 230, 395, 1985.
49. **Elliott, N. G., Swain, R., and Ritz, D. A.,** Metal interaction during accumulation by the mussel *Mytilus edulis planulatus, Mar. Biol.,* 93, 395, 1986.
50. **Hemalraad, J., Kleinveld, H. A., de Roos, A. M., Holwerda, D. A., and Zandee, D. I.,** Cadmium kinetics in freshwater Clams. III. Effects of zinc on uptake and distribution of cadmium in *Anodonta cygnea, Arch. Environ. Contam. Toxicol.,* 16, 95, 1987.
51. **Cosson, R. P., Amiard-Triquet, C., and Amiard, J.-C.,** Metallothioneins and detoxification. Is the use of detoxification protein for MTs a language abuse? *Water, Air, Soil Pollut.,* 57/58, 555, 1991.
52. **Hogstrand, C. and Haux, C.,** Binding and detoxification of heavy metals in lower vertebrates with reference to metallothionein, *Comp. Biochem. Physiol.,* 100C, 137, 1991.
53. **Zeng, J., Vallee, B. L., and Kägi, J. H. R.,** Zinc transfer from transcription factor IIIA fingers to thionein clusters, *Proc. Natl. Acad. Sci. U.S.A.,* 88, 9984, 1991.

54. **Bremner, I. and Beattie, J. H.**, Metallothionein and the trace minerals, *Annu. Rev. Nutr.*, 10, 63, 1990.

CHAPTER 16

Quantitative Aspects of Zinc and Cadmium Binding in *Helix pomatia*: Differences Between an Essential and a Nonessential Trace Element

Reinhard Dallinger, Burkhard Berger, and Alexandra Gruber

TABLE OF CONTENTS

0-87371-734-1/93/$0.00 + $.50
© 1993 by Lewis Publishers

315

I. INTRODUCTION

Many species of soil invertebrates are strong concentrators of cadmium and zinc (see Chapter 14, this volume).[1] Examples for this are provided by terrestrial gastropods: cadmium concentrations of more than 100 μg/g and zinc concentrations as high as 2000 μg/g (dry weight) were reported in tissues of snails and slugs from contaminated environments.[2] At the same time these organisms seem to be relatively tolerant against the elevated metal concentrations in their bodies.[3-5] Thus from an ecological point of view, snails and slugs represent biological sinks for these metals. Terrestrial gastropods may contribute considerably to the cycling of nutrients and trace elements in soil habitats.[6] Hence the question arises if and by how much these organisms account for the fixation or the mobilization of cadmium and zinc in terrestrial foodwebs. One may ask, for example, if the binding of cadmium to metallothioneins in terrestrial snails[7] renders the metal more available to potential predators. The answer to such questions cannot be generalized, since it appears that even closely related species of terrestrial gastropods show considerable differences in metal accumulation patterns (see Chapter 15, this volume).[5]

The species *Helix pomatia*, for instance, accumulates cadmium and zinc predominantly in the midgut gland.[8] However, no interactions between cadmium and zinc in terms of metal uptake, assimilation, and elimination could be found in this species after combined exposure to both metals (see Chapter 15, this volume).[5] This is not trivial since several examples of interference between the two metals have been reported for many vertebrate and invertebrate species.[9,10] One reason for this may be that zinc and cadmium share several physicochemical properties: both elements belong to the borderline metals showing ambivalent affinities for metal-binding donor atoms in ligands.[11] From a biological point of view, however, zinc is an essential trace element,[12] while cadmium is a non-essential trace metal which adversely interacts with many biological molecules.[13] In mammals, for instance, cadmium exposure caused changes in zinc concentrations of different tissues.[14] In the freshwater clam *Anodonta cygnaea*, the uptake of cadmium via the gills was found to be significantly reduced in the presence of zinc.[15] In terrestrial invertebrates too, several examples of interaction between cadmium and zinc have been reported. In the flour beetle *Tribolium confusum*, for instance, supplemental zinc concentrations in the diet reduced the toxicity of cadmium.[16]

In the light of these findings the apparent lack of interferences between cadmium and zinc in *Helix pomatia* on accumulation of both metals is significant. In the present study we show, by means of chromatographic techniques, why cadmium and zinc accumulate in the same target organ of *Helix pomatia* without displaying any sign of reciprocal interference. It appears that in this species the two metals are treated by compartmentalization and binding to different ligands.

II. MATERIAL AND METHODS

A. Animals and Feeding Conditions

Specimens of *Helix pomatia* were obtained from a commercial dealer (Fa. Stein, Lauingen, Germany). The animals were reared in plastic containers with soil substrate, and fed lettuce *(Lactuca sativa)* until the beginning of the experiment.

The feeding experiment was carried out at constant temperature (18°C) and photoperiod (12:12). A total of 24 snails were reared in 12 plastic boxes (18 × 15 × 9 cm), each box containing two individuals. During a feeding period of 30 days agar plates of known metal concentrations were offered to the animals. The two snails of each box were fed three times a week with one agar plate containing a combination of cadmium (105.5 μg/g dry weight $\pm$ 11.9, n = 10) and zinc (698.5 μg/g dry weight $\pm$ 82, n = 10). The preparation of agar plates has been described by Berger et al. in Chapter 15.[5]

B. Protein Purification

1. Purification Protocol

Cadmium- and zinc-binding ligands of pooled hepatopancreas tissues were separated by a series of fractionation steps: (1) homogenization and centrifugation, (2) gel filtration on Sephacryl S-200, (3) anion exchange chromatography (DEAE cellulose), (4) gel filtration on Sephacryl S-100, and (5) HPLC on a reverse-phase column (RP-18). In order to quantify the absolute (μg) and relative (%) amounts of metals associated with different ligands, absorption at 254 nm as well as concentrations of cadmium and zinc were measured in all fractions. Metal-containing fractions were pooled and stored for further separation or for amino acid analysis.

2. Sample Preparation Prior to Chromatography

At the end of the feeding experiment the 16 snails were sacrificed and dissected, the hepatopancreas tissues pooled in ice-cold centrifuge tubes. The tissues were weighed and homogenized with an Ultraturrax in a twofold volume of Tris-HCl buffer (50 mM, pH 7.5) containing 5 mM sodium azide, 10 mM 2-mercaptoethanol, 0.2 mM antipain, and 0.05 mM pepstatin. The homogenates were centrifuged for 20 min at 27,000 g (Sorvall RC2-B) and subsequently for 1 h at 100,000 g (Sorvall OTD-2). Aliquots of supernatants (3 × 1 ml) and pellets were prepared for analyses of cadmium and zinc. The remaining supernatant was filtered (Sartorius, Minisart NML, 0.2 μm) and stored in portions of 6 ml at 4°C.

3. Gel Filtration Chromatography

Gel filtration was carried out on either a Sephacryl S-200 HR (Pharmacia) column (25 mm × 85 cm) or a Sephacryl S-100 HR (Pharmacia) column (18 mm × 90 cm). The mobile phase was 50 mM Tris-HCl buffer, pH 7.5, containing 5 mM sodium azide. Fractions of 4.5 (S-200) or 6 ml (S-100) were collected and analyzed for absorption at 280 and 254 nm (Beckman DU-6), as well as for cadmium and zinc. Cadmium- or zinc-containing fractions were pooled and stored at 4°C.

The columns had been calibrated with substances of known molecular weight: blue dextran (2,000,000 Da), bovine albumin (68,000 Da), ovalbumin (45,000 Da), myoglobin (17,800 Da), cytochrome C (12,500 Da), rabbit metallothionein (6500 Da), and vitamin B12 (1350 Da).

4. Anion Exchange Chromatography

Metal-containing fractions derived from gel filtration chromatography (S-200) were applied to an ion exchange column (26 mm × 40 cm) filled with DEAE cellulose (Whatman) and eluted on a gradient of 0 to 0.5 M sodium chloride (dissolved in 50 mM Tris-HCl buffer, pH 7.5) at a flow rate of 32.5 ml/h. Absorption (280/254 nm), salinity (refraction index), and cadmium and zinc concentrations were measured in fractions of 12 ml.

5. Reverse-Phase High Performance Liquid Chromatography

The system used for reverse-phase chromatography consisted of a pair of HPLC pumps (501, Waters, U.S.), a Rheodyne injector (model 7125) with either 25 μl or 200 μl sample loops, and a Waters 490E programmable wavelength detector. The gradient was controlled by a personal computer equipped with

Baseline 810 chromatography software. Chromatography was performed on 3.9 mm × 300 mm μBondapak C18 columns (10 μm particle size, 125 Å pore size). The composition of the buffers used was as follows: buffer A — 25 mM Tris-HCl, 5 mM sodium azide, pH 7.5; buffer B — 25 mM Tris-HCl, 5 mM sodium azide, pH 7.5, containing 60% (v/v) acetonitrile. The samples were eluted by a linear gradient during 25 min from 0% to 40% buffer B at a flow rate of 1 ml/min.

Absorption was recorded at 220 and 254 nm. Fractions of 1 ml were collected and concentrations of cadmium, zinc, and copper measured by flame atomic absorption spectrophotometry.

C. Amino Acid Analysis

Purified proteins were oxidized and hydrolyzed as described by Dallinger et al.[7] or derivatized by S-methylation according to Hunziker.[17] Amino acid analyses were carried out using a Liquimat 3 apparatus (Labotron, Germany) or an Applied Biosystems analyzer (model 420) equipped with on-line hydrolysis.

D. Atomic Absorption Spectrophotometry

Tissue samples, aliquots of supernatants, and pellets were transferred to 10 ml polyethylene tubes (Greiner, Austria) and dried at 60°C. Digestion was performed for several days at 70°C with a mixture of 1 to 2 ml of nitric acid (Suprapure, Merck, Germany) and distilled water (1:1) and with a few drops of hydrogen peroxide.

Metal concentrations in the fractions collected during the chromatographic separations were measured, without prior digestion, by flame atomic absorption spectrophotometry with deuterium background correction (Perkin Elmer model 3280).

III. RESULTS AND DISCUSSION

A. Subcellular Distribution of Cadmium and Zinc in the Midgut Gland

A large proportion of cadmium and zinc in metal-exposed *Helix pomatia* accumulated in the midgut gland which contained, on average, 4916 μg zinc and 619 μg cadmium (Table 1). After subcellular fractionation most of the zinc (79%) was found in the pellet while a predominant fraction of cadmium (86%) was associated with the supernatant (Table 1).

The cytosolic components of cadmium and zinc were fractionated by gel chromatography (Sephacryl S-200), yielding different distribution patterns for each metal (Figure 1). All the cadmium (Figure 1b), but only 30% of supernatant zinc (Figure 1c) — corresponding to only 6% of the metal in the whole hepa-

Table 1. Absolute (μg) and Relative (%) Contents of
Cadmium and Zinc in Pooled Hepatopancreas
Fractions of Metal-Fed Specimens of *Helix
pomatia* After Centrifugation. The Data Represent
a Tissue Pool of 16 Individuals

Metal	Hepatopancreas (homogenate)	Supernatant	Pellet
Zinc μg	4619	973	3647
%	100	21	79
Cadmium μg	619	531	88
%	100	86	14

topancreas — coeluted within a single peak, exhibiting an apparent molecular weight of 11,500 Da. This peak also displayed an elevated absorption at 254 nm (Figure 1a) which indicates that the metals were bound to metallothionein fractions. Most of the cytosolic zinc, however, was associated with fractions of a molecular weight of less than 4000 Da (Figure 1c) which are likely to represent a specific zinc-binding ligand, while a minor proportion of the metal (19%) was bound to fractions of a molecular weight higher than 200,000 Da.

The predominant role of the midgut gland in accumulating zinc and cadmium is a common characteristic of terrestrial gastropods.[8,18-21] Distribution patterns for cadmium in the midgut gland similar to those obtained in the present study have also been reported for the slug *Arion ater*.[20]

Although in the present study both metals were concentrated in the same target organ, they were largely diverted into different compartments: the prevalent association of cadmium with supernatant fractions indicates that most of this metal is bound to cytosolic components. On the other hand, the occurrence of a predominant proportion of zinc in the pellet means that the metal must be associated with cellular particles. Among these, two possible candidates may be envisaged: the lipofuscin-containing residual bodies within the digestive cells, and the calcium phosphate granules within the calcium cells. Electron microscopic studies have shown that in the midgut gland of zinc-exposed snails, electron-dense metal precipitations were present within the calcium granules, but not within the lipofuscin vesicles (see Chapter 14, this volume).[1] This indicates that in the midgut gland of *Helix pomatia* a large proportion of zinc is represented by calcium granules in which the metal coprecipitates with calcium pyrophosphate.[22,23] Similar findings involving radioactive zinc were reported for the species *Arion rufus*.[24]

The different fractionation patterns of supernatant cadmium and zinc in the hepatopancreas of *Helix pomatia* show that apart from differences in subcellular distribution, each metal was predominantly bound to different ligands even in the cytosol. The prevalent appearance of cadmium within one single peak displaying features of metallothioneins represents a characteristic pattern in metal-

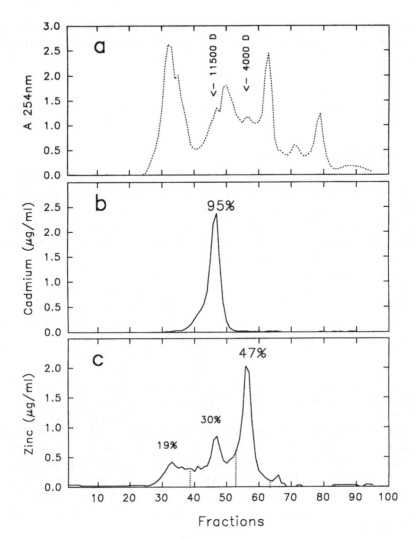

FIGURE 1. Gel chromatography (Sephacryl S-200) of supernatants derived from hepatopancreas homogenates of cadmium/zinc-fed snails.

a: Absorption at 254 nm. Arrows and molecular weights refer to calibration substances.

b: Elution profile of cadmium. Percentage amounts refer to total cadmium contained in the whole sample.

c: Elution profile of zinc. Percentage amounts refer to total zinc contained in the whole sample.

exposed terrestrial gastropods,[7,25,26] but may also depend on the use of reducing agents during fractionation,[27,28] and on the degree of cadmium-loading in exposed individuals.[29] In contrast, the preferential association of zinc with low molecular weight ligands indicates that this metal competes only marginally with cadmium for binding sites in metallothioneins. This is corroborated by the fact that even in uncontaminated and in zinc-exposed snails, with low cadmium concentrations in their tissues, zinc was always associated with low molecular weight fractions.[25]

B. Speciation of Cadmium

Cadmium-containing fractions derived from Sephacryl S-200 chromatrography (Figure 1b) were further separated by ion exchange chromatography (DEAE cellulose), displaying two major cadmium peaks (Figure 2a): one pronounced and narrow peak containing 30% of the metal (peak A) and a second, broader peak containing 31% of cadmium (peak B) (Figure 2a). Fractions of both peaks were pooled and stored for subsequent separation steps.

After gel chromatography on Sephacryl S-100, most of the cadmium was eluted in a pronounced peak (called peak A) with a molecular weight of more than 6000 Da (Figure 3a). This peak also proved to be homogeneous after reverse-phase HPLC (Figures 3b and c). Fractions of this peak (Figure 3b) were pooled and characterized by amino acid analysis, showing that it consisted of metal-lothionein, as suggested by the high content of cysteine residues (17.5%) and the total absence of aromatic amino acids (Table 2, peak A).

Peak B from ion exchange chromatography (Figure 2a) seemed to be homogeneous after Sephacryl S-100 gel chromatography (Figure 4a). After reverse-phase HPLC, however, the same peak was split into two separate peaks, called B1 and B2, respectively (Figure 4b). Subsequent metal analyses showed that peak B1 contained cadmium, whereas peak B2 proved to be rich in copper (Figure 4c). Fractions of the cadmium-containing peak (B1) were pooled and analyzed for amino acid composition (Table 2, peak B1). The results show that the peak consisted of a protein with a rather low content of cysteine residues (5%), a high proportion of threonine (24.6%) and serine (16.6%), but no aromatic amino acids.

The association of cadmium with metallothionein seems to be a general characteristic in terrestrial pulmonates and has also been shown for the species *Arianta arbustroum*,[7] *Cepaea hortensis*,[7] and *Arion lusitanicus*.[26] Fractions of the metallothionein peak derived from reverse-phase HPLC were concentrated and further prepared for amino acid sequence analysis,[30] preliminary results of which are shown in Chapter 14, this volume.[1]

The amino acid composition of the second cadmium binding protein derived from HPLC peak B1 (Figures 4b and c, Table 2) is surprising: the relatively low cysteine content of this ligand (5%), its exceptionally high amount of threonine (24.6%), as well as the presence of methionine (4.2%) suggest that this component is not a true metallothionein. However, its cadmium binding capacity may be due to the sulfur atoms provided by the cysteine and methionine residues.

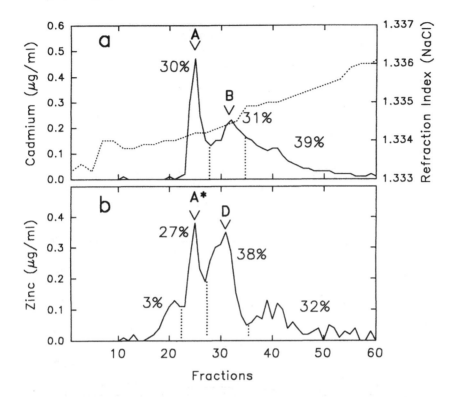

FIGURE 2. Ion exchange chromatography (DEAE-cellulose) of pooled fractions derived from the cadmium peak shown in Figure 1b and from the 30% zinc peak shown in Figure 1c.
a: Elution profile of cadmium (solid line) and refraction index (dotted line). Percentage amounts refer to total cadmium contained in the whole sample. Fractions of peak A and B were pooled and stored for further separation.
b: Elution profile of zinc. Percentage amounts refer to total zinc contained in the whole sample. Fractions of peaks A* and D were pooled and stored for further separation.

C. Speciation of Zinc

Fractions of the main zinc peak from Sephacryl S-200 chromatography, containing 47% of the metal (Figure 1c), were pooled and separated by ion exchange chromatography, yielding one main zinc peak, called peak C (Figure 5a). Fractions of this peak were further separated by Sephacryl S-100 chromatography (Figure 5b), the predominant fraction of zinc (70%) appearing in a pronounced peak with a molecular weight of approximately 3000 Da. This compound was identified as a unique, low molecular weight, zinc-binding ligand, the structure of which remains to be elucidated.

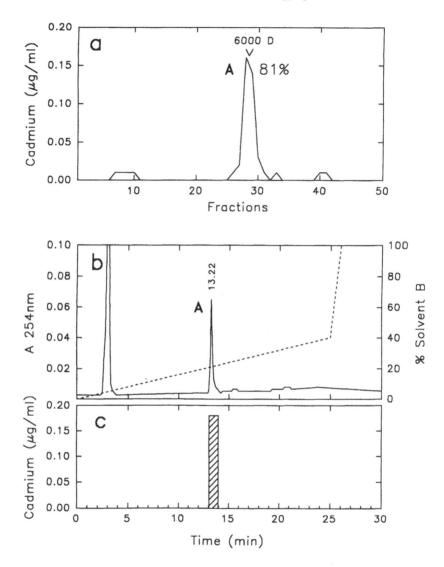

FIGURE 3. Gel permeation chromatography (Sephacryl S-100) and reverse-phase HPLC of pooled fractions derived from cadmium peak A shown in Figure 2a.

a: Elution profile of cadmium after Sephacryl S-100 chromatography. Percentage amounts refer to total cadmium contained in the whole sample. The molecular weight of 6000 Da is marked.

b: Elution profile after HPLC, showing the absorption at 254 nm (solid line) in fractions of peak A and elution time in a gradient of solvent B (dashed line).

c: Concentration of cadmium (bar) in HPLC fractions of peak A.

Table 2. Amino Acid Analyses of Purified Proteins from *Helix pomatia*

Amino acid	Peak A (Res.%)	Peak B1 (Res.%)
Cys	17.5[a]	5.0[b]
Asx	3.2	8.3
Thr	13.2	24.6
Ser	13.2	16.6
Glx	7.7	8.0
Pro	5.0	5.4
Gly	18.5	9.9
Ala	6.3	8.7
Val	0.3	3.2
Met	0[c]	4.2
Ile	0	3.0
Leu	0.3	0
Phe	0	0
Lys	12.4	3.2
His	0	0
Arg	0.8	0

Note: Amino acid analyses of purified proteins from Peak A (see Figures 3b and 5b) and Peak B1 (see Figure 4b) after gel filtration chromatography on Sephacryl S-100 (peak A) or after reverse-phase HPLC (peak B1). Mean percentage residues (Res.%) of 4 repeated analyses are reported.

[a] Determined as cysteic acid.
[b] Determined as S-methyl-cysteine.
[c] Determined as methionine sulfone.

Fractions of the second zinc peak from Sephacryl S-200 chromatography, containing 30% of the metal (Figure 1c), coeluted with the cadmium peak (Figure 1b) and were pooled for further separation by ion exchange chromatography (Figure 2). After this fractionation step, two pronounced zinc peaks appeared (Figure 2b). The first was eluted at the same ionic strength as the main cadmium peak (Figure 2a) and was called peak A* (Figure 2b). Further separation of this

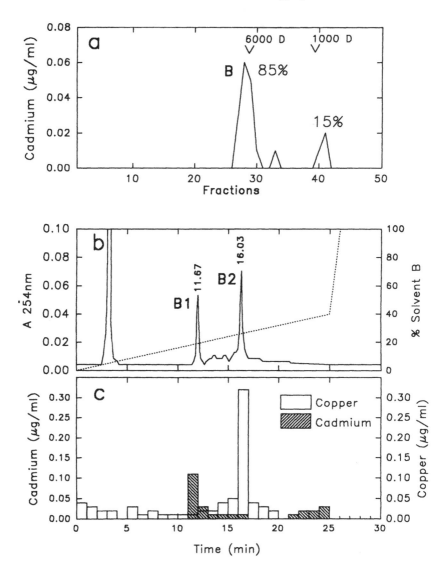

FIGURE 4. Gel permeation chromatography (Sephacryl S-100) and reverse-phase HPLC of pooled fractions derived from cadmium peak B shown in Figure 2a.

a: Elution profile of cadmium after Sephacryl S-100 chromatography. Percentage amounts refer to total cadmium contained in the whole sample. Molecular weights of 6000 and 1000 Da are marked.

b: Elution profile after HPLC, showing the absorption at 254 nm (solid line) in fractions of peaks B1 and B2, as well as elution times in a gradient of solvent B (dotted line).

c: Concentrations of cadmium and copper (bars) in HPLC fractions of peaks B1 and B2.

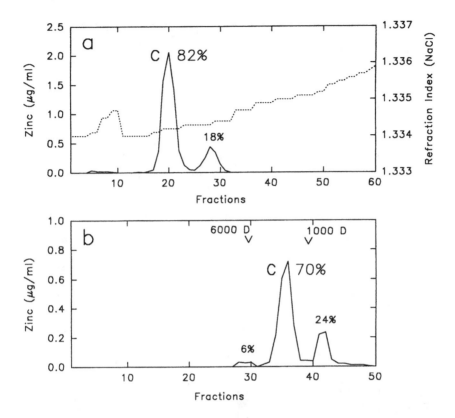

FIGURE 5. a: Ion exchange chromatography (DEAE-cellulose) of pooled fractions derived from the main zinc peak (containing 47% of zinc) shown in Figure 1c: elution profile of zinc (solid line) and refraction index (dotted line). Percentage amounts refer to total zinc contained in the whole sample. Fractions of peak C were pooled and further fractionated.
b: Gel permeation chromatography (Sephacryl S-100) of pooled fractions derived from zinc peak C shown in Figure 5a: elution profile of zinc. Percentage amounts refer to total zinc contained in the whole sample.

peak by Sephacryl S-100 chromatography (Figure 6) revealed that it consisted of two zinc compounds, the first of which (peak A) contained 59% of the metal, eluting at a molecular weight of more than 6000 Da (Figure 6a); the second peak (peak D), with 41% of the metal, eluted at a molecular weight of less than 1000 Da (Figure 6a). This peak consisted mainly of ionic zinc and proved to be identical with peak D in Figure 7.

Zinc peak A proved to be homogeneous after reverse-phase HPLC (Figures 6b and c), eluting together with cadmium peak A (Figure 3b). It was thus recognized as the metallothionein already characterized by its amino acid composition (Table 2, peak A).

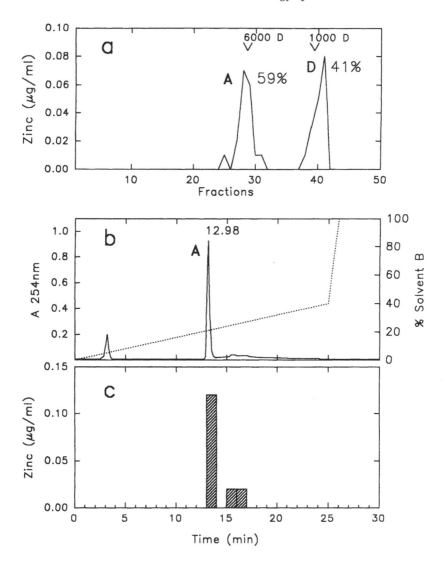

FIGURE 6. Gel permeation chromatography (Sephacryl S-100) and reverse-phase HPLC of pooled fractions derived from zinc peak A* shown in Figure 2b.

a: Elution profile of zinc after Sephacryl S-100 chromatography. Percentage amounts refer to total zinc contained in the whole sample. Molecular weights of 6000 and 1000 Da are marked.

b: HPLC elution profile of pooled fractions from peak A* in Figure 2b, showing the absorption at 254 nm (solid line) in fractions of peak A with the elution time in the gradient of solvent B (dotted line).

c: Concentrations of zinc (bars) in HPLC fractions.

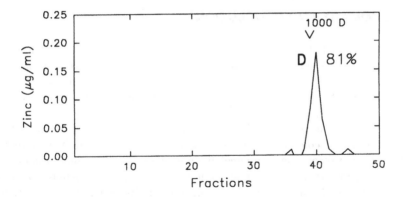

FIGURE 7. Gel permeation chromatography (Sephacryl S-100) of pooled fractions derived from zinc peak D shown in Figure 2b: elution profile of zinc. Percentage amounts refer to total zinc contained in the whole sample. Molecular weight 1000 Da is marked.

The second zinc peak appearing after the first ion exchange chromatography, called peak D (Figure 2b), was rechromatographed on a Sephacryl S-100 column (Figure 7) and exhibited an apparent molecular weight of less than 1000 Da. It was identified as free, ionic zinc.

Most of the cytosolic zinc in the hepatopancreas of *Helix pomatia* was associated with a low molecular weight, zinc-binding ligand. It is interesting that in the kidney cytosol of the mussel *Mytilus edulis* zinc was also found to be complexed by a low molecular weight ligand of approximately 1000 Da.[28] Since the reducing agents used in chromatography may interfere with these compounds, their true nature has to be verified using buffers without reducing chemicals.[28] In the case of *Helix pomatia* the low molecular weight zinc complex was also observed in the absence of such agents (not shown). It is concluded that the ligand concerned does play a role in zinc metabolism of this species.

On the other hand, the metallothionein of *Helix pomatia* was found to bind only 6% of total hepatopancreas zinc, the ratio of cadmium to zinc in this protein usually being 10:1. Since the synthesis of metallothionein is induced rapidly by cadmium exposure (Chapter 14, this volume),[1] it is suggested that the newly produced metallothionein binds predominantly cadmium, while the small amounts of zinc already bound to this protein remain unaffected. Thus metallothionein might play an ambivalent role, binding and storing small proportions of essential zinc on the one hand, and detoxifying nonessential cadmium on the other hand. In addition, however, a low molecular weight zinc-binding ligand is likely to play an important role in the metabolism of this essential trace element in *Helix pomatia*.

IV. SUMMARY AND CONCLUSIONS

Individuals of *Helix pomatia* were fed on a cadmium and zinc enriched diet. After homogenization and centrifugation of midgut gland, the main proportion of cadmium (86%), but only 21% of zinc were found in the supernatant.

After gel chromatography, 95% of the cadmium were bound to fractions showing some properties of metallothioneins. These fractions were further separated by ion exchange and gel chromatography, and by reverse-phase HPLC. They appeared to consist of two main components, one of them being a metallothionein, the second one a nonmetallothionein cadmium binding protein.

The supernatant of zinc was shown to consist of different ligands belonging to mainly two molecular weight classes: the greater proportion of zinc was bound to fractions with a molecular weight of 3000 Da and was shown to be a homogeneous ligand the nature of which is unknown so far. The second proportion of zinc was bound to metallothionein, but at much lower absolute amounts in comparison to cadmium (Cd:Zn = 10:1).

The results may explain why in *Helix pomatia* no interactions between cadmium and zinc were observed (see Chapter 15, this volume).[5] It is speculated that one of the adverse effects exerted by cadmium in most animals is the disturbance of zinc metabolism. Such disturbances are not likely to occur in *Helix pomatia* which is able to channel the two metals into different cellular compartments.

From an ecotoxicological point of view this may also explain why *Helix pomatia* suffers little from cadmium exposure alone or from combined exposure of cadmium and zinc (see Chapter 15, this volume).[5] The two metals often occur together on contaminated sites in the field.[31] The extraordinary capacity of *Helix pomatia* in handling both trace elements may render this species one of the most efficient metal concentrators in polluted soil habitats (see Chapter 14, this volume).[1] It remains to be elucidated if this capacity is significant in terms of increased bioavailability of cadmium and zinc in terrestrial foodwebs.

ACKNOWLEDGMENTS

The authors thank Professor W. Wieser, University of Innsbruck, for reviewing the manuscript and for helpful discussion. The study was supported by the Austrian "Fonds zur Förderung der wissenschaftlichen Forschung in Österreich", project Nr. P7815.

REFERENCES

1. **Dallinger, R.,** Strategies of metal detoxification in terrestrial invertebrates, this volume, chap. 14.
2. **Berger, B. and Dallinger, R.,** Terrestrial snails as quantitative indicators of environmental metal pollution, *Biol. Monit. Assess.,* in press.
3. **Russell, L. K., Dehaven, J. I., and Botts, R. P.,** Toxic effects of cadmium on the garden snail *Helix aspersa, Bull. Environ. Contam. Toxicol.,* 26, 634, 1981.
4. **Marigómez, J. A., Angulo, E., Saez, V.,** Feeding and growth responses to copper, zinc, mercury and lead in the terrestrial gastropod *Arion ater* (Linné), *J. Mollusc Stud.,* 52, 68, 1986.
5. **Berger, B., Dallinger, R., Felder, E., and Moser, J.,** Budgeting the flow of cadmium and zinc through the terrestrial gastropod, *Helix pomatia* L., this volume, chap. 15.
6. **Beeby, A.,** The role of *Helix aspersa* as a major herbivore in the transfer of lead through a polluted ecosystem, *J. Appl. Ecol.,* 22, 267, 1985.
7. **Dallinger, R., Berger, B., and Bauer-Hilty, A.,** Purification of cadmium-binding proteins from related species of terrestrial Helicidae (Gastropoda, Mollusca): A comparative study, *Mol. Cell. Biochem.,* 85, 135, 1989.
8. **Dallinger, R. and Wieser, W.,** Patterns of accumulation, distribution and liberation of Zn, Cu, Cd, and Pb in different organs of the land snail *Helix pomatia* L., *Comp. Biochem. Physiol.,* 79C, 117, 1984.
9. **Elinder, C.-G. and Piscator, M.,** Cadmium and zinc relationships, *Environ. Health Perspect.,* 25, 129, 1978.
10. **Petering, H. G.,** Some observations on the interaction of zinc, copper, and iron metabolism in lead and cadmium toxicity, *Environ. Health Perspect.,* 25, 141, 1978.
11. **Niboer, E. and Richardson, D. H. S.,** The replacement of the nondescript term "heavy metals" by a biologically and chemically significant classification of metal ions, *Environ. Pollut.,* 1B, 3, 1980.
12. **Ohnesorge, F. K. and Wilhelm, M.,** Zinc, in *Metals and Their Compounds in the Environment. Occurrence, Analysis and Biological Relevance,* VCH Publishers, Weinheim, Germany, 1991.
13. **Wood, J. M., Fanchiang, Y.-T., and Ridley, W. P.,** The biochemistry of toxic elements, *Q. Rev. Biophys. II,* 4, 467, 1978.
14. **Chmielnicka, J., Bem, E. M., Brzeźnicka, E. A., and Kasperek, M.,** The tissue disposition of zinc and copper following repeated administration of cadmium and selenium in rats, *Environ. Res.,* 37, 419, 1985.
15. **Hemelraad, J., Kleinveld, H. A., de Roos, A. M., Holwerda, D. A., and Zandee, D. I.,** Cadmium kinetics in freshwater clams. III. Effects of zinc on uptake and distribution of cadmium in *Anodonta cygnea, Arch. Environ. Contam. Toxicol.,* 16, 95, 1987.
16. **Medici, J. C. and Taylor, M. W.,** Interrelationships among copper, zinc and cadmium in the diet of the confused beetle, *J. Nutr.,* 38, 307, 1967.
17. **Hunziker, P. E.,** Cystein modification of metallothionein, in *Methods of Enzymology,* Vol. 205, Riordan, J. S. and Vallee, B. L., Eds., Academic Press, San Diego, 1991, 399.

18. **Coughtrey, P. J. and Martin, M. H.,** The distribution of Pb, Zn, Cd, and Cu within the pulmonate mollusc *Helix aspersa* Müller, *Oecologia,* 23, 315, 1976.

19. **Cooke, M., Jackson, A., Nickless, G., and Roberts, D. J.,** Distribution and speciation of cadmium in the terrestrial snail *Helix aspersa, Bull. Environ. Contam. Toxicol.,* 23, 445, 1979.

20. **Ireland, M. P.,** Uptake and distribution of cadmium in the terrestrial slug *Arion ater* (L)., *Comp. Biochem. Physiol.,* 68A, 37, 1981.

21. **Berger, B. and Dallinger, R.,** Accumulation of cadmium and copper by the terrestrial snail *Arianta arbustorum* L.: kinetics and budgets, *Oecologia,* 79, 60, 1989.

22. **Simkiss, K. and Taylor, M. G.,** Convergence of cellular systems of metal detoxification, *Mar. Environ. Res.,* 28, 211, 1989.

23. **Taylor, M. G., Graves, G. N., and Simkiss, K.,** Biotransformation of intracellular minerals by zinc ions *in vivo* and *in vitro, Eur. J. Biochem.,* 192, 783, 1990.

24. **Schötti, G. and Seiler, H. G.,** Uptake and localisation of radioactive zinc in the visceral complex of the land pulmonate *Arion rufus, Experientia,* 26, 1212, 1970.

25. **Dallinger, R. and Wieser, W.,** Molecular fractionation of Zn, Cu, Cd, and Pb in the midgut gland of *Helix pomatia* L., *Comp. Biochem. Physiol.,* 79C, 125, 1984.

26. **Dallinger, R., Janssen, H. H., Bauer-Hilty, A., and Berger, B.,** Characterization of an inducible cadmium-binding protein from hepatopancreas of metal-exposed slugs (Arionidae, Mollusca), *Comp. Biochem. Physiol.,* 92C, 355, 1989.

27. **Minkel, D. T., Poulsen, K., Wielgus, S., Shaw, C. F., and Petering, D. H.,** On the sensitivity of metallothionein during isolation, *Biochem. J.,* 191, 475, 1980.

28. **Lobel, P. B.,** The effect of dithiothreitol on the subcellular distribution of zinc in the cytosol of mussel kidney *(Mytilus edulis)*: isolation of metallothionein and a unique low molecular weight zinc-binding ligand, *Comp. Biochem. Physiol.,* 92C, 189, 1989.

29. **Janssen, H. H. and Dallinger, R.,** Diversification of cadmium binding proteins due to different levels of contamination in *Arion lusitanicus, Arch. Environ. Contam. Toxicol.,* 20, 132, 1991.

30. **Dallinger, R., Berger, B., Hunziker, E. P., Hauer, Ch., and Kägi, J. H. R.,** in preparation.

31. **Dallinger, R.,** unpublished data.

CHAPTER 17

Metal Relationships of Earthworms

A. John Morgan, John E. Morgan, Michael Turner, Carole Winters, and Andrew Yarwood

TABLE OF CONTENTS

0-87371-734-1/93/$0.00 + $.50
© 1993 by Lewis Publishers

I. GENERAL INTRODUCTION

Earthworms are relatively large, soft-bodied detritivores: they usually feed on decomposing plant materials, but may also graze on roots and consume seeds, algae, fungi, and protozoa.[1] Despite the surprisingly low taxonomic and morphological variability within such a well-established and functionally important animal group, earthworms can be subdivided into at least three ecophysiological categories:[2]

1. Epigeic Species — relatively small, usually pigmented, litter-inhabiting species;
2. Endogeic species — small or large, weakly pigmented worms living in horizontal burrows in the organomineral soil horizon; and
3. Anecic species — large worms living in deep vertical burrows that may also feed on the surface.

Other schemes of dividing earthworms have been adopted which complement rather than contradict the scheme proposed by Bouche.[2] For example, some earthworm species have active, mineralizing, calciferous glands that appear to be associated with the ability to tolerate a fairly wide soil pH range; other species have inactive or nonmineralizing glands.[3,4] Some species enter a discrete resting state, 'diapause', during periods of adverse climatic conditions; other species remain active or experience spells of reduced activity, 'quiescence', during certain periods of the year.[5] Earthworms are in fairly intimate contact with litter and soil, although they may be much more discriminate consumers than is often appreciated.[6] It has been convincingly shown that earthworms do accumulate several different metals from soils contaminated from diverse anthropogenic sources,[7,8] with tissue concentrations often exceeding those in organisms occupying higher trophic levels and inhabiting the same polluted location.[9] While these observations are consistent with those made on oligochaetes in freshwater food chains,[10] some caution needs to be exercised in the interpretation of these findings.

Terrestrial organisms such as geophilid centipedes (Myriapoda; Chilopoda) and slugs (Mollusca; Gastropoda; Pulmonata) spend relatively large proportions of their lives beneath the soil surface,[11] and yet they do not accumulate metals to the same extent as earthworms. A subterranean lifestyle *per se* does not lead to high accumulated metal concentrations, even in animals with soft, highly permeable integuments. Why, therefore, can earthworms accumulate high metal burdens? Answering this deceptively simple question is difficult, but it appears that it reflects the detritivorous habit, itself promoted by a resident gut microflora and the functional anatomy of the oligochaete alimentary system. This hypothesis can probably be extended to explain why terrestrial isopods are also prodigious metal accumulators.[9,12,13]

Attempting to explain metal accumulation goes to the crux of the issue of exposure. Exposure is not simply consumption. Metal accumulators must not only consume metal-contaminated materials, either as selected dietary constituents, or as ingesta indiscriminately consumed during burrowing; the released metals must also be rendered available for transport across the absorptive epithelia. (Of course, the existence of these processes does not preclude the possibility that a proportion of the body burden derives from metal transport across integumentary and/or respiratory surfaces.) Finally, the metal accumulator must possess the capacity to sequester metals within its cells and tissues, either by binding to existing or induced ligands.

Overgeneralizations may conceal important information. For example:

1. Tissue metal accumulation is the net effect of uptake and excretion. An organism is a nonaccumulator for one of several reasons:
    - it is not exposed, in either the ecological or physiological senses, to the given metal;
    - it is exposed, but effectively excludes the metal by blocking access to transepithelial pathways;
    - it has a highly discriminating uptake pathway(s);
    - it has a high excretory rate;
    - it has a high uptake and low excretory rates, but poor intracellular sequestration/detoxification capacity, so that low body burdens are highly debilitating or lethal.
2. Accumulators may not accumulate all metals with comparable efficiencies. This is a principle that applies to both essential and nonessential metals. Earthworms and isopods, for example, are both metal accumulating groups,[9] but isopods are very avid Cu accumulators,[12] while worms do not accumulate high Cu concentrations even from relatively heavily contaminated soils.[13-15]
3. Even nonaccumulators can accumulate high concentrations of certain essential metals. For example, centipedes do not assimilate Pb; but even under noncontaminated conditions they contain high Zn burdens, the large proportion of which is restricted to a subcuticular tissue.[16]

Metal relationships, including the impact of exposure to toxic metals, can be examined at several different levels of functional organization, namely: ecological, physiological, cellular, and molecular. The present paper is confined to a brief review of recent findings on the ecological and cellular relationships between earthworms and certain essential (Zn) and nonessential (Pb, Cd) metals. We have two modest heuristic objectives: (1) to highlight information that may contribute to the establishment of general principles regarding metal accumulation, compartmentation, and detoxification by these large terrestrial detritivores; and (2) to draw attention to areas where further information is required.

II. EARTHWORM AND MONITORING SOIL METAL POLLUTION

Aquatic organisms have been used relatively extensively compared with terrestrial species for monitoring the relative degrees of pollution at different sites. There appear to be two pragmatic motivations for their use: (1) efforts to safeguard human health by measuring accumulated metal concentrations in key food organisms, and subsequently comparing the observed data with standards established by toxicity tests; and (2) detecting spatial and temporal trends in metal contamination, so that emission control practices can be mobilized at the source(s). It is pertinent to ask why it is deemed desirable, if not necessary, to measure pollutants in living tissues rather than directly in an abiotic component of the study site(s).

The general concept of measuring the tissue concentrations of metals in a selected species to yield information concerning 'bioavailable pollution fractions' in a given habitat is theoretically attractive, but is fraught with practical diffi- culties. The fact that bioavailability data are species specific leads to interpre- tative problems. Differences have been recorded in the concentrations of metals accumulated by closely related species of slugs[9,17,18] and isopods[9,19-21] living, respectively, in the same contaminated habitats. In the case of isopods, the observed differences between the species *Porcellio scaber* and *Oniscus asellus* are attributed largely to differences in the turnover of certain metal-accumulating cells in their hepatopancreases.[19] It is not entirely unexpected that significant differences have also been found in the tissue metal concentrations of ecophys- iologically different earthworm species: the endogeic species, *Apporrectodea caliginosa* and *Octolasion cyaneum*, accumulate higher in Pb and Zn concen- trations than the epigeic species, *Lumbricus rubellus*.[22-25] However, detailed analyses of the metal concentrations in the crop contents, feces, and tissues of epigeic and endogeic species sampled from the same metalliferous site have failed to identify the basis of the observed species differences (Figure 1), but concluded that there are a number of interactive, mainly biotic, determinants.[24] Even less easy to explain are the striking differences that occur between two epigeic earthworms, *L. rubellus* and *Dendrodrilus rubidus*,[26-28] where the smaller *D. rubidus* accumulates higher concentrations of Pb (Figure 2) and Cd, but lower concentrations of Zn and Ca than *L. rubellus*.[28] Clearly the 'bioavailable metal fraction' estimated by analyzing tissue concentrations is not equivalent or anal- ogous to any 'soil extractable fraction' estimated by standard physicochemical procedures. Because the measured accumulated metal concentrations in the tis- sues of a given biomonitoring organism are not absolute measures of either 'bioavailability' per se, or of metal 'speciation' within a soil, it is only valid to use such data as indices for comparing the relative and potential bioavailabilities of specific metals between different sites.

Having defined the objectives of the biomonitoring exercise and selected the biomonitoring species, other fundamental issues must be addressed. Unambig- uous biomonitoring, according to Phillips,[29] can best be achieved if there is a relatively simple statistical relationship (i.e., linear relationship) between metal concentrations in the biomonitoring organism and in a relevant abiotic component of its environment. Even if this proposal were acceptable, a number of factors can detract from such a relationship, occasionally in unpredictable ways. Take, for example, the earthworm species *A. caliginosa* and *L. rubellus*. *A. caliginosa*, but not *L. rubellus*, enters a distinctive resting state during dry spells in the temperate summer. During diapause, the tissue concentrations of Cd and Zn are significantly lower than in 'active' *A. caliginosa*, while Pb concentrations are significantly elevated during diapause.[25] We saw earlier that *A. caliginosa* is a more efficient accumulator of metals than *L. rubellus*, a fact that recommends it for biomonitoring purposes. However, the relationship between the metal concentrations in the tissues of *A. caliginosa* and in the soil that it inhabits is subject to change dictated by local climatic, and perhaps edaphic, conditions.

A linear relationship between metal concentrations in earthworm tissues and in soil suggests that the same information concerning relative pollution amounts at different sites could probably be derived more readily by the direct analysis of soils. However, the recognition of 'outliers', that deviate from the linear relationship, can reveal important information that integrates in space and time the interactive influences of abiotic and biotic factors.[30,31]

Two 'outlier' populations can be identified in the data plotted in Figure 3, each showing nonconformity for a different metal: (1) the Pb concentration in *L. rubellus* sampled from site "g" exceeded the prediction of the regression analysis, such that the concentration factor for this population was 3.9, compared with values well below unity for all other populations; and (2) the Cd concentrations in worms from site "f" were well below statistical prediction such that, of all the populations analyzed, this was the only one yielding a concentration factor below unity for this particular metal. It is of more than passing interest that site "g" soil is both acidic and relatively Ca poor, especially since it has been shown that both edaphic factors exert strong codeterminant influences on the availability of Pb to earthworms.[32-35] Moreover, the site "f" soil was notable because it had a Zn concentration (= 150,000 μg/g dry weight) about twice that of the next most contaminated soil, and it also had by far the highest organic matter content (loss on ignition = 35%). One or both of these factors may render the Cd at site "f" relatively unavailable. Certainly, the immobilizing influence of soil macromolecules is well known.[32,34]

'Outlier' populations could be explained not only by the interaction of edaphic synergists and antagonists, but also by the existence of intrinsic factors whereby earthworm ecotypes have evolved the capacity to either exclude or hyperaccumulate certain metals, due to the heavy selection pressures extant in their polluted environments. If such local adaptations are shown to exist, then it may be prudent to biomonitor by exposing genotypically uniform test organisms to the soils to be tested under field-enclosure or laboratory conditions.[36,37]

III. EFFECTS OF METALS ON EARTHWORMS IN THE FIELD: METAL INTERACTIONS ON A BIOCHEMICAL LEVEL OF ORGANIZATION

Earthworms are used extensively to evaluate the toxicities of xenobiotics by standardized acute toxicity tests.[38] A number of behavioral, reproductive, bio-

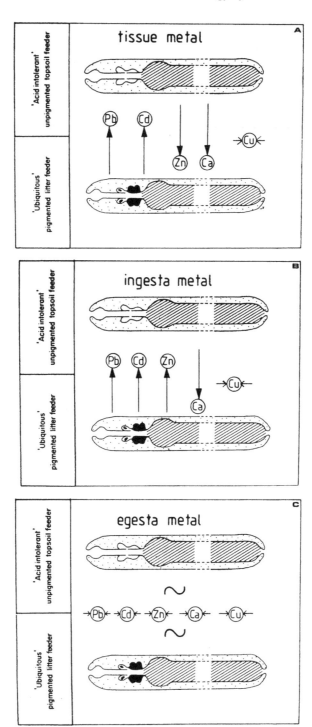

FIGURE 1. Schematic diagrams summarizing analytical data on the metal concentrations in whole-worm tissue, ingesta (crop contents), and egesta (feces), of two very broad ecophysiological categories of earthworms sampled from five 'stations' at a disused Pb/Zn mine (Llantrisant: O.S. Map Ref. = ST 048822). All worms were analyzed individually; for smaller species, crop contents, and feces, respectively, from more than one worm were pooled to provide a sample. The 'ubiquitous' ecophysiological group includes *L. rubellus* (epigeic) and *L. terrestris* (anecic), both litter feeders if not litter inhabitors, and both possessing active mineralizing calciferous glands. The 'acid intolerant' group includes *A. caliginosa* (endogeic) and *A. longa* (anecic), both apparently consuming more mineralized materials, and both possessing inactive nonmineralizing calciferous glands. Upward pointing (↑) arrows indicate that the 'acid intolerant' species have significantly higher concentrations of a given metal; downward pointing (↓) arrows indicate that the 'ubiquitous' species have the higher concentrations; horizontal (→ ←) arrows indicate that there are no significant differences between the species groups. (These diagrams were constructed from data presented in tabular form by Morgan, J. E. and Morgan, A. J., *Soil Biol. Biochem.*, in press. With permission.)

chemical, and immunological performance indicators have also been developed for evaluating the sublethal toxicities of chemicals, including metals, to earthworms chronically exposed under laboratory and field conditions.[39-41] However, very few studies have been conducted to examine the effects of toxic metals on earthworms inhabiting heavily contaminated metalliferous soils. Such populations are not only chronically exposed to complex metal mixtures, and diverse edaphic factors, but have also had the opportunity of evolving over periods that could accommodate several tens, if not hundreds, of earthworm generations.

In earthworms, very high proportions of the total accumulated body burdens of Pb, Cd, and Zn are confined to the posterior alimentary tissues,[15] with the chloragogenous tissue separating the absorptive intestinal epithelium from the coelom being a major metal depository (see Figure 4).[26,27,42-44] Although the morphology of individual chloragocytes changes significantly due to the accumulation of toxic metals (see Figure 5),[26,42,45] these changes have not been accurately described nor measured morphometrically. Furthermore, it is not known whether the cellular effects are a reflection of an increased turnover of intracellular metal-sequestering organelles and/or the direct toxic effects of the extraneous metals. However, it appears that earthworm chloragocytes respond to Pb accumulation in a way that is analogous to the response of isopod 'B' cells[46] (see also Chapter 18). Lead accumulation reduces energy stores, in the form of glycogen, in the earthworm cells (Figure 5).[26,45] But the direct effects of toxic metals on the biochemical pathways involved in energy storage should not be dismissed.[47]

It is well known that low concentrations of Pb inhibit hemoglobin biosynthesis in vertebrates.[48] The metal can interfere with several steps in the haem

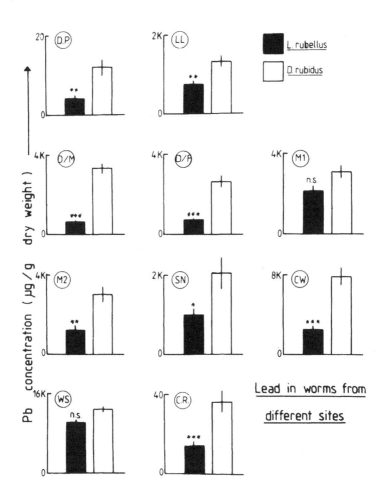

FIGURE 2. Histograms comparing the Pb concentrations in two epigeic earth-worms (*L. rubellus* and *D. rubidus*) sampled from an uncontaminated site (D.P.) and nine disused metalliferous sites (the others). Data are presented as means ± S.E.; number of observations vary from 7 to 42. Symbols: * = $p < 0.05$, ** = $p < 0.01$, *** = $p < 0.001$, and n.s. = non significant. Capital letters in circles are site codes. (Figures are drawn from data presented in tabular form by Morgan, J. E. and Morgan, A. J., *Bull. Environ. Contam. Toxicol.*, 47, 296, 1991. With permission.)

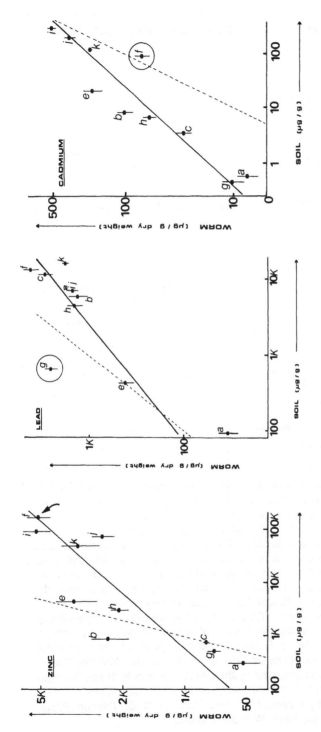

FIGURE 3. Relationships between Pb ($\log_{10}$ Y = 0.801 + 0.642 $\log_{10}$ X; r^2 = 66.5), and Cd ($\log_{10}$ Y = 1.23 + 0.553 $\log_{10}$ X; r^2 = 85.6) concentrations in whole-worm (*L. rubellus*) tissues and the soils that they inhabited. Vertical lines represent standard errors about geometric means; letters (a to l) refer to the different sampling sites; "a" is an uncontaminated site, the others are disused metalliferous sites. Broken lines are unity lines. Circled data points ("g" in the case of Pb; "f" in the case of Cd) are subjectively considered to be outlier populations. Note that the soil at site "f" has the highest Zn concentration. (These figures are redrawn from figures published by Corp, N. and Morgan, A. J., *Environ. Pollut.*, 74, 39, 1991. With permission.)

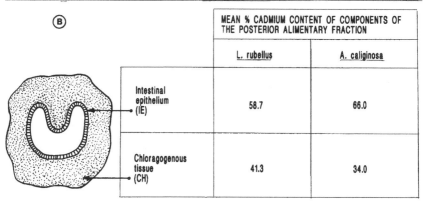

	MEAN % DISTRIBUTION IN TISSUE FRACTIONS					
(A)	Dry weight	Cu	Cd	Pb	Zn	Ca
•Anterior alimentary fraction (AF)	10.8	12.3	8.8	4.4	10.5	67.0
•Posterior alimentary fraction (PF)	22.0	57.3	72.8 (66)	87.7	78.8	21.3
•"Rest" fraction (RF)	67.0	30.4	18.4	7.9	10.7	11.7

(B)	MEAN % CADMIUM CONTENT OF COMPONENTS OF THE POSTERIOR ALIMENTARY FRACTION	
	L. rubellus	A. caliginosa
Intestinal epithelium (IE)	58.7	66.0
Chloragogenous tissue (CH)	41.3	34.0

FIGURE 4. A. Mean percentage distribution of dry weight, and metal (Cu, Cd, Pb, Zn, Ca) contents (μg) in three major tissue fractions of the earthworm *L. rubellus* sampled from a Pb/Zn mine site at Llantrisant, South Wales. The AF includes the esophagus, calciferous glands, crop, and gizzard; PF includes the intestinal epithelium and chloragogenous tissue; RF consists mainly of the body wall. (The data are derived from Morgan and Morgan;[15] the Cd value in parentheses was estimated from data presented by Ireland and Richards,[42] and was included for comparative purposes.)
B. Mean percentage Cd content in the two major components of the posterior alimentary fraction of two ecophysiologically different earthworm species, *L. rubellus* and *A. caliginosa*, maintained under laboratory conditions on Cd-soaked filter paper. Note that the Cd content of the intestinal epithelium is higher in both species than the Cd content of the chloragog. (Values were calculated from data presented by Ireland, M. P. and Richards, K. S., *Environ. Pollut. Ser. A*, 26, 69, 1981. With permission.)

pathway, for example: (1) it inhibits the enzyme delta-aminolevulinic acid dehydratase (ALAD), leading to the accumulation of delta-aminolevulinic acid; (2) it inhibits the enzyme coproporphyrinogen oxidase, leading to the excretion of coproporphyrin; and (3) in inhibits ferrochelatase, leading to the accumulation of protoporphyrin IX. Hemoglobin is the respiratory pigment of earthworms.[49] Available evidence suggests that the tissue site of haem synthesis in earthworms is the chloragog,[50] the very tissue that accumulates Pb to concentrations as high as >20,000 µg/g dry mass in the intracellular chloragosom granules.[44] Not surprisingly, Ireland and Fischer[51] were able to reduce the activity of the enzyme ALAD in the chloragog tissue of *L. terrestris* maintained on a Pb-spiked filter paper for several days under laboratory conditions. Our own provisional observations on *L. rubellus* sampled from an uncontaminated and two different Pb-contaminated sites indicated that the effect of Pb on haem synthesis under field conditions may be rather more complicated (see Figure 6).

The major conclusions of our analytical study, involving analysis by High Pressure Liquid Chromatography (HPLC) of chloroform extracts of whole *L. rubellus*, can be summarized as follows:

1. The dominant porphyrin in the earthworm tissues was protoporphyrin IX.
2. The concentration of protoporphyrin IX in the tissues of worms inhabiting one contaminated site (Cwmystwyth) was twice as high as in 'control' worms (Dinas Powys).
3. The protoporphyrin IX concentration in the tissues of worms inhabiting a second contaminated site (Llantrisant) was only half that measured in the Dinas Powys 'controls'.
4. The extremely dark, violet-brown pigmentation of the Cwmystwyth worms indicated that the protoporphyrin IX accumulates primarily in the body wall.

It has been shown that Pb accumulation inhibits ALAD activity, while Zn activates the enzyme and protects it from Pb-inhibition.[52,53] Ferrochelatase, the enzyme catalyzing the insertion of ferrous iron into protoporphyrin IX to produce hemoglobin, may be similarly activated by Zn and inhibited by Pb. Our observations suggest that the elevated protoporphyrin levels in Cwmystwyth worms may be due to two causative factors: (1) high tissue Pb concentrations accumulated from the moderately Pb-contaminated soil due to the low pH of the soil matrix,[36] and (2) tissue Zn concentrations that are too low to compensate fully for Pb-inhibition by reactivating ferrochelatase. The lower protoporphyrin IX concentration in the worms inhabiting the relatively heavily Pb- and Zn-contaminated, but neutral soil at Llantrisant may reflect the relatively low Pb and high Zn tissue concentrations in this population.

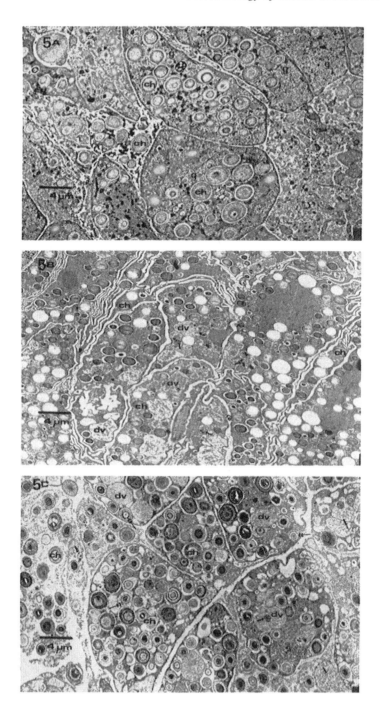

FIGURE 5. Transmission electron micrographs (low magnification) providing an overview of the morphology of the chloragocytes of *L. rubellus* inhabiting: an uncontaminated site (5A: Dinas Powys); an acidic site with relatively low total, but high 'available' soil Pb concentration (5B: Cwmystwyth, site "g" in Figure 3; see also Figure 6); a calcareous site heavily contaminated with Pb, Zn, and Cd; (5C: Draethen; see Figure 7). Note that there are a number of obvious differences between the 'normal' and the 'stressed' cells. Apart from the characteristic chloragosome granules (ch), the control chloragocytes contain vast quantities of glycogen (g). The stressed cells contain very little glycogen; they do, however, contain large numbers of debris vesicles (dv) packed between the chloragosomes. The debris vesicles often contain fine dense particulates (arrows) dispersed within more lucent material; occasionally, the chloragosomes also contain small dense particles, especially near their peripheries. (The origins and fate of these structures cannot be established merely from micrographs such as these; in addition, the appearance of some of the structures, including the distribution of the microparticulates, may be the result of the leaching and redistribution of materials during chemical fixation, dehydration, and resin embedding.)

IV. SUBCELLULAR COMPARTMENTATION OF METALS WITHIN EARTHWORM CHLORAGOCYTES

Three main intracellular pathways have evolved for the binding and immobilization of metals accumulated in terrestrial invertebrates.[54-56] Two of these pathways coexist in earthworm chloragocytes (Figure 7).

The type A pathway is exemplified by phosphate-rich chloragosome granules that bind Ca and Zn.[3,21,43,57,58] It should be noted, however, that the multifunctional chloragosomes differ from the more inorganic phosphate-rich calcospherites that typically occur in most other invertebrates. Earthworm chloragosomes contain a complex mixture of organic macromolecules,[59] and they may have the capacity to release ions for buffering the coelomic fluid and blood.[3,60] Lead appears to bind preferentially to the matrix of chloragosomes (Figure 7), with some evidence suggesting that it does so by effectively displacing Ca^{2+} ions.[44,61] Given the high concentration of Pb that can be bound by the chloragosomes of worms inhabiting heavily polluted soils,[44] it is important to evaluate the efficiency of Pb transport, binding and extrusion mechanisms, and their relationship to cellular Ca status. Lead ions can, as we saw above, inhibit enzymes involved in haem synthesis. Lead may also interfere with cellular function by disturbing Ca homeostasis in one of two general ways. First, it has been claimed,[62] but subsequently challenged,[63] that picomolar concentrations of Pb can mimic Ca in the activation of protein kinase C, an enzyme intimately involved in cellular growth and differentiation. Second, by either binding to SH-containing enzymes such as Ca^{2+} transport ATPases,[64] or by displacing Ca from

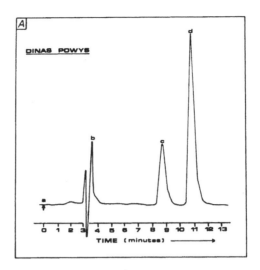

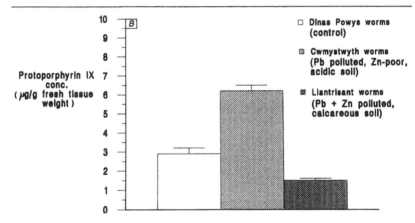

FIGURE 6. The effects of Pb and Zn on haem biosynthesis in *L. rubellus* sampled from an uncontaminated site (Dinas Powys) and two contaminated sites, one (Cwmystwyth) with an acidic soil containing a high proportion of bioavailable Pb but low Zn, and the other (Llantrisant) with a calcareous soil containing high absolute concentrations of both Pb and Zn. After a 3-day period on filter paper to clear their guts, individual worms were weighed, frozen in liquid N$_2$, thawed, and homogenized. The homogenates were then extracted in 5% H$_2$SO$_4$ in methanol, and the porphyrin methyl esters were extracted into chloroform. Porphyrin analysis was performed in a Gilson HPLC, using a Hypersil silica-packed column (25 cm × 4.6 cm i.d.) and a 100 μl injection loop; the wavelength was set at 400 nm, and the solvent system used was 20% ethyl acetate and 80% hexane. (To check the identity of the worm porphyrin, standard protoporphyrin IX, mesoporphyrin IX, and deuteroporphyrin IX were also injected under identical analytical conditions.)

intracellular binding sites, Pb may cause a rise in the concentration of cytosolic free Ca^{2+} leading to a loss of cellular viability due to the activation of phospholipases and endonucleases (see Nicotera et al.[65]). It is not known whether any of these cytotoxic effects are manifest in earthworm chloragocytes exposed to Pb, or whether the populations of earthworms inhabiting polluted sites have evolved mechanisms that safeguard against Pb toxicity or confer resistance to Pb toxicity. An example of an effective resistance mechanism would be the induction of the synthesis of more chloragosomal material by the displaced Ca, so that the capacity to bind Ca (and Pb) is maintained, thus reducing the chance of Ca overload within sensitive organelles.

The type B pathway is characterized by the ability of sulfur-donating ligands to bind Cd. Low molecular weight Cd-binding proteins, exhibiting many of the characteristics of metallothioneins, have been identified in earthworms experimentally exposed to Cd or exposed to the metal in their native soils.[7,66,67] Electron microprobe studies suggest, but cannot confirm, that the Cd-metallothionein complex is confined within vacuoles (Figure 7) possibly of lysosomal origin. A number of authors have in fact suggested that toxic metals bound to metallothioneins are compartmentalized within elements of the lysosomal systems of certain terrestrial and aquatic invertebrates.[68-76] In some cases the accumulation of Cd, and the expression of Cd-resistance, has been shown,[77,78] or suggested,[79,80] to be determined by duplications of metallothionein genes.

Because our knowledge of the mechanisms of metal binding within the earthworm chloragocyte is so relatively poor, it is proposed to highlight certain aspects of the data presented in Figure 7 hopefully, to, stimulate further research on this cell and other metal-accumulating cell types.

The X-ray maps of freeze-dried, cryosectioned, chloragocytes from *L. rubellus* inhabiting a Pb, Zn, and Cd-contaminated mine site indicate that Pb, Zn, and Ca are preferentially bound to the Pb-rich matrix of chloragosomes, and that Cd is confined to S-rich compartments called, perhaps misleadingly, "cadmosomes".[27,81] X-ray microanalysis is an analytical tool of high spatial resolution, but poor sensitivity. The minimum detectable 'local' concentration of an element varies with its atomic number, but even under the most favorable conditions, is not better than about 200 μg/g dry mass.[82] Thus it is possible that the chloragosomes could contain significant, albeit nondetectable, amounts of Cd, and the cadmosomes significant amounts of Pb and Zn. If this were the case, the metal accumulating pathways in the chloragocyte may not be as discriminatory as the maps suggest. Earlier 'static' microprobe analysis of the chloragocytes of Draethen worms indicated that at least three distinct metal-

A: A typical analytical spectrum of porphyrins from Dinas Powys ('control') worms, a = sample injection point; b = solvent front; c = unidentified peak at 8 min; d = major protoporphyrin IX peak at 11 min.

B: Histogram (means $\pm$ S.D.) comparing the concentration of protoporphyrin IX in whole earthworms from the three study sites.

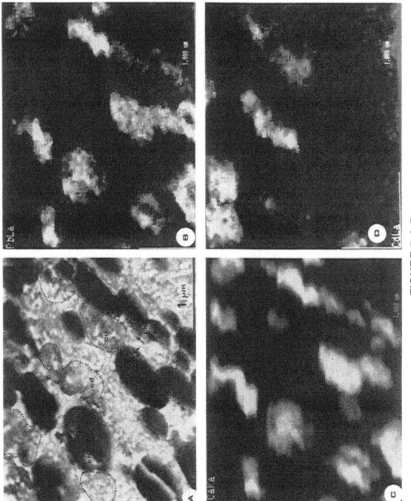

FIGURE 7 A-D.

FIGURE 7.   A: Scanning transmission electron micrographs (bright field image) of a freeze-dried, unfixed, thin cryosection of a chloragocyte from an earthworm, *L. rubellus*, sampled from a site (Draethen) contaminated with Pb, Zn, and Cd. The fresh tissue was mounted on a metal pin and plunged into stirred liquid propane, sectioned on a glass knife at −120°C in a Reichert Ultracut E fitted with an FC4 cryochamber, and the thin (∼100 nm) sections mounted on coated titanium grids and transferred in a liquid N_2-cooled aluminum lidded pot to a vacuum coating unit, where they were freeze dried and coated with carbon. The electron opaque chloragosome granules (ch) are prominent in the micrographs (compare with Figure 5); the cadmosomes (cd) are more difficult to discern morphologically and have been delineated with dotted lines after reference to the Cd X-ray map (Figure 7D).

accumulating compartments could be compositionally, if not morphologically, identified.[61] These were the typical chloragosomes and cadmosomes, and a 'hybrid' compartment containing Pb, Zn, and Cd. One interpretation of these observations is that the chloragosomes and cadmosomes are distinct lysosomal lineages that 'converge' to form a tertiary lysosome, or residual body, which may then be disposed of in a yet unknown manner. This hypothesis, derived in part from that suggested by George[83,84] to describe the compartmentation of Cd in *Mytilus* kidney, requires further testing.

Metallothioneins and similar proteins have long been recognized to play important roles in the intracellular homeostasis of essential metals, such as Zn and Cu, by perhaps storing and donating these metals to meet metabolic needs.[85] Goering and Fowler[86] demonstrated *in vitro* that Zn-metallothionein could donate Zn to, and thus activate, the enzyme delta-aminolevulinic acid dehydratase. More recently, Sunderman[87] showed that Cd could substitute for Zn in the DNA-binding "zinc finger" proteins that regulate gene expression (see Klug and Rhodes[88] and Evans and Hollenberg[89]). It is therefore possible that not only are the Zn-dependent transcription factors the molecular targets for Cd toxicity (it is of more than passing interest that Cd is a potent genotoxin in the Zn-rich reproductive tissues of mammals), but that metallothioneins may have the capability of transporting and donating Cd to the vulnerable intracellular targets. On the other hand, the confinement of Cd-metallothionein complexes within vesicular structures in at least some invertebrate cells may be a mechanism of harnessing the capacity of metallothionein to sequester toxic metals without the freedom to transport and redistribute them freely within the cell. Indeed, George[83] showed that the tertiary lysosomes in the kidney cells of *Mytilus* contained Cd but no metallothionein, suggesting that the protein had been hydrolyzed, and perhaps that the released amino acids had been recycled for further synthetic purposes.

Copper causes mortality and sublethal toxic injury to earthworms at environmental concentrations much lower than do Pb and Zn.[90] Furthermore, earthworms do not appear to accumulate high Cu concentrations even from heavily

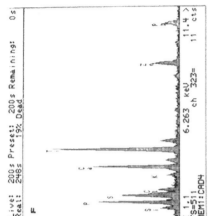

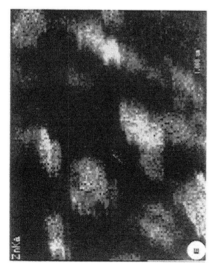

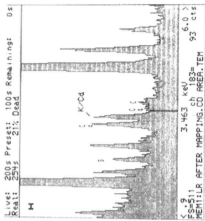

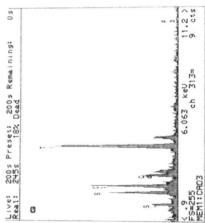

FIGURE 7 E-H.

FIGURE 7. B to E: Electron generated X-ray maps for Pb (B), Ca (C), Cd (D), and Zn (E) in the area depicted in the micrograph. Note that the Pb, Ca, and Zn are codistributed in the chloragosomes and that Cd is equally discretely distributed but is confined to a second metal-sequestering compartment, the cadmosome. In this cellular field, there appears to be no overlap in the distributions of Pb and Cd — this was more convincingly apparent in the colored maps from which these figures were derived. With color it is possible to superimpose maps for two or more elements so that their relative distributions can be appreciated readily at high spatial resolution. However, the specificity of intracellular compartmentation is confirmed by the X-ray spectra, obtained with 'stationary probes' positioned on a chloragosome (F) and a cadmosome (G and H), respectively. Note the presence of Zn in spectrum F, but not in G. Conditions for mapping and/or spectral acquisition were JEOL JEM-2000 FXII STEM operated at 120 kV, with 12 microamps heating current on a LaB_6 emitter. Spectra were acquired with 200 s 'live' time using a horizontal Link Pentafer ultra-thin-window detector fitted with a 30 mm² silicon crystal (FWHM resolution = 133 eV at 5.9 keV); Link Systems eXL X-ray analyzer; screen resolution = 128 × 128 pixels; dwell time on each pixel = 5 s; total map acquisition time = $22^3/_4$ h; condenser aperture = 200 μm. Background was subtracted from all spectra to exclude the effects of local specimen topographical effects on the maps; most maps (all except Zn) were median filtered to eliminate 'noise' (single random black or white points), but test points were always made without filtering to ensure that no data were lost or artificially created by this process. Drift correction was not available on the equipment used for mapping, but this was not a serious handicap because no significant drift was observed even though we analyzed a thin cryosection for a protracted period. Micrographs of the analyzed areas were taken immediately before and after analysis, either directly from the STEM unit onto Polaroid film or via the X-ray microanalyzer. Colored X-ray maps were recorded using a Mitsubishi CP200B color printer connected to the video output of the analyzer.

contaminated soils.[14,15] The high toxicity and low accumulation of Cu may be a function of the inability to induce the metallothionein gene in earthworms, despite the fact that Cu probably would form a more stable complex with the protein than does Cd (see Bremner[91]). Thus, we hypothesize that earthworms inhabiting a Cd-contaminated soil, and whose metallothionein gene(s) are activated, would not only be more Cu resistant, but would accumulate relatively high Cu concentrations in the so-called cadmosome vesicle.

There is no doubt that transintestinal transport is a major mode of entry of metals into earthworm tissues,[4] although transintegumentary uptake may not be insignificant. In earthworms, unlike terrestrial macroinvertebrates with relatively inpermeable exoskeletons, such as isopods and spiders,[56] the metal storage mechanism of the alimentary epithelium has been largely neglected. Ireland and Richards[42] (see Figure 4) showed that the intestine accumulated high Cd concentrations in worms exposed to filter paper soaked with a 5 ppm solution of Cd for 26 days. This suggests that the epithelium acts as a major barrier limiting the entry of Cd into the body fluids. The relationship between the Cd burden of the intestine and that of the chloragog has not been studied. Does the chloragog merely accumulate Cd that cannot be sequestered by the intestine? If so, it would be expected that the proportion of Cd bound by the intestine relative to the chloragog would decrease as the body burden increased. Are metallothionein gene(s) expressed in the intestinal cells? If so, is the Cd-metallothionein complex confined to "cadmosomes", or other components of the lysosomal system? Is metallothionein implicated in the transfer of Cd from the intestinal cells to the chloragog? Morgan et al.[67] showed that the posterior alimentary tissue fraction of *L. rubellus* contained two slightly different Cd-binding metallothionein-like proteins. Is one confined to the chloragog and the other to the intestine, or do they coexist in both tissues? Does Cd accumulation alter the structure and function of the intestinal epithelium? It is less likely that the integumentary cells act as metal storage and transit sites, because a very high proportion of the body wall Cd burden appears to be associated with surface mucus and not internalized.[42]

ACKNOWLEDGMENTS

We would like to thank Pam Ayres and Janet Davies for typing the manuscript, and Vyv Williams for his assistance with the illustrations. A. John Morgan would also like to thank Nadia Corp and Islam Choudhury for the discussions that led to some of the ideas expressed in this contribution.

REFERENCES

1. **Wallwork, J. A.,** *Earthworm Biology,* Inst. Biol. Stud. Biol., No. 161, Edward Arnold, London, 1983.
2. **Bouché, M.,** Strategies lombriciennes, *Ecol. Bull. (Stockholm),* 25, 122, 1977.
3. **Morgan, A. J.,** The elemental composition of the chloragosomes of nine species of British earthworms in relation to calciferous gland activity, *Comp. Biochem. Physiol.,* 73A, 207, 1982.
4. **Piearce, T. G.,** The calcium relations of selected Lumbricidae, *J. Anim. Ecol.,* 41, 167, 1972.
5. **Olive, P. J. W. and Clark, R. B.,** Physiology of reproduction, in *Physiology of Annelids,* Mill, P. J., Ed., Academic Press, London, 1978, 271.
6. **Piearce, T. G.,** Gut contents of some lumbricid earthworms, *Pedobiologia,* 18, 153, 1978.

7. **Ireland, M. P.,** Heavy metal uptake and tissue distribution in earthworms, in *Earthworm Ecology from Darwin to Vermiculture,* Satchell, J. E., Ed., Chapman and Hall, London, 1983, 247.

8. **Ireland, M. P.,** Heavy metal sources — uptake and distribution in terrestrial macroinvertebrates, in *Biological Monitoring of Exposure to Chemicals: Methods,* Dillion, H. K. and Ho, M. H., Eds., John Wiley & Sons, New York, 1991, 263.

9. **Morgan, A. J., Morris, B., James, N., Morgan, J. E., and Leyshon, K.,** Heavy metals in terrestrial macroinvertebrates: species differences within and between trophic levels, *Chem. Ecol.,* 2, 319, 1986.

10. **Prosi, F.,** Heavy metals in aquatic organisms, in *Metal Pollution in the Aquatic Environment,* Förstner, U. and Whittman, G. T. W., Eds., Springer-Verlag, Berlin, 1981, 271.

11. **Brown, A. L.,** *Ecology of Soil Organisms,* Heinemann, London, 1978.

12. **Hopkin, S. P. and Martin, M. H.,** The distribution of zinc, cadmium, lead and copper within the woodlouse *Oniscus asellus* (Crustacea, Isopoda), *Oecologia (Berlin),* 54, 227, 1982.

13. **Hunter, B. A., Johnson, M. S., and Thompson, D. J.,** Ecotoxicology of copper and cadmium in a contaminated grassland ecosystem, *J. Appl. Ecol.,* 24, 587, 1987.

14. **Streit, B.,** Effects of high copper concentrations on soil invertebrates (earthworms and oribatid mites): Experimental results and a model, *Oecologia (Berlin),* 64, 381, 1984.

15. **Morgan, J. E. and Morgan, A. J.,** The distribution of cadmium, copper, lead, zinc and calcium in the tissues of the earthworm, *Lumbricus rubellus,* sampled from one uncontaminated and four polluted soils, *Oecologia (Berlin),* 84, 559, 1990.

16. **Hopkin, S. P. and Martin, M. H.,** Assimilation of zinc, cadmium, lead and copper by the centipede *Lithobius variegatus* (Chilopoda), *J. Appl. Ecol.,* 21, 535, 1984.

17. **Greville, R. W. and Morgan, A. J.,** Seasonal changes in metal levels (Cu, Pb, Cd, Zn, and Ca) within the grey field slug, *Deroceras reticulatum,* living in a highly polluted habitat, *Environ. Pollut.,* 59, 287, 1989.

18. **Greville, R. W. and Morgan, A. J.,** The influence of size on the accumulated amounts of metals (Cu, Pb, Cd, Zn, and Ca) in six species of slug sampled from a contaminated woodland site, *J. Mol. Stud.,* 56, 355, 1990.

19. **Hopkin, S. P.,** Species-specific differences in the net assimilation of zinc, cadmium, lead, copper, and iron in the terrestrial isopods *Oniscus asellus* and *Porcellio scaber, J. Appl. Ecol.,* 27, 460, 1990.

20. **Hopkin, S. P., Martin, M. H., and Moss, S. M.,** Heavy metals in isopods from the supra-littoral zone on the southern shore of the Severn estuary, U.K., *Environ. Pollut. Ser. A.,* 9, 239, 1985.

21. **Morgan, A. J. and Winters, C.,** The contribution of electron probe X-ray microanalysis (EPXMA) to pollution studies, *Scanning Electron Microsc.,* 1, 133, 1987.

22. **Morris, B. and Morgan, A. J.,** Calcium-lead interactions in earthworms: observations on *Lumbricus terrestris* L. sampled from a calcareous abandoned lead mine site, *Bull. Environ. Contam. Toxicol.,* 37, 226, 1986.

23. **Morgan, J. E. and Morgan, A. J.,** Calcium-lead interactions involving earthworms. Part 1: The effect of exogenous calcium on lead accumulation by earthworms under field and laboratory conditions, *Environ. Pollut.,* 54, 41, 1988.

24. **Morgan, J. E. and Morgan, A. J.,** A comparison of heavy metal concentrations in the tissues, ingesta and faeces of ecophysiologically different earthworm species, *Soil. Biol. Biochem.,* in press.

25. **Morgan, J. E. and Morgan, A. J.,** Seasonal changes in the tissue metal (Cd, Zn, and Pb) concentrations in two ecophysiologically dissimilar earthworm species: pollution monitoring implications, *Environ. Pollut.,* (in press).

26. **Ireland, M. P. and Richards, K. S.,** The occurrence and localization of heavy metals and glycogen in the earthworms *Lumbricus rubellus* and *Dendrobaena rubida* from a heavy metal site, *Histochemistry,* 51, 153, 1977.

27. **Morgan, A. J. and Morris, B.,** The accumulation and intracellular compartmentation of cadmium, lead, zinc, and calcium in two earthworm species (*Dendrobaena rubida* and *Lumbricus rubellus*) living in highly contaminated soil, *Histochemistry,* 75, 269, 1982.

28. **Morgan, J. E. and Morgan, A. J.,** Differences in the accumulated metal concentrations in two epigeic earthworm species (*Lumbricus rubellus* and *Dendrodrilus rubidus*) living in contaminated soils, *Bull. Environ. Contam. Toxicol.,* 47, 296, 1991.

29. **Phillips, D. J. H.,** *Quantitative Aquatic Biological Indicators: Their Use to Monitor Trace Metal and Organochlorine Pollution,* Applied Science Publishers, London, 1980.

30. **Morgan, J. E.,** Earthworms as biological monitors of cadmium and zinc in highly contaminated metalliferous soils, *SEESoil J.,* 5, 24, 1987.

31. **Morgan, J. E., Morgan, A. J., and Corp, N.,** Assessing soil metal pollution with earthworms: indices derived from regression analyses, in *Earthworm Ecotoxicology,* Greig-Smith, P. W., Becker, H., Edwards, P. J., Heimbach, F., Eds., Intercept Ltd., Andover, U.K., 1992, 233.

32. **Ma, W.-C.,** The influence of soil properties and worm-related factors on the concentration of heavy metals in earthworms, *Pedobiologia,* 24, 109, 1982.

33. **Beyer, W. N., Hensler, G., and Moore, J.,** Relation of pH and other soil variables on concentrations of Pb, Zn, Cd, and Se in earthworms, *Pedobiologia,* 30, 167, 1987.

34. **Morgan, J. E. and Morgan, A. J.,** Earthworms as biological monitors of cadmium, copper, lead, and zinc in metalliferous soils, *Environ. Pollut.,* 54, 123, 1988.

35. **Kiewiet, A. T. and Ma, W.-C.,** Effect of pH and calcium on lead and cadmium uptake by earthworms in water, *Ecotoxicol. Environ. Saf.,* 21, 32, 1991.

36. **Corp, N. and Morgan, A. J.,** Accumulation of heavy metals from polluted soils by the earthworm, *Lumbricus rubellus*: can laboratory exposure of 'control' worms reduce biomonitoring problems?, *Environ. Pollut.,* 74, 39, 1991.

37. **Greville, R. W. and Morgan, A. J.,** A comparison of (Pb, Cd, and Zn) accumulation in terrestrial slugs maintained in microcosms: evidence for metal tolerance, *Environ. Pollut.,* 74, 115, 1991.

38. **Greig-Smith, P. W.,** *Recommendations from the International Workshop on Ecotoxicology of Earthworms,* Intercept Ltd., Andover, U.K., 1992.

39. **Andersen, C.,** Cadmium, lead and calcium content, number and biomass in earthworms (Lumbricidae) from sewage sludge treated soil, *Pedobiologia,* 19, 309, 1979.

40. **Ma, W.-C.,** Sublethal toxic effects of copper on growth, reproduction and litter breakdown activity in the earthworm *Lumbricus rubellus,* with observations on the influence of temperature and soil pH, *Environ. Pollut. Ser. A,* 33, 207, 1984.

41. **Chen, S. C., Fitzpatrick, L. C., Goven, A. J., Venables, B. J., and Cooper, E. L.,** Nitroblue tetrazolium dye reduction by earthworm *(Lumbricus terrestris)* coelomocytes: an enzyme assay for nonspecific immunotoxicity of xenobiotics, *Environ. Toxicol. Chem.,* 10, 1037, 1991.

42. **Ireland, M. P. and Richards, K. S.,** Metal content, after exposure to cadmium, of two species of earthworms of known differing calcium metabolic activity, *Environ. Pollut. Ser. A,* 26, 69, 1981.

43. **Morgan, J. E. and Morgan, A. J.,** Zinc sequestration by earthworm (Annelida: Oligochaeta) chloragocytes: An *in vivo* investigation using fully quantitative electron probe X-ray micro-analysis, *Histochemistry,* 90, 405, 1989.

44. **Morgan, J. E. and Morgan, A. J.,** The effect of lead incorporation on the elemental composition of earthworm (Annelida, Oligochaeta) chloragosome granules, *Histochemistry,* 92, 237, 1989.

45. **Richards, K. S. and Ireland, M. P.,** Glycogen-lead relationship in the earthworm *Dendrobaena rubida* from a heavy metal site, *Histochemistry,* 56, 55, 1978.

46. **Morgan, A. J., Gregory, Z. D. E., and Winters, C.,** Responses of the hepatopancreatic 'B' cells of a terrestrial isopod, *Oniscus asellus,* to metals accumulated from a contaminated habitat: a morphometric analysis, *Bull. Environ. Contam. Toxicol.,* 44, 363, 1990.

47. **Hazelhoff-Roelfzema, W., Tohyama, C., Nishimura, H., Nishimura, N., and Morselt, A. F. W.,** Quantitative immunohistochemistry of metallothionein in rat placenta, *Histochemistry,* 90, 365, 1989.

48. **Moore, M. R., Meredith, P. A., and Goldberg, A.,** Lead and haem biosynthesis, in *Lead Toxicity,* Singhal, R. L. and Thomas, B., Eds., Urban and Shwarzenberg, Baltimore, MD, 1980, 79.

49. **Laverack, M. S.,** *The Physiology of Earthworms,* Pergamon Press, Oxford, 1963.

50. **Delkeskamp, E.,** Über den Eisenstoffwechsel bei *Lumbricus terrestris* L., *Z. Vgl. Physiol.,* 48, 400, 1964.

51. **Ireland, M. P. and Fischer, E.,** Effect of Pb^{++} on Fe^{+++} tissue concentration and delta-aminolaevulinic acid dehydratase activity in *Lumbricus terrestris, Acta Biol. Acad. Sci. Hung.,* 29, 395, 1978.

52. **Baxter, C. S., Wey, H. E., and Cardin, A. D.,** Evidence for specific lead-delta-aminolevulinate complex formation by carbon 13 nuclear magnetic spectroscopy, *Toxicol. Appl. Pharmacol.,* 47, 477, 1979.

53. **Hutton, M.,** The effects of environmental lead exposure and *in vitro* zinc on tissue delta-aminolevulinic acid dehydratase in urban pigeons, *Comp. Biochem. Physiol.,* 74C, 441, 1983.

54. **Morgan, A. J.,** The localization of heavy metals in the tissues of terrestrial invertebrates by electron microprobe X-ray analysis, *Scanning Electron Microsc.,* 4, 1847, 1984.

55. **Hopkin, S. P.,** *Ecophysiology of Metals in Terrestrial Invertebrates,* Elsevier Applied Science, Barking, U.K., 1989.

56. **Hopkin, S. P.,** Critical concentrations, pathways of detoxification and cellular ecotoxicology of metals in terrestrial arthropods, *Functional Ecol.,* 4, 321, 1990.

57. **Morgan, A. J. and Winters, C.,** The elemental composition of the chloragosomes of two earthworm species (*Lumbricus terrestris* and *Allolobophora longa*) determined by electron probe X-ray microanalysis of freeze-dried cryosections, *Histochemistry,* 73, 589, 1982.

58. **Winters, C. and Morgan, A. J.,** Quantitative electron probe X-ray microanalysis of lead-sequestering organelles in earthworms: technical appraisal of air-dried smears and air-dried cryosections, *Scanning Electron Microsc.,* 2, 947, 1988.

59. **Jamieson, B. G. M.,** *The Ultrastructure of the Oligochaeta,* Academic Press, New York, 1981.

60. **Prentø, P.,** Metals and phosphate in the chloragosomes of *Lumbricus terrestris* and their possible physiological significance, *Cell Tissue Res.,* 196, 123, 1979.

61. **Morgan, A. J., Roos, N., Morgan, J. E., and Winters, C.,** The subcellular accumulation of toxic heavy metals: qualitative and quantitative X-ray microanalysis, in *Electron Probe Microanalysis. Applications in Biology and Medicine,* Zierold, K. and Hagler, H. K., Eds., Springer-Verlag, New York, 1989, 59.

62. **Markovac, J. and Goldstein, G. W.,** Picomolar concentrations of lead stimulate brain protein kinase C, *Nature,* 334, 71, 1988.

63. **Simons, T. J. B.,** Lead contamination, *Nature,* 337, 514, 1989.

64. **Viarengo, A. and Nicotera, P.,** Possible role of Ca^{2+} in heavy metal cytotoxicity, *Comp. Biochem. Physiol.,* 110C, 81, 1991.

65. **Nicotera, P., McConkey, D. J., Svensson, S. A., Bellomo, G., and Orrenius, S.,** Correlation between cytosolic Ca^{2+} concentration and cytotoxicity in hepatocytes exposed to oxidative stress, *Toxicology,* 52, 55, 1988.

66. **Suzuki, K. T., Yamamura, M., and Mori, T.,** Cadmium-binding proteins induced in the earthworm, *Arch. Environ. Contam. Toxicol.,* 9, 415, 1980.

67. **Morgan, J. E., Norey, C. G., Morgan, A. J., and Kay, J.,** A comparison of the cadmium-binding proteins isolated from the posterior alimentary canal of the earthworms *Dendrodrilus rubidus* and *Lumbricus rubellus*, *Comp. Biochem. Physiol.,* 92C, 15, 1989.

68. **Ballan-Dufrancais, C., Ruste, J., and Jeantet, A. Y.,** Quantitative electron probe microanalysis on insects exposed to mercury. I. Methods. An approach on the molecular form of the stored mercury. Possible occurrence of metallothionein-like proteins, *Biol. Cell,* 39, 317, 1980.

69. **Jeantet, A. Y., Ballan-Dufrancais, C., and Ruste, J.,** Quantitative electron probe microanalysis on insects exposed to mercury. II. Involvement of the lysosomal system in detoxification processes, *Biol. Cell,* 39, 325, 1980.

70. **George, S. G. and Pirie, B. J. S.,** The occurrence of cadmium in sub-cellular particles in the kidney of the marine mussel, *Mytilus edulis*, exposed to cadmium. The use of electron microprobe analysis, *Biochem. Biophys. Acta,* 580, 234, 1979.

71. **Viarengo, A., Pertica, M., Mancinelli, G., and Orunesu, M.,** Possible role of lysosomes in the detoxification of copper in the digestive gland cells of metal-exposed mussels, *Mar. Environ. Res.,* 14, 469, 1984.

72. **Viarengo, A., Moore, M. N., Mancinelli, G., Mazzucotelli, A., Pipe, R. K., and Farrar, S. V.,** Metallothioneins and lysosomes in metal toxicity and accumulation in marine mussels: the effect of cadmium in the presence and absence of phenanthrene, *Mar. Biol.,* 94, 251, 1987.

73. **Dallinger, R. and Prosi, F.,** Heavy metals in the terrestrial isopod *Porcellio scaber* Latr. II. Subcellular fractionation of metal accumulating lysosomes from hepato-pancreas, *Cell Biol. Toxicol.,* 4, 97, 1988.

74. **Prosi, F. and Dallinger, R.,** Heavy metals in the terrestrial isopod *Porcellio scaber* Latr. I. Histochemical and ultrastructural characterization of metal-containing ly-sosomes, *Cell Biol. Toxicol.,* 4, 81, 1988.

75. **Laverjat, S., Ballan-Dufrancais, C., and Wegnez, M.,** Detoxication of cad-mium. Ultrastructural study and electron-probe microanalysis of the midgut in a cadmium-resistant strain of *Drosophila melanogaster, Biol. Metals,* 2, 97, 1989.

76. **Vovelle, J. and Grasset, M.,** Experimental silver bioaccumulation in the poly-chaete *Pomatoceros triqueter* (L.), *Biol. Metals,* 4, 107, 1991.

77. **Otto, E., Young, J. E., and Maroni, G.,** Structure and expression of a tandem duplication of the *Drosophila* metallothionein gene, *Proc. Natl. Acad. Sci. U.S.A.,* 83, 6025, 1986.

78. **Maroni, G., Wise, J., Young, J. E., and Otto, E.,** Metallothionein gene du-plications and metal tolerance in natural populations of *Drosophila melanogaster, Genetics,* 117, 739, 1987.

79. **Klerks, P. L. and Levinton, J. S.,** Rapid evolution of metal resistance in a benthic oligochaete inhabiting a metal-polluted site, *Biol. Bull.,* 176, 135, 1989.

80. **Klerks, P. L. and Bartholomew, P. R.,** Cadmium accumulation and detoxification in a Cd-resistant population of the oligochaete *Limnodrilus hoffmeisteri, Aquat. Toxicol.,* 19, 97, 1991.

81. **Morgan, A. J., Morgan, J. E., and Winters, C.,** Subcellular cadmium seques-tration by the chloragocytes of earthworms living in highly contaminated soil, *Mar. Environ. Res.,* 28, 221, 1989.

82. **Morgan, A. J.,** *X-ray Microanalysis in Electron Microscopy for Biologists,* Oxford University Press, U.K., 1985.

83. **George, S. G.,** Heavy metal detoxification in the mussel *Mytilus edulis* — com-position of Cd-containing kidney granules (tertiary lysosomes), *Comp. Biochem. Physiol.,* 96C, 53, 1983.

84. **George, S. G.,** Heavy metal detoxication in *Mytilus* kidney — an *in vitro* study of Cd- and Zn-binding to isolated tertiary lysosomes, *Comp. Biochem. Physiol.,* 76C, 59, 1983.

85. **Hamer, D. H.,** Metallothionein, *Annu. Rev. Biochem.,* 55, 913, 1986.

86. **Goering, P. L. and Fowler, B. A.,** Activation of δ-aminolevulinic acid dehy-dratase following donation of zinc from kidney metallothionein, in *Metallothionein. Proc. 2nd Int. Meet. Metallothionein and Other Low Mol. Weight Metal-Binding Proteins,* Kägi, J. H. R. and Kojima, Y., Eds., Birkhäuser, Verlag, Boston, MA, 1987.

87. **Sunderman, F. W.,** Cadmium substitution for zinc in finger-loop domains of gene-regulating proteins as a possible mechanism for the genotoxicity and carcinogenicity of cadmium compounds, *Toxicol. Environ. Chem.,* 27, 131, 1990.

88. **Klug, A. and Rhodes, D.,** "Zinc fingers": a novel protein motif for nucleic acid recognition, *T.I.B.S.,* 12, 464, 1987.

89. **Evans, R. M. and Hollenberg, S. M.,** Zinc fingers: guilt by association, *Cell,* 52, 1, 1988.

90. **Neuhauser, E. F., Malecki, M. R., and Loehr, R. C.,** Growth and reproduction of the earthworm *Eisenia foetida* after exposure to sublethal concentrations of metals, *Pedobiologia,* 27, 89, 1984.
91. **Bremner, I.,** Factors influencing the occurrence of copper thioneins in tissues, in *Metallothionein,* Kägi, J. H. R. and Nordberg, M., Eds., Birkhäuser-Verlag, Basel, 1979, 273.

CHAPTER 18

Deficiency and Excess of Copper in Terrestrial Isopods

Stephen P. Hopkin

TABLE OF CONTENTS

0-87371-734-1/93/$0.00 + $.50

I. INTRODUCTION

The importance of essential and nonessential metals in the biology of terrestrial isopods (woodlice) has fascinated researchers for the past 30 years. Wieser[1] first drew attention to the remarkable ability of woodlice to accumulate copper. Since then, terrestrial isopods have been shown to store other metals such as zinc, cadmium, and lead at extremely high concentrations, especially in metal-contaminated sites.[2,3]

The importance of Wieser and colleagues' theories is often overlooked. While many of the suggestions as to the importance of coprophagy for copper balance[4,5] have been modified in the light of recent research, these ideas stimulated many researchers to examine essential trace elements in invertebrates at a time when most work concentrated on metals as pollutants rather than potentially limiting nutrients.

The purpose of this review is to reexamine these theories in the light of research on the nutrition of terrestrial isopods which has been conducted in the last few years. Four main areas will be covered. First, the role of copper in isopod nutrition and the disputed significance of coprophagy (Section II). Second, the influence of the structure and function of the digestive system on uptake and loss of metals (Section III). Third, copper requirements of woodlice, and an examination of whether they are likely to suffer from copper deficiency in the wild (Section IV). Fourth, to ask whether isopods are ever poisoned by excessive intake of copper in polluted sites (Section V). The concluding remarks (Section VI) include suggestions for further research.

II. COPPER AND THE COLONIZATION OF THE LAND

A. Copper as an Essential Nutrient

Copper is an essential component of the diet of all animals. There are at least 12 major proteins that require copper as an integral part of their structure. These include the respiratory enzyme cytochrome oxidase, without which no animal can survive.[6]

Most crustaceans (and molluscs) possess hemocyanin as their main oxygen-carrying blood protein. This doubles their need for copper, in comparison to invertebrates which do not contain hemocyanin.[7] Terrestrial isopods must extract sufficient quantities of this metal from their food to supply these needs. Estimates of their copper requirements have been made and are described in more detail in Section IV.

B. Colonization of the Land

Terrestrial isopods constitute the Suborder Oniscidea which is almost certainly a monophyletic group.[8] Although the earliest fossil Oniscidea are from the Quaternary, consideration of the distribution of present-day taxa in the light of theories of continental drift, makes it likely that they first appeared in the Carboniferous over 300 million years ago.[9]

The first isopods to colonize the land were probably similar to present-day species of *Ligia*. These possess several 'primitive' features including antennal flagella with numerous segments, an open water-conducting system, and a hepatopancreas with six lobes instead of the four lobes of more 'advanced' oniscideans.[10,11]

Seawater in uncontaminated coastal areas contains about 10 μg l^{-1} copper.[12] Marine isopods can assimilate this copper directly across their respiratory epithelia. This route is of course not available to terrestrial isopods, which have to obtain all their copper requirements from the food.

Marine Crustacea are able to permanently reduce the rate at which copper is assimilated by the gut. This guards against a sudden influx of copper following ingestion of food containing high concentrations.[13] This strategy can not be adopted by woodlice as it would render them vulnerable to deficiency. However, maximizing the permeability of the gut to copper may lead to excessive uptake and eventual copper poisoning. The ways in which woodlice obtain enough copper, but detoxify that which is assimilated in excess of requirements, are described in the remaining parts of this review.

C. Feeding Strategies

A glance at the leaf litter layer of a mature woodland might lead to the belief that food for terrestrial isopods and other primary decomposers is abundant. However, much of this material is unpalatable. It is either too tough, or contains high levels of secondary defensive chemicals. These substances repel potential consumers until sufficient time has elapsed for them to break down.[14]

Numerous studies have shown that woodlice, and other terrestrial arthropods, can discriminate quite subtle differences between similar foods.[15-17] *Porcellio scaber*, for example, can discern the difference between 'sun' and 'shade' leaves from the same tree,[18] and avoid ingesting leaves contaminated with lead salts.[19]

Woodlice are attracted strongly to leaves on which large numbers of fungal hyphae are growing. When presented with a heavily infected leaf, they always eat the hyphae before tackling the plant tissue.[20-22] Collembola detect the odor of fungus[23] and will migrate toward the surface of the litter layer where fungal growth is at a maximum after heavy rain.[24] Woodlice are probably attracted toward fungus by following concentration gradients of volatile chemicals given off by the hyphae.

Most early invertebrate colonizers of the land relied on microorganisms to supply digestive enzymes for the breakdown of plant tissues.[25] Microorganisms may perform this role in woodlice, although isopods do not contain a permanent symbiotic gut microflora.[26] Other early colonizers such as Collembola, relied on fungal hyphae to supply them with a source of nutrients that was digested more easily than plant litter.

The earliest terrestrial habitat to be colonized by woodlice was probably moist litter of pteridophytes on base-rich soils.[27] A major proportion of their diet would have been fungal hyphae,[25] which grow in profusion on dead leaves of horsetails and ferns.[27] Although isopod feces are poorer nutritionally than the food from which they are derived,[28] deposition of the pellets in the moist depths of the litter probably stimulated the growth of hyphae on which they could graze and in which nutrients were more readily available.[29] Fungal grazing may have been especially important before the full development of the hindgut typhlosole (see Section IV,A) which allows more effective circulation of digestive fluids in the digestive system.[10]

Woodlice have flexible feeding strategies. If food is of 'poor' quality, or is in short supply, it is retained in the gut for a long period to allow the maximum amounts of nutrients to be extracted. If, on the other hand, there is abundant food of 'high' quality, woodlice become hyperphagous and pass material through the gut as rapidly as possible.[14]

Of course, assimilation rates of certain substances will differ markedly from assimilation measured by subtracting the weight of feces from the weight of food consumed.[30] Some nutrients such as simple carbohydrates (and perhaps copper) may be extracted with efficiencies approaching 100%, whereas others (e.g., cellulose) may hardly be assimilated at all. Thus woodlice could obtain more of these easily assimilated substances by passing food through the gut as rapidly as possible. Individual *Oniscus asellus* are able to consume up to 30% of their dry body weight in leaves of field maple *(Acer campestre)* on a dry weight basis every 24 h. This represents filling and emptying of the gut some six times. However, in overall terms (weight of food consumed/weight of feces voided), experiments have shown that the *weight* of ingested food which is assimilated is remarkably consistent.[22]

Copper has a very high affinity for organic matter and is bound tightly within plant tissue.[31] Furthermore, concentrations in leaf litter of angiosperms are often <10 μg Cu g^{-1} dry weight.[6] In lower plants (the litter of which woodlice would have been consuming during their early evolution), levels of 5 μg Cu g^{-1} dry weight are typical.[32,33] However, some simple calculations (given in detail in Section IV) show that even if woodlice consumed food at the maximum rate possible, and extracted 100% of the copper, leaf litter alone would not supply their copper needs.

The answer to this apparent contradiction lies in their choice of food. Woodlice ingest fungal hyphae in which concentrations of copper are very much higher than the leaves on which they are growing (Table 1). Indeed, concentrations of copper in fungus in excess of 1% of the dry weight have been recorded on numerous occasions.[34-38,80] Fungi isolated from contaminated soils can even be used to extract metals from industrial effluents.[39] We shall return to this topic in Section IV,B.

D. Coprophagy

Wieser[5] argued that during periods of hyperphagy, food passed through the gut so rapidly that the minimum copper requirement was not extracted from the food. Unless the isopods could find some other source of copper, they would become deficient. This source, he suggested, was the feces. Copper was made more available than in the original food due to the activities of microorganisms.[5]

Wieser's theory rested on the assumption that the only source of copper to woodlice is leaf litter. It is clear from the above discussion, and the data presented in Table 1, that isopods can supply all their copper needs by grazing fungal hyphae from the surfaces of leaves. They need to consume leaf litter to supply their major nutrients, but relatively small quantities of fungi can satisfy their trace element requirements, and possibly those for other essential trace substances. In other words, fungal hyphae are the woodlouse equivalent of 'vitamin pills'!

Some invertebrates must eat their own feces if they are to grow and reproduce successfully (e.g., certain species of millipede[40]). So, is it essential for woodlice to consume their own feces? When the only food available is of poor quality, consumption of feces may improve growth and reproduction.[41,42] However, if fed on shredded carrot root, *Porcellio scaber* grows faster than when fed on recently fallen leaf litter and their own feces. Hassall and Rushton[41] concluded that enhanced microbial activity in the feces increases the nutrient status is such a way that some coprophagy is necessary in order to optimize overall nutrient uptake of woodlice. They suggested also that the ability to vary the extent to which feces are recycled in response to differences in food quality is important in that it introduces greater flexibility into the feeding strategies of terrestrial isopods.

Several observations have supported the hypothesis that isopods are attracted to fungal hyphae growing on the feces rather than to the material within the

Table 1. Comparison of Uncontaminated and Cu-Contaminated Leaves of Field Maple (*Acer campestre*) Collected From the Litter Layer of a Site in Reading and Maintained for 1 Month at 25°C at 100% Relative Humidity in Complete Darkness

	Leaf			Fungus		
	Weight (mg)	Cu (µg)	Cu (µg g^{-1})	Weight (mg)	Cu (µg)	Cu (µg g^{-1})
Control (n = 5)						
Mean	61.6	0.65	10.5	0.057	0.037	610
Range	(54.1–73.0)	(0.51–0.82)	(8.2–14.8)	(0.029–0.093)	(0.010–0.060)	(212–916)
S.D.	6.7	0.11	2.2	0.021	0.021	246
Cu-treated (n = 5)						
Mean	64.6	6.9	108	0.149	0.51	3960
Range	(52.3–80.9)	(6.1–7.9)	(98–118)	(0.099–0.209)	(0.20–0.76)	(956–6720)
S.D.	11.4	0.66	9	0.039	0.18	2080

Note: Five leaves were untreated (Control). A solution of copper nitrate (Cu(NO$_3$)$_2$·3H$_2$O) was applied topically as tiny droplets to the upper surfaces of another five leaves (Copper-treated), which were then air dried to give a whole leaf concentration of approximately 100 µg Cu g^{-1} dry weight. All values are given on a dry weight basis (S.D. = standard deviation). At the end of the experiment, a rich growth of white fungal hyphae (unidentified) was present on the surfaces of all the leaves. This was more dense on the treated leaves, due probably to the stimulatory effect of the nitrate in the applied solution. The fungus was scraped from the individual leaves onto preweighed pieces of Millipore filter paper using forceps. Nitric acid digests of all samples were analyzed by flame or flameless atomic absorption spectrometry. For further details of analytical techniques, see Ref. 2 and 46.

pellet. Woodlice prefer to consume feces of two to three weeks in age, on which fungal growth is at a maximum.[43] When access to feces is restricted, consumption of fungal hyphae on leaves increases.[44] Fungal populations on leaf litter are reduced considerably after passage of food through the gut, whereas numbers of bacteria increase by two orders of magnitude.[45]

Thus, if isopods are prevented from eating fungal hyphae, *and* prevented from consuming their own feces, it might be possible to induce copper deficiency (and deficiency of other micronutrients) in the laboratory. In consequence, they should grow more slowly than a population with access to fungus.

Of course the copper contained in the fungus is derived from the leaf on which it is growing. If the isopod is forced to eat the whole leaf before a new one is supplied, consumption of fungus from the surface will not increase the net amount of copper ingested. Thus, isopods must graze fungal hyphae from the surfaces of a much larger number of leaves than they will eventually eat.

If woodlice are forced to consume all of a leaf before they are supplied with a new one, then they may become copper-deficient. Such experiments should be performed using juveniles, which do not possess significant reserves of copper in the hepatopancreas. If adults are maintained in containers with a large excess of food, concentrations of copper in the hepatopancreas rise rapidly, due almost certainly to increased ingestion of fungal hyphae in which copper levels are high.[46]

Consumption of fecal material can not provide more copper than consumption of the leaf litter from which it is derived. However, there may be periods when fungus growing on the feces extracts copper so that concentrations in the hyphae are higher than in the material in the pellets. These hyphae could provide a copper-rich food source for isopods to graze when all the fungus has been grazed from leaf litter. However, it has now been accepted that consumption of entire fecal pellets is not *essential* for the maintenance of adequate reserves of copper in woodlice.[41,47]

III. THE IMPORTANCE OF THE DIGESTIVE SYSTEM

A. Structure and Function of the Gut

The digestive system of terrestrial isopods has been studied extensively (for reviews see Hopkin[2] and Hopkin and Martin[3]). It consists essentially of a straight tube lined with cuticle from mouth to anus, and a four-lobed hepatopancreas (six in *Ligia*) which secretes enzymes and absorbs products of digestion. The hepatopancreas is connected to a complicated sorting and filtering stomach at the anterior end of the hindgut.[48] The dorsal wall of the hindgut is folded to form a typhlosole and a pair of channels. However, it is only recently that the detailed structure and function of the component parts of the gut have been elucidated.[10]

One of the early suggestions of Wieser to support the essentiality of copro-phagy was that once food was posterior to the openings of the hepatopancreas, the only way in which nutrients released by subsequent digestion could enter the hepatopancreas was by reingestion of feces. It is now known that products of digestion are carried *anteriorly* along the typhlosole channels from the posterior ('papillate') region of the gut into the lumen of the hepatopancreas via the stomach.[10] Thus, whether or not coprophagy is required for other reasons, it is not necessary for absorption of nutrients released by digestive enzymes posterior to the openings of the hepatopancreas since there is effective mixing of fluids throughout the lumen of the digestive system.[10]

The role of bacteria in digestion has still to be elucidated. These microorganisms do not seem to supply much in the way of nutrients, but may be important in producing enzymes which isopods can not manufacture themselves.[49] When the gut is full, there are large populations of bacteria in the lumen of the hindgut and hepatopancreas.[50,51] However, when the gut is empty, bacteria are few and there is no evidence for the presence of a permanent symbiotic microflora in the lumen.[52] Fungal hyphae are digested in the gut but the spores are more resistant and may germinate in the rectum prior to voiding of a fecal pellet.[10]

B. The Hepatopancreas

1. Ultrastructure

The epithelium of the hepatopancreas is one cell in thickness and consists of two cell types. The large 'B' cells are arranged evenly throughout the length of the tubules and are each surrounded by the smaller 'S' cells. At the blind end, there is a region of cells which differentiate into new B and S cells as the woodlouse grows and the tubules of the hepatopancreas increase in length.

The B cells undergo profound changes in their ultrastructural appearance during a digestive cycle. At the beginning of a digestive cycle, the apical contents of each B cell are voided into the lumen of the hepatopancreas. This material contains digestive enzymes which are forced into the gut by contraction of the hepatopancreas. This material contains digestive enzymes which are forced into the gut by contraction of the hepatopancreas. Subsequently, fluids are returned to the hepatopancreas via the typhlosole channels and the B cells swell in size as they accumulate products of digestion.[53]

The S cells, in contrast, do not change their appearance during a digestive cycle. This difference in the responses of B and S cells to feeding (following starvation) has also been observed by Storch.[54]

2. Pathways of Metal Accumulation

Material in the food must pass through S and B cells (or across the cuticular lining of the hindgut) before entering the blood. Thus, there is the opportunity for these cells to 'filter out' unwanted metals such as cadmium and lead, or essential metals which are in excess of requirements that include zinc, copper, and iron. Material in the cells of the epithelium of the hindgut may not necessarily have passed through the cuticular lining, but may have arrived there from the basal regions of the hepatopancreas via the blood.[55] The large amounts of iron detected in the hindgut epithelia of *Porcellio scaber* and *Oniscus asellus* in the feeding experiments of Hopkin[46] may have arrived via such a route.

The B and S cells in the hepatopancreas accumulate metals in modified lysosomes which are formed into residual bodies.[56,57] These are quite variable in hardness and appearance in thin section and are usually referred to as 'granules'.[2] When homogenates of the hepatopancreas are centrifuged, delicate granules may rupture. Their metal-rich contents may then occur in the supernatant rather than the pellet.[58]

The granules in the S cells of isopods from uncontaminated areas always contain predominantly copper and sulfur together with a much smaller amount of calcium (Figure 1a). If the food contains cadmium, this is stored in the S cell granules and appears to substitute for copper (Figure 1b), probably during their formation. Zinc and lead may also occur in these granules, but if present, are usually found in diffuse deposits associated with phosphorus. This material may surround existing granules, be dispersed throughout the cell, or be associated with the membranes.[59,60]

The granules in the B cells always contain iron,[61] probably as hemosiderin. In contaminated woodlice, zinc and lead may also be present but cadmium and copper have not so far been detected in B cells.[59,60]

Thus, there appear to be three main pathways of metal accumulation into the cells of the hepatopancreas of terrestrial isopods.[59,60] In the 'type A' pathway, zinc and/or lead enter B or S cells, are bound to phosphate, and are precipitated throughout the cytoplasm, or in and around the periphery of existing granules of other types. In the 'type B' pathway, cadmium and copper are bound to a sulfur-containing ligand (possibly metallothionein) and deposited in granules in the S cells. In the 'type C' pathway, iron assimilated in excess of requirements is deposited as hemosiderin in granules in the B cells.

Evidence from studies on vertebrates suggests that assimilation of zinc and iron into cells is regulated metabolically and involves specific carrier molecules.[62-64] In contrast, copper transport is strictly passive. A variety of metabolic inhibitors have no effect on uptake rates of copper into rat hepatocytes, which are dependent solely on extracellular-intracellular concentration gradients.[65]

If this is true for terrestrial isopods, storage of copper in granules may maintain a permanent concentration gradient between the S cells and the digestive fluids and result in all copper released from the food being assimilated. Once inside the cell, the copper is bound to a storage protein from which some is released into the blood for essential metabolic requirements (hemocyanin, etc.), the remainder being deposited in granules. These intracellular metal-binding proteins may be metallothioneins. However, intracellular binding of metals in isopods is an area of active research and it is not possible at present to provide a definitive description of the precise pathways involved.

The storage of copper and cadmium in insoluble granules in the S cells explains why contaminated isopods do not readily lose these metals when fed

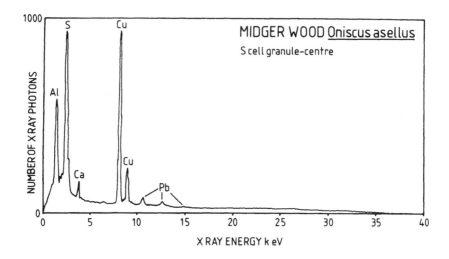

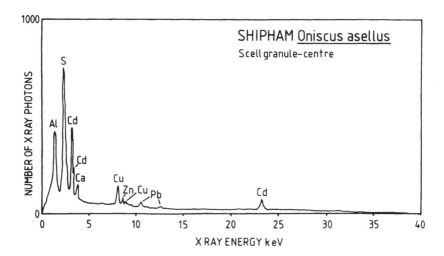

FIGURE 1. X-ray microanalytical spectra of the center of 'type B' granules in the hepatopancreas of *Oniscus asellus* collected from an uncontaminated site (A, Midger Wood) and the spoil tips of a disused zinc/cadmium mine (B, Shipham). (For further details of the analytical techniques employed, see References 2 and 60).

an uncontaminated diet.[46] It has been suggested[52,66] that the greater rate of excretion of zinc by *Oniscus asellus* compared to *Porcellio scaber* is because *P. scaber* stores a greater proportion of assimilated zinc in the S cells than *O. asellus*. Metals stored in B cells would be lost more rapidly due to the rapid turnover of cytoplasm. Studies using autoradiography are in progress to examine this possibility.

Depledge and Rainbow[67] have argued that rather than define 'regulators' or 'nonregulators', it is more useful to define the extent to which organisms use (a) uptake impairment, (b) enhanced excretion, and (c) intraorganism storage of metals in an insoluble form when exposed to elevated concentrations in the environment or food. Terrestrial isopods appear to rely almost entirely on option (c), using storage-detoxification in the hepatopancreas to regulate concentrations in the other tissues.

IV. DO ISOPODS SUFFER FROM COPPER DEFICIENCY?

A. Copper Requirements

Carefoot[68] could find no significant differences in growth and survival of *Ligia pallasii* fed on an artificial diet containing 4 μg Cu g^{-1}, and diets containing 8.5 or 17.5 μg Cu g^{-1}. However, the concentrations of copper in the animals were not determined at the end of the experiment. Since adults were used, it is possible that copper was mobilized from the hepatopancreas to supplement any deficiencies from the hepatopancreas to supplement any deficiencies in the diet. It would be interesting to repeat such experiments using woodlice released recently from a brood pouch. These juveniles will not have had the opportunity to accumulate significant reserves of copper in the hepatopancreas.

Depledge[7] has calculated that Crustacea require a whole body concentration of at least 82.8 μg g^{-1} dry weight (including exoskeleton) to satisfy their copper requirements for hemocyanin and other proteins. While extrapolation of his proposals to terrestrial isopods is open to criticism, it is clear from Figure 2 that *Oniscus asellus* regulates the concentration of copper in tissues other than the gut and hepatopancreas (the 'rest') at a similar level (about 50 μg g^{-1}). This is achieved by storage of excess copper in the hepatopancreas. Even when the hepatopancreas contains more than 1% copper (the most contaminated isopods from Caradon), the concentrations in the rest are maintained below 100 μg g^{-1}.

The rest fraction includes the exoskeleton which does not contain detectable amounts of copper when molted.[46] Therefore, tissues other than the hepatopancreas and gut probably require about 100 μg Cu g^{-1} dry weight. We would not expect a terrestrial isopod to be vulnerable to copper deficiency unless its hepatopancreas had a concentration of less than 100 μg g^{-1}. Is there any evidence that wild populations of woodlice contain individuals that are copper deficient?

B. Copper Deficiency in the Wild

If we assume that terrestrial isopods need a minimum whole body concentration of 50 μg Cu g^{-1} dry weight (including exoskeleton), it is possible to perform some simple calculations to determine how much food they need to eat to satisfy this requirement.

A woodlouse of 20 mg dry weight would need to contain 1 μg of copper to have a whole body concentration of 50 μg g^{-1}. Uncontaminated leaf litter

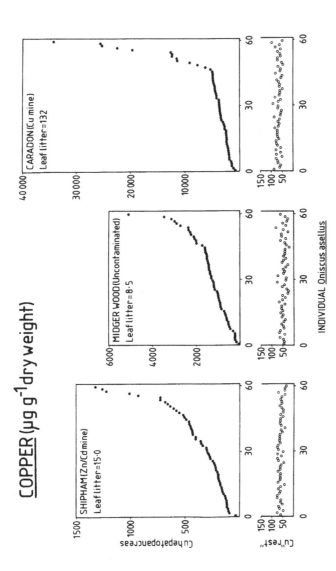

FIGURE 2. Concentrations of copper (μg g⁻¹ dry weight) in the hepatopancreas of individual *Oniscus asellus* from three sites in southwestern England (n = 60 for each site, using data originally collected for Reference 76). Individuals are ranked in order of increasing copper concentration in the hepatopancreas (●). The highest concentrations recorded in the hindgut (not shown) were 78 μg g⁻¹ (Shipham), 182 μg g⁻¹ (Midger) and 402 μg g⁻¹ (Caradon). Concentrations in tissues other than the hindgut and hepatopancreas (the 'rest' (○) — including exoskeleton) are regulated at about 50 μg g⁻¹, irrespective of the concentration in the hepatopancreas.

typically contains 10 μg Cu g^{-1} dry weight (Table 1). Assuming that woodlice can extract 100% of the copper from leaves, and that it took the isopod 200 days to grow to 20 mg dry weight, it would have to eat 100 mg of leaf to obtain 1 μg of copper. This represents a consumption rate of 5% of its dry body weight in dry leaf per day and is equivalent to filling and emptying the gut once every 24 h.

However, isopods are unlikely to be able to extract all the copper from leaf material. In addition, there are areas in which leaf litter has considerably less than 10 μg Cu g^{-1}. In a study where samples of leaf litter and *Porcellio scaber* were collected from 89 sites in Southwestern England,[69] there were 10 sites where the concentration of copper in leaf litter was below 5 μg g^{-1} dry weight. At none of these sites was the mean concentrations of copper in whole *Porcellio scaber* less than 100 μg g^{-1}.

Furthermore, when individuals are analyzed in uncontaminated sites, concentrations of copper in the hepatopancreas may reach several thousand μg g^{-1} (Midger Wood, Figure 2). It is clear that these individuals could not have obtained such large amounts of copper from eating leaf litter alone. Coprophagy would not provide the answer since the feces are poorer nutritionally than the leaf litter from which they are derived[70] and must contain even less copper. The most probable source of this copper is fungal hyphae. Indeed, Table 1 shows that in an uncontaminated site, they would only have to eat about 1.6 mg of hyphae to obtain 1 μg of copper, whereas 95 mg of leaf would have to be eaten to obtain the same amount.

If copper-deficient isopods do exist in the wild, this is most likely to be detected by analysis of individuals, a fact that is often overlooked.[71] There were two *Oniscus asellus* from Midger Wood in which the concentrations of copper in the hepatopancreas were less than 200 μg g^{-1} (Figure 3), one with 162 and the other with only 56 μg Cu g^{-1}. The latter animal had no surplus copper if we accept Depledge's estimate[71] for the requirements of Crustacea. This individual may have been very near to being deficient. Thus, individual woodlice in Midger Wood (and by implication many other uncontaminated sites) may be dying prematurely from copper deficiency although of course this is impossible to prove without further detailed field and laboratory experiments.

C. Antagonism

Where in the wild are copper-deficient populations of isopods likely to exist? Possible sites are regions where concentrations of copper in the soil have been shown to be below the critical level for successful grazing of livestock.[72] Midger Wood is an oak-hazel woodland on calcareous soil where a pooled sample of leaf litter contains only 8.5 μg Cu g^{-1} dry weight (Figure 2). The woodlice were collected from under stones and there were apparently healthy populations of isopods at this site. If copper-deficient isopods were occurring in Midger Wood, their early mortality did not seem to be having a significant effect on the population. However, there may be areas yet to be recognized where isopods are absent or scarce through limited availability of copper.

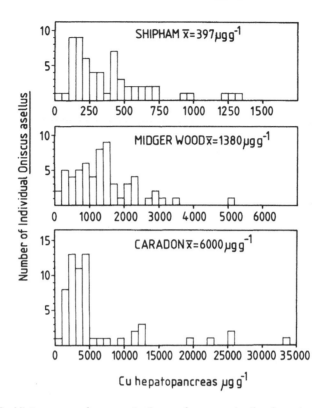

FIGURE 3. Histograms of concentrations of copper in the hepatopancreas of individual *Oniscus asellus* from three sites (same data as Figure 2). Note the very large individual variation which has important implications for predators (see text).

One other possibility is that a substance present in the food of woodlice may compete with sites of copper uptake in the hepatopancreas. Of the 89 sites surveyed by Hopkin et al.,[69] there was only one site in which the mean concentration of copper in whole *Porcellio scaber* was significantly less than 100 $\mu g\ g^{-1}$. This was on the spoil tips of a disused zinc mine at Shipham, North Somerset, where concentrations of cadmium and zinc in the soils are among the highest in the world.[73]

Numerous studies on vertebrates have shown that a high level of cadmium in the diet depresses copper uptake. Zinc may also be a copper antagonist.[65] In the light of these observations, it is interesting to note the large number of *Oniscus asellus* from Shipham which had very low concentrations of copper in the hepatopancreas, relative to uncontaminated animals from Midger Wood. One possible explanation for this phenomenon is that the high levels of zinc and cadmium in leaf litter at this site (2250 and 23.7 $\mu g\ g^{-1}$ dry weight, respectively) were inhibiting the assimilation of copper into the hepatopancreas (Figure 4).

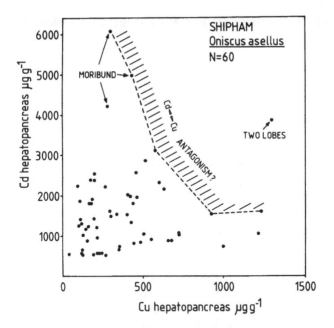

FIGURE 4. Plot of concentration of cadmium against concentration of copper in the hepatopancreas of individual *Oniscus asellus* (n = 60) from Shipham (disused zinc/cadmium mine). These results suggest that cadmium (and possibly zinc — not shown, but also very high) may be competing with copper at uptake sites in the hepatopancreas. The hepatopancreas of one isopod had only two lobes instead of the usual four, but it is not known why this individual had such a relatively high concentration of copper.

The moribund animals with concentrations of 4960, 4190, and 6080 μg Cd g^{-1} in the hepatopancreas (Figure 4) had 19,200, 21,400, and 19,778 μg Zn g^{-1}, respectively. These three individuals were among those woodlice which had the lowest concentrations of copper in the hepatopancreas (Figure 4). This high Cd:Cu ratio was reflected in the composition of the type B granules in the S cells of the hepatopancreas. The granules were extremely rich in cadmium (and poor in copper) in comparison to isopods from Midger Wood (Figure 1).

Most of the zinc in the hepatopancreas of isopods from Shipham was associated with phosphorus in 'type A' material surrounding existing type B granules in the S cells, mixed with the iron-rich type C granules in the B cells, and dispersed throughout the cytoplasm and on the membranes of both cell types.

Woodlice near the Avonmouth zinc, lead, and cadmium smelting works to the north of Shipham are unlikely to suffer from copper deficiency. Despite being a major source of cadmium and zinc,[69,74] the factory also emits substantial amounts of copper which supply the needs of the isopods many times over.

V. DO ISOPODS SUFFER FROM COPPER POISONING?

Caradon is a site contaminated heavily with copper from past mining activity. The concentration of 132 μg copper g^{-1} in leaf litter exceeds the 'critical concentration' of Bengtsson and Tranvik[75] of 100 μg g^{-1}, the 'maximum allowable level' above which adverse effects on the biota are likely to occur. *Oniscus asellus* from this site contain substantially more copper than those from Midger (Figures 2 and 3) although populations at the disused mine were apparently healthy.

Caradon woodlice are unlikely to suffer from copper deficiency! However, there were ten individuals with more than 1% copper in the hepatopancreas and one with more than 3% (Figure 3). These isopods showed no sign of being moribund. However, the possibility can not be excluded that some individuals in this population were dying when detoxification sites in the hepatopancreas became saturated.

In laboratory experiments, isopods are reluctant to ingest food which is heavily contaminated with copper salts.[41] They may die through starvation rather than excessive copper assimilation if the food contains more than 200 μg Cu g^{-1} dry weight. The results presented in Table 1 show that their reluctance to feed may be due to an aversion to fungal hyphae on leaf surfaces in which the concentration of copper may exceed 0.5% on a dry weight basis.

Again, the importance of analyzing individuals needs to be stressed. For example, there is overlap between Midger and Caradon populations in the concentrations of copper in the hepatopancreas. Despite highly significant differences between the mean concentrations,[76] 13 isopods from Midger had more than 2000 μg Cu g^{-1} in the hepatopancreas whereas 9 woodlice from Caradon had less than 2000 μg g^{-1}.

Thus a predator of woodlice such as the spider *Dysdera crocata*,[60] during feeding on a single woodlouse, could ingest less copper at Caradon if it attacked one of the least-contaminated individuals than at Midger if it consumed one of the most-contaminated isopods. However, if the distribution of *D. crocata* is limited by copper contamination of its food, this will be by ingestion of the most contaminated individuals. The nature of the digestive system and feeding methods of the spider render it vulnerable to acute rather than chronic poisoning. Most of the *O. asellus* that *Dysdera* consumes at Caradon have less than 5000 μg copper g^{-1} in the hepatopancreas, but 1 in 60 isopods has over 30,000 μg g^{-1}.

VI. CONCLUDING REMARKS

A. 'Assimilation' and 'Concentration' of Metals

When discussing the concepts of deficiency and excess of essential elements, it is important to define clearly the terms *assimilation* and *concentration*. An organism exhibits net assimilation of copper if the number of atoms in its tissues

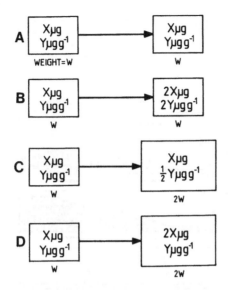

FIGURE 5. Schematic diagrams showing the relationships between the weight of an organism (W), the amount of metal it contains (μg) and the concentration (μg g^{-1}) between the start (left boxes) and end (right boxes) of a period of time. A: No change in parameters. B: Metal content doubles (2X), weight stays the same (W) so concentration also doubles (2Y), possibly resulting in toxicity symptoms if the critical level = 2Y. C: Metal content unchanged (X), weight doubles (2W) so concentration is halved (½Y), possibly resulting in deficiency symptoms if minimum tissue requirement = Y. D: Metal content doubles (2X), weight doubles (2W) so concentration remains the same (Y). This model can also be applied to individual organs, or cells within organs.

increases during a period of time. However, the *concentration* may fall, stay the same, or rise over the same period, depending on whether the weight decreases, remains constant, or increases. The implications of this concept are demonstrated schematically in Figure 5.

For example, the isopods at Caradon can be split into two groups, those with less than, and those with more than 5000 μg Cu g^{-1} in the hepatopancreas (Figure 2). Hopkin and Martin[76] suggested that these two groups might represent different year classes, with the more contaminated isopods being those in their second year of growth. Similar arguments have been presented to explain why older isopods in sites contaminated with zinc are more vulnerable to zinc poisoning than younger ones.[46]

The model shown in Figure 5 can also be applied to organs within organisms. However, the overall response of an organism to a metal 'insult', in terms of changes in concentrations, can disguise variations in responses of organs and cells (shown schematically in Figure 6). Thus, response B (Figure 6) may be

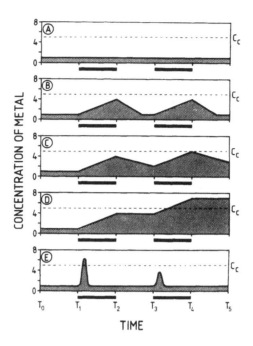

FIGURE 6. Schematic diagrams showing possible responses of organisms to two periods of dietary exposure to a metal (T_1-T_2, T_3-T_4) in terms of concentration measured on six occasions (T_0-T_5). *Response A* — no change. *Response B* — increase in concentration during exposure (T_1-T_2) but return to preexposure level before the subsequent exposure period (T_3-T_4). *Response C* — increase in concentration during exposure (T_1-T_2) but insufficient length of recovery period (T_2-T_3) for concentration to return to preexposure level. Subsequent exposure for a second time (T_3-T_4) results in the critical concentration (C_c) being reached at T_4. *Response D* — no decrease in concentration after exposure period so critical concentration is exceeded before the end of the second exposure period. *Response E* — initial exposure (T_1-T_2) 'switches on' detoxification mechanisms which are able to regulate the concentration at a lower level during subsequent exposure. This model can also be applied to individual organs or cells within organs (e.g., S and B cells of isopods — see text).

followed by zinc in the B cells of the hepatopancreas of *O. asellus* since the metal is probably assimilated and lost during each digestive cycle.[53,66] Response D is probably followed by zinc in the S cells. Thus the overall response of the hepatopancreas is a stepwise increase in loading represented by response C. For elements such as cadmium and copper, which apparently are accumulated only in the S cells, the B cells may follow response A and the S cells (and the hepatopancreas and isopod as a whole) response D.

Response E is a special case where concentration of a metal above a certain critical level (C_c) switches on a detoxification mechanism. This mechanism is able to prevent such rapid uptake on a second insult, or engage carrier systems that remove the element at a faster rate so that concentrations do not reach such high values. This may be the switching on of a gene that codes for a detoxifying protein. There is evidence for genetically based tolerance to cadmium in isopods,[77] but further research is required before the exact molecular basis of this can be established.

B. Suggestions for Further Research

Further work is required on the role of microorganisms in the feeding ecology of terrestrial isopods. It is still not clear what many species eat. In polluted sites, isopods (and other fungivores[78]) are clearly being exposed to much greater concentrations of metals in their diet than analysis of pooled samples of leaf litter would suggest (Table 1).

Although the concentrations of metals have been measured in the sporophores of a wide range of fungi, and have been shown to be comparable to those in leaf litter,[79] there is very little information on the levels in hyphae before the fruiting body is produced. During the experiments of Hopkin,[46] in which isopods were confined to plastic tanks containing leaf litter, sporophores were never seen. Presumably the grazing activities of the isopods prevented the hyphae from producing a fruiting body. However, after removal of all the woodlice, a thick white growth of hyphae developed on the leaves, and toadstools (unidentified) would often appear in the tanks in 24 to 48 h. The reason for the hyperaccumulation of copper in fungal hyphae of basiodomycetes growing on uncontaminated leaves (Table 1) now becomes clear. Apparently, the hyphae need such relatively large amounts of copper since production of the fruiting body involves massive growth-dilution of nutrients. If copper (and presumably other essential elements) were not stored, the fungus would become copper-deficient. This is clearly an area in need of further research.[80]

There are at least 7000 species of terrestrial isopods in the world, only half of which have been described.[8] However, only a tiny proportion of these (<10) have been studied with regard to metal dynamics. The vast majority of experiments have been conducted on *Oniscus asellus, Porcellio scaber* and to a lesser extent *Philoscia muscorum* and *Armadillidium vulgare*. These are the most common and widespread species of the temperate zones.[11] However, it would be interesting to examine other species which are less cosmopolitan in their choice of habitat. Some species may be rare, or have a restricted distribution, because they have a highly specialized diet. It is possible that this could include a requirement for an essential metal, but at present there is almost nothing known about the role of trace elements in the ecology of terrestrial invertebrates.

VII. SUMMARY

Terrestrial isopods probably colonized the land via leaf litter of pteridophytes during the Carboniferous period. They require copper as integral parts of proteins that are common to all animals, and for hemocyanin, their oxygen-carrying blood pigment. This copper must be obtained from their food since direct uptake from seawater, an option available to their marine ancestors, is no longer possible.

Concentrations of copper in leaf litter of most plants are too low to provide isopods with the copper they need at rates of food ingestion that are likely to occur in the field. Woodlice appear to have solved this problem by supplementing their diet with fungal hyphae, which they graze from leaf surfaces. These hyphae contain concentrations of copper at least an order of magnitude greater than the leaves on which they are growing.

Consumption of fecal pellets is not essential to provide the copper requirements of terrestrial isopods. Nevertheless, in times of food shortage, it may be favorable for woodlice to graze fungal hyphae that are growing on the pellets.

Analysis of concentrations of copper in individual isopods from an uncontaminated site suggests that copper deficiency may occur in a few individuals in the wild. However, this is unlikely to have a significant impact in most populations.

Isopods have evolved a storage-excretion strategy to detoxify copper that is assimilated in excess of their needs. The copper is stored in the S cells of the hepatopancreas in granules which are retained until death.

In sites contaminated heavily with copper, some older individuals may be killed by copper poisoning. However, the storage capacity of the hepatopancreas for copper is remarkable. Individual *Oniscus asellus* can tolerate a concentration of copper in the hepatopancreas of 3.4% of the dry weight with no apparent ill effects.

ACKNOWLEDGMENT

I am grateful to the U.K. Natural Environment Research Council who provided financial support for this study.

REFERENCES

1. **Wieser, W.**, Copper in isopods, *Nature,* 191, 1020, 1961.
2. **Hopkin, S. P.**, *Ecophysiology of Metals in Terrestrial Invertebrates,* Elsevier Applied Science, New York, 1989.
3. **Hopkin, S. P. and Martin, M. H.**, Heavy metals in woodlice, *Symp. Zool. Soc. London,* 53, 143, 1984.

4. **Dallinger, R. and Wieser, W.**, The flow of copper through a terrestrial food chain. 1. Copper and nutrition in isopods, *Oecologia (Berlin)*, 30, 253, 1977.

5. **Wieser, W.**, Conquering *terra firma*: the copper problem from the isopod's point of view. *Helgol. Wiss. Meeresunters*, 15, 282, 1967.

6. **Scheinberg, H.**, Copper, in *Metals and Their Compounds in the Environment*, Merian, E., Ed., VCH Publishers, Weinheim, Germany, 1991, 893.

7. **Depledge, M. H.**, Re-evaluation of metabolic requirements for copper and zinc in decapod crustaceans, *Mar. Environ. Res.*, 27, 115, 1989.

8. **Schmalfuss, H.**, Phylogenetics in Oniscidea, *Monit. Zool. Ital.*, Monogr. 4, 3, 1989.

9. **Little, C.**, *The Terrestrial Invasion: An Ecophysiological Approach to the Origins of the Land Animals*, Cambridge University Press, U.K., 1990.

10. **Hames, C. A. C. and Hopkin, S. P.**, The structure and function of the digestive system of terrestrial isopods, *J. Zool. London*, 217, 599, 1989.

11. **Hopkin, S. P.**, A key to the woodlice of Britain and Ireland, *Field Stud.*, 7, 599, 1991.

12. **Flemming, C. A. and Trevors, J. T.**, Copper toxicity and chemistry in the environment: a review. *Water, Air, Soil Pollut.*, 44, 143, 1989.

13. **Rainbow, P. S., Phillips, D. J. H., and Depledge, M. H.**, The significance of trace metal concentrations in marine invertebrates. A need for laboratory investigation of accumulation strategies, *Mar. Pollut. Bull.*, 21, 321, 1990.

14. **Hassall, M. and Rushton, S. P.**, Feeding behaviour of terrestrial isopods in relation to plant defences and microbial activity, *Symp. Zool. Soc. London*, 53, 487, 1984.

15. **Bengtsson, G. and Rundgren, S.**, The Gusum case: a brass mill and the distribution of soil Collembola, *Can. J. Zool.*, 66, 1518, 1988.

16. **Tyler, G., Balsberg-Pahlsson, A. M., Bengtsson, G., Baath, E., and Tranvik, L.**, Heavy-metal ecology of terrestrial plants, microorganisms and invertebrates, *Water, Air, Soil Pollut.*, 47, 189, 1989.

17. **Waldbauer, G. P. and Friedman, S.**, Self-selection of optimal diets by insects, *Annu. Rev. Entomol.*, 36, 43, 1991.

18. **Stockli, H.**, Das Unterscheidungsvermögen von *Porcellio scaber* (Crustacea, Isopoda) zwischen Blättern einer Baumart, unter Berücksichtigung der makroskopisch sichtbaren Verpilzung, *Pedobiologia*, 34, 191, 1990.

19. **Van Capelleveen, H. E., Van Straalen, N. M., Van Den Berg, M., and Van Wachem, E.**, Avoidance as a mechanism of tolerance for lead in terrestrial arthropods, *Proc. 3rd Europ. Congr. Entomol.*, Velthuis, H. H. W., Ed., Ned. Entomol. Vereniging, 1986, 251.

20. **Gunnarsson, T.**, Selective feeding on a maple leaf by *Oniscus asellus* (Isopoda), *Pedobiologia*, 30, 161, 1987.

21. **Hanlon, R. D. G. and Anderson, J. M.**, Influence of macroarthropod feeding activities on microflora in decomposing oak leaves, *Soil Biol. Biochem.*, 12, 255, 1980.

22. **Soma, K. and Saito, T.**, Ecological studies of soil organisms with references to the decomposition of pine needles. II. Litter feeding and breakdown by the woodlouse, *Porcellio scaber, Plant Soil*, 75, 139, 1983.

23. **Bengtsson, G., Erlandsson, A., and Rundgren, S.**, Fungal odour attracts soil Collembola, *Soil Biol. Biochem.*, 20, 25, 1988.

24. **Hassall, M., Visser, S., and Parkinson, D.,** Vertical migration of *Onychiurus subtenuis* (Collembola) in relation to rainfall and microbial activity, *Pedobiologia,* 29, 175, 1986.

25. **Price, P. W.,** An overview of organismal interactions in ecosystems in evolutionary and ecological time, *Agric. Ecosyst. Environ.,* 24, 369, 1988.

26. **Kukor, J. J. and Martin, M. M.,** The effect of acquired microbial enzymes on assimilation efficiency in the common woodlouse *Tracheoniscus rathkei, Oecologia (Berlin),* 69, 360, 1986.

27. **Piearce, T. G.,** Acceptability of pteridophyte litters to *Lumbricus terrestris* and *Oniscus asellus* and implications for the nature of ancient soils, *Pedobiologia,* 33, 91, 1989.

28. **Jambu, P., Juchault, P., and Mocquard, J. P.,** Étude expérimentale de la contribution du crustacé isopode *Oniscus asellus* à la transformation des litières forestières sous chêne sessile, *Pedobiologia,* 32, 147, 1988.

29. **Hassall, M., Turner, J. G., and Rands, M. R. W.,** Effects of terrestrial isopods on the decomposition of woodland leaf litter, *Oecologia (Berlin),* 72, 597, 1987.

30. **Kohler, H. R., Ullrich, B., Storch, V., Schairer, H. U., and Alberti, G.,** Massen- und Energiefluss bei Diplopoden und Isopoden, *Mitt. Dtsch. Ges. Allg. Angew. Ent.,* 7, 263, 1989.

31. **Bergkvist, B., Folkeson, L., and Berggren, D.,** Fluxes of Cu, Zn, Pb, Cd, Cr, and Ni in temperate forest ecosystems, *Water, Air, Soil Pollut.,* 47, 217, 1989.

32. **Burton, M. A. S.,** *Biological Monitoring of Environmental Contaminants (Plants),* Monit. Assess. Res. Centre, Kings College, London, 1986.

33. **Klein, R. M., Badger, G. J., and Novak, K.,** Metal ion concentrations in red spruce foliage over time, *Environ. Exp. Bot.,* 31, 141, 1991.

34. **Bengtsson, G., Gunnarsson, T., and Rundgren, S.,** Growth changes caused by metal uptake in a population of *Onychiurus armatus* (Collembola) feeding on metal polluted fungi, *Oikos,* 40, 216, 1983.

35. **Gadd, G. M.,** Accumulation of metals by microorganisms and algae, *Biotechnology,* Vol. 6b, Rehm, H. J. and Reed, G., Eds., VCH Publishers Weinheim, Germany, 1988, 401.

36. **Gadd, G. M.,** Heavy metal accumulation by bacteria and other microorganisms, *Experentia,* 46, 834, 1990.

37. **Gadd, G. M. and White, C.,** Heavy metal and radionuclide accumulation and toxicity in fungi and yeasts, in *Metal-Microbe Interactions,* Poole, R. K. and Gadd, G. M., Eds., IRL Press, Oxford, 1989, 19.

38. **Hughes, M. N. and Poole, R. K.,** *Metals and Micro-organisms,* Chapman and Hall, New York, 1989.

39. **Campbell, R. and Martin, M. H.,** Continuous flow fermentation to purify waste water by the removal of cadmium, *Water, Air, Soil Pollut.,* 50, 397, 1990.

40. **Hopkin, S. P. and Read, H. J.,** *Biology of Millipedes,* Oxford University Press, 1992.

41. **Hassall, M. and Rushton, S. P.,** The role of coprophagy in the feeding strategies of terrestrial isopods, *Oecologia (Berlin),* 53, 374, 1982.

42. **Mead, F. and Gabouriaut, D.,** Belated and decreased reproduction in isolated females of *Helleria brevicornis* Ebner (Crustacea, Oniscoidea). Recuperation after the addition of faeces to the female environment, *Int. J. Invertebr. Reprod. Dev.,* 14, 95, 1988.

43. **Hassall, M. and Rushton, S. P.,** The adaptive significance of coprophagous behaviour in the terrestrial isopod *Porcellio scaber, Pedobiologia,* 28, 169, 1985.

44. **Coughtrey, P. J., Martin, M. H., Chard, J., and Shales, S. W.,** Micro-organisms and metal retention in the woodlouse *Oniscus asellus, Soil Biol. Biochem.,* 12, 23, 1980.

45. **Ineson, P. and Anderson, J. M.,** Aerobically isolated bacteria associated with the gut and faeces of the litter feeding macroarthropods *Oniscus asellus* and *Glomeris marginata, Soil Biol. Biochem.,* 17, 843, 1985.

46. **Hopkin, S. P.,** Species-specific differences in the net assimilation of zinc, cadmium, lead, copper and iron by the terrestrial isopods *Oniscus asellus* and *Porcellio scaber, J. Appl. Ecol.,* 27, 460, 1990.

47. **Wieser, W.,** Ecophysiological adaptations of terrestrial isopods: a brief review, *Symp. Zool. Soc. London,* 53, 247, 1984.

48. **Storch, V.,** Microscopic anatomy and ultrastructure of the stomach of *Porcellio scaber* (Crustacea, Isopoda), *Zoomorphology,* 106, 301, 1987.

49. **Ullrich, B., Storch, V., and Schairer, H.,** Bacteria on the food, in the intestine and on the faeces of the woodlouse *Oniscus asellus* (Crustacea, Isopoda), *Pedobiologia,* 35, 41, 1991.

50. **Watkins, B. and Simkiss, K.,** Interactions between soil bacteria and the molluscan alimentary tract, *J. Mol. Stud.,* 56, 267, 1990.

51. **Wood, S. and Griffiths, B. S.,** Bacteria associated with the hepatopancreas of the woodlice *Oniscus asellus* and *Porcellio scaber, Pedobiologia,* 31, 89, 1988.

52. **Griffiths, B. S. and Wood, S.,** Microorganisms associated with the hindgut of *Oniscus asellus* (Crustacea, Isopoda), *Pedobiologia,* 28, 377, 1985.

53. **Hames, C. A. C. and Hopkin, S. P.,** A daily cycle of apocrine secretion by the B cells in the hepatopancreas of terrestrial isopods, *Can. J. Zool.,* 69, 1931, 1991.

54. **Storch, V.,** The influence of nutritional stress on the ultrastructure of the hepatopancreas of terrestrial isopods, *Symp. Zool. Soc. London,* 53, 167, 1984.

55. **Hryniewiecka-Szyfter, Z. and Storch, V.,** The influence of starvation and different diets on the hindgut of Isopoda *(Mesidotea entomon, Oniscus asellus, Porcellio scaber), Protoplasma,* 134, 53, 1986.

56. **Dallinger, R. and Prosi, F.,** Heavy metals in the terrestrial isopod *Porcellio scaber* Latreille. II. Subcellular fractionation of metal-accumulating lysosomes from hepatopancreas, *Cell Biol. Toxicol.,* 4, 97, 1988.

57. **Prosi, F. and Dallinger, R.,** Heavy metals in the terrestrial isopod *Porcellio scaber* Latreille. I. Histochemical and ultrastructural characterization of metal-containing lysosomes, *Cell Biol. Toxicol.,* 4, 81, 1988.

58. **Donker, M. H., Koevoets, P., Verkleij, J. A. C., and Van Straalen, N. M.,** Metal binding compounds in hepatopancreas and haemolymph of *Porcellio scaber* (Isopoda) from contaminated and reference areas, *Comp. Biochem. Physiol.,* 97C, 119, 1990.

59. **Hopkin, S. P.,** Critical concentrations, pathways of detoxification and cellular ecotoxicology of metals in terrestrial arthropods, *Functional Ecol.,* 4, 321, 1990.

60. **Hopkin, S. P., Hames, C. A. C., and Dray, A.,** X-ray microanalytical mapping of the intracellular distribution of pollutant metals, *Microsc. Anal.,* 14, 23, 1989.

61. **Morgan, A. J., Gregory, Z. D. E., and Winters, C.,** Responses of the hepatopancreatic 'B' cells of a terrestrial isopod *Oniscus asellus,* to metals accumulated from a contaminated habitat: a morphometric analysis, *Bull. Environ. Contam. Toxicol.,* 44, 363, 1990.

62. **Morgan, E. H. and Baker, E.,** Iron uptake and metabolism by hepatocytes, *Fed. Proc.,* 45, 2810, 1986.

63. **Pattison, S. E. and Cousins, R. J.,** Zinc uptake and metabolism by hepatocytes, *Fed. Proc.,* 45, 2805, 1986.

64. **Song, M. K.,** Low molecular-weight zinc-binding ligand: a regulatory modulator for intestinal zinc transport, *Comp. Biochem. Physiol.,* 87A, 223, 1987.

65. **Ettinger, M. J., Darwish, H. M., and Schmitt, R. C.,** Mechanism of copper transport from plasma to hepatocytes, *Fed. Proc.,* 45, 2800, 1986.

66. **Hames, C. A. C. and Hopkin, S. P.,** Assimilation and loss of [109]Cd and [65]Zn by the terrestrial isopods *Oniscus asellus* and *Porcellio scaber, Bull. Environ. Contam. Toxicol.,* 47, 440, 1991.

67. **Depledge, M. H. and Rainbow, P. S.,** Models of regulations and accumulation of trace metals in marine invertebrates, *Comp. Biochem. Physiol.,* 97C, 1, 1990.

68. **Carefoot, T. H.,** Studies on the nutrition of the supralittoral isopod *Ligia pallasii* using chemically defined artificial diets: assessment of vitamin, carbohydrate, fatty acid, cholesterol and mineral requirements, *Comp. Biochem. Physiol.,* 79A, 655, 1984.

69. **Hopkin, S. P., Hardisty, G. N., and Martin, M. H.,** The woodlouse *Porcellio scaber* as a 'biological indicator' of zinc, cadmium, lead and copper pollution, *Environ. Pollut.,* 11B, 271, 1986.

70. **Gunnarsson, T. and Tunlid, A.,** Recycling of fecal pellets in isopods: micro-organisms and nitrogen compounds as potential food for *Oniscus asellus* L., *Soil Biol. Biochem.,* 18, 595, 1986.

71. **Depledge, M. H.,** New approaches in ectoxicology: can inter-individual physio-logical variability be used as a tool to investigate pollution effects?, *Ambio,* 19, 251, 1990.

72. **Webb, J. S., Thornton, I., Thompson, M., Howarth, R. J., and Lowenstein, P. L.,** *The Wolfson Geochemical Atlas of England and Wales,* Oxford University Press, 1978.

73. **Davies, B. E. and Ballinger, R. C.,** Heavy metals in soils in north Somerset, England, with special reference to contamination from base metal mining in the Mendips, *Environ. Geochem. Health,* 12, 291, 1990.

74. **Jones, D. T. and Hopkin, S. P.,** Biological monitoring of metal pollution in terrestrial ecosystems, in *Terrestrial and Aquatic Ecosystems: Perturbation and Recovery,* Ravera, O., Ed., Ellis Horwood, London, 1991, 148.

75. **Bengtsson, G. and Tranvik, L.,** Critical metal concentrations for forest soil invertebrates, *Water, Air, Soil Pollut.,* 47, 381, 1989.

76. **Hopkin, S. P. and Martin, M. H.,** The distribution of zinc, cadmium, lead and copper within the woodlouse *Oniscus asellus* (Crustacea, Isopoda), *Oecologia (Berlin),* 54, 227, 1982.

77. **Donker, M. H. and Bogert, C. G.,** Adaptation to cadmium in three populations of the isopod *Porcellio scaber, Comp. Biochem. Physiol.,* 100C, 143, 1991.

78. **Hove, K., Pedersen,O., Garmo, T. H., Hansen, H. S., and Staaland, H.,** Fungi: a major source of radiocesium contamination of grazing ruminants in Nor-way, *Health Phys.,* 59, 189, 1990.

79. **Tyler, G.,** Metal accumulation by wood-decaying fungi, *Chemosphere,* 11, 1141, 1982.

80. **Mitani, T. and Misic, D. M.,** Copper accumulation by *Penicillium* sp. isolated from soil, *Soil Sci. Plant Nutr.,* 37, 347, 1991.

CHAPTER 19

Metal Contamination Affects Size-Structure and Life-History Dynamics in Isopod Field Populations

Marianne H. Donker, H. Erik van Capelleveen, and Nico M. van Straalen

TABLE OF CONTENTS

I. INTRODUCTION

When soils become contaminated with metals, susceptible species may disappear from the area. Several field studies have documented changes in species composition among soil invertebrate fauna found around sources of metal pollution.[1] Laboratory studies have clearly demonstrated that significantly different responses to metals exists among the various invertebrate groups.[2] Change in species diversity of soil organisms may be used as an indication of serious environmental effects but, when such effects are observed, it is often too late for remedial measures. Thus, there is a need for better parameters than species abundance in ecotoxicological field studies. Demographic population attributes, such as size distribution and brood size, will respond to environmental pollution at an earlier stage. Through effects on the fitness of individual organisms in ecological time, certain combinations for reproduction and survival are selected. These combinations arise through trade-offs between different tactics (i.e., extent of investment in reproductive activity or investment in physiological adaptations) and depend on the frequency of disturbances and the general level of adversity.[3] The pattern of trade-offs, though, being constrained by available genetic variability. In a study of two populations of the freshwater isopod *Asellus aquaticus*, separated by an effluent discharge, it was shown that isopods below the discharge had a lower reproductive effort and allocated these investments into fewer and larger offspring.[4]

Demographic studies of terrestrial isopods have identified various environmental factors influencing population dynamics.[5,6] For isopods, of course, the most important factors are suitable shelter sites and availability and quality of food. Metals may be a third important factor, both as trace elements for mineral nutrition[7] and as toxic agents when present in high concentrations. The isopod *Porcellio scaber* (Latreille) is a common woodlouse inhabiting unpolluted as well as highly metal polluted sites. The species is rather resistant to metals in terms of acute toxicity, but sublethal effects on growth and reproduction have been observed under longer exposure time.[8] The presence of *Porcellio scaber* in metal polluted sites may relate to the unique ability of isopods to concentrate large amounts of metals in the hepatopancreas.[9] This continuous accumulation of metals, associated with complexation inside granules,[10] is only lethal to the isopod when a certain concentration is exceeded. For example, the lethal zinc concentration for *P. scaber* is 27 μmol g^{-1}[11] and for cadmium, the LC_{50} decreases with increasing exposure time, the ultimate LC_{50} approaches zero.[35]

As accumulation of metals results in increased mortality in older isopods, mortality may act as a selection pressure on the life-history characteristics of a population, and early maturity or increased reproduction may be favored. This presupposition is confirmed in a study performed with populations of the crustacean *Daphnia magna*; when large individuals were harvested for about ten generations, reproduction occurred at a smaller size.[12] The aim of the present study was to determine demographic patterns of a field population of *P. scaber*

Table 1. Metal Concentrations in Isopods *(Porcellio scaber)* and in the Organic Part of the Soil Profile (A-horizon) at the Two Sampling Sites. Means are Given With Standard Errors, Based on 15 Replicate Samples

Metal	Santpoort (reference site)		Budel (smelter site)	
	A-horizon (μg g^{-1})	Isopods (μg g^{-1})	A-horizon (μg g^{-1})	Isopods (μg g^{-1})
Zinc	399 $\pm$ 59	287 $\pm$ 46	864 $\pm$ 275	1347 $\pm$ 183
Cadmium	1.12 $\pm$ 0.22	6.7 $\pm$ 2.2	9.0 $\pm$ 1.1	80.9 $\pm$ 13.5
Lead	22.7 $\pm$ 4.14	6.2 $\pm$ 2.0	606 $\pm$ 93	50 $\pm$ 6.2
Copper	26.0 $\pm$ 3.2	362 $\pm$ 42	83.8 $\pm$ 7.6	250 $\pm$ 29.0

in the surroundings of a zinc smelter in Budel, the Netherlands. The population in this metal contaminated area was compared to a reference population living in a similar, but unpolluted area.

II. MATERIAL AND METHODS

The sampling site of the metal-exposed population is located approximately 1 km southeast of a zinc smelter at Budel-Dorplein, the Netherlands. The smelter has been in operation since 1892; emissions were reduced after 1974 following a change in process technology. The Budel site is a woodland consisting mainly of poplar *(Populus* spec.), birch *(Betula pendula)*, and oak *(Quercus robur)* on sandy soil. The reference site is in a similar woodland vegetation close to Santpoort, 25 km to the west of Amsterdam. Table 1 presents the metal concentrations in isopods and in the A-horizon of the soil profile; the results are based on 15 replicates. Zinc, lead, copper, and cadmium were analyzed for, using the methods described by Van Straalen and Van Wensem.[13]

Porcellio scaber Latr. populations were sampled during the years 1983 and 1984 at intervals of one month. A sample of 50 to 100 individuals was collected from shelter microhabitats, no attempt was made to estimate absolute population densities. The sample was transferred live to the laboratory in plastic pots with litter from the site.

Each individual was fresh weighed and its head width was determined using an ocular micrometer in a microscope with stereo view. Sex determination was possible for individuals with a head width larger than 0.8 mm, or heavier than 2.4 mg fresh weight, by screening the endopodites of the first and second pleopods. Animals with a head width less than 0.8 mm were not considered in the analysis. The presence of brood pouches in females was noted; the brood was removed from the pouch with forceps and counted.

III. RESULTS

Size-structure of isopods from the reference and smelter site are presented in Figures 1 and 2. The numbers of males and females in each size-class are given as a percentage of the total number sampled on each date. Although there was a clear correlation between head width and fresh weight, the latter seems to be more variable; head width is probably a better indicator of age, and was used to define population structure in Figures 1 and 2.

The populations appeared to consist of two size-groups during most of the year; occasionally, in summer (July to September), three size-groups could be discerned due to the recruitment of young adults in the population. The size frequency patterns show that *Porcellio scaber* has overlapping generations. The overlap is more pronounced in the males, which are smaller and grow more slowly than females at all stages of development. The overlap between generations seemed more pronounced in the smelter population (Figures 1 and 2) and is probably due to reduced growth in the smelter isopods. New adults entered both populations during July, August, and September.

Although head width is a good indicator of age, weight of females is a good indicator of brood size. Several studies have shown a relationship between a female's fresh weight and the number of eggs produced.[14,15] Thus, female weight is an important population dynamic characteristic which will be analyzed below.

Figure 3 shows the weight frequency distribution of the female isopods, sampled monthly and cumulated over the two sampling years at the reference and smelter sites (24 time intervals). The weight distribution of the reference population has two peaks, one peak consists of the first-year class and the second peak contains the second- and third-year classes. A similar pattern, shifted to the left, is shown for the smelter population. Clearly, the smelter females grew more slowly, which resulted in a more extensive overlap between the two generations, and a lower average weight. The mean weight of the smelter females was 20.7% in 1983 and 24.8% in 1984 less than the mean weight of the reference females.

In Figure 4 the number of females is expressed as a percentage of the total population for each month, composed of the pooled data for the corresponding months from the two sampling years. It appears that in both populations the sex ratio fluctuated throughout the year, with a total average of 55.6%. However, during spring (April, May, and June) females clearly outnumbered males in the smelter population. This can be explained by a difference in the time of main mortality: large smelter males mostly died in April and May, as also can be seen from Figure 2; while females mostly died in July, August, and December. The mortality of females hardly influences the sex ratio, as in August the new year class enters the population. This ratio is largely determined by the new year class and thus returns to 50%. From the difference in sex ratio between reference and smelter isopods in April, it can be concluded that life expectancy of male isopods is reduced at the smelter site.

With respect to reproduction, the data provide information on four parameters: (1) the period in which broods are carried, (2) the size distribution of gravid females, (3) the percentage of females participating in reproduction, and (4) the size of the brood in relation to the size of the female.

Figure 5 presents the phenology of gravid females, expressed as a percentage of the total female population collected on each sampling date. In both populations, the first females with a brood pouch were observed in March and the last ones in October. On the average, the percentage of gravid females was lower in the smelter isopods than in the reference isopods. This is mainly due to the summer period: in the reference population a second peak of reproductive females in August/September occurred, this peak was hardly present in 1983 and effectively absent in 1984 in the smelter population.

The relationship between brood size and female size was analyzed separately for the spring and the summer periods. Regression analysis produced a linear relationship between brood size and female fresh weight in all cases. The regression equations are given in Table 2. The weight dependence of brood size was more pronounced during spring; compared to the summer breeding period, summer broods were smaller. The weight of the females reproducing in August and September is about 25 mg; these are probably the females born in autumn of the previous year or in early spring. In relation to their size, smelter females produce more eggs per brood than reference females. This can also be inferred from Figure 6, which presents the time course of average brood sizes for each year separately; this figure clearly shows that, although smelter females are smaller and are less likely to reproduce during summer, their average brood size is larger both in spring and in summer.

IV. DISCUSSION

The *Porcellio scaber* population inhabiting a metal contaminated site was shown to differ from a reference population in several demographic parameters such as size distribution, sex ratio, and reproduction. In our interpretation, these differences reflect effects of cadmium and zinc, leading to advanced mortality, especially in male isopods, and to a smaller body size in both sexes. Nevertheless, the population persists through increased reproductive effort with larger broods for a given female size, this being predicted by life-history theories of Southwood[3] and Sibly and Calow.[16] Similar life-history responses have been observed in *Asellus aquaticus*,[4] in *Daphnia magna*,[12] in Collembola from contaminated sites,[17,18] and in fish.[19]

Although the differences between *Porcellio* populations have been consistent over two sampling years, and are in accordance with life-history theory, it is difficult to prove that this is causally related to increased metal levels in the litter layer. The same is true for gradient studies around a single source of pollution, where various soil factors such as pH, litter depth, and vegetation

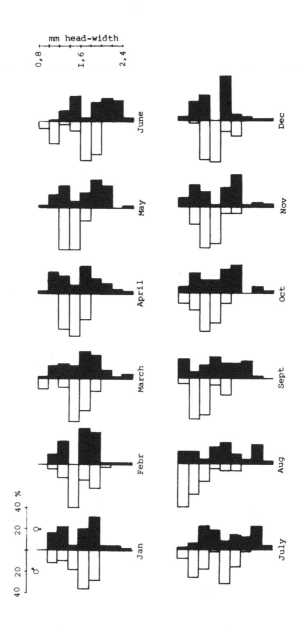

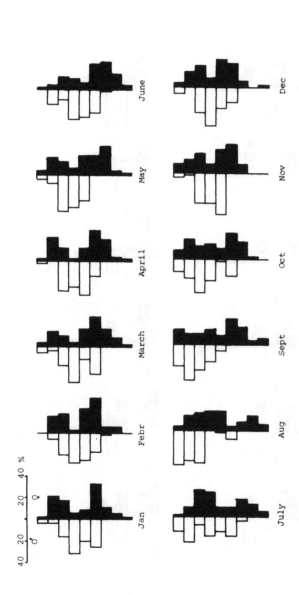

FIGURE 1. Frequency distributions for *Porcellio scaber* population structure at the reference site (Santpoort). Males (left) and females (right) are classified separately according to head width, for each monthly sampling occasion during a two-year period.

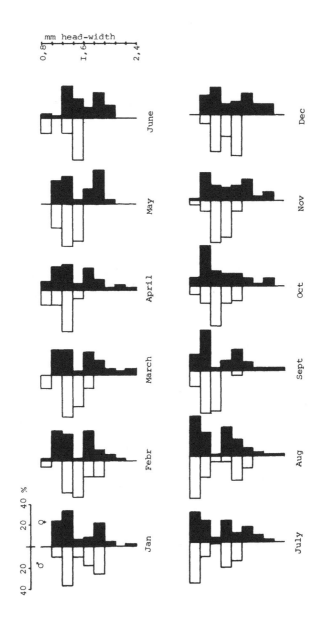

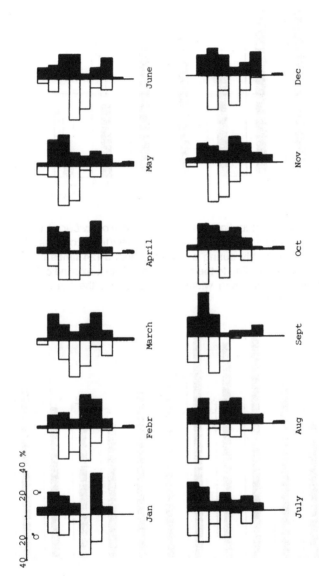

FIGURE 2. Frequency distributions for *Porcellio scaber* population structure at the smelter site (Budel). Males (left) and females (right) are classified separately according to head width, for each monthly sampling occasion during a two-year period.

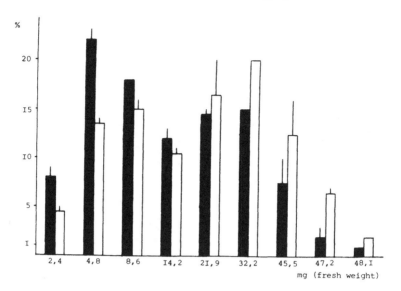

FIGURE 3. Frequency distributions of female *Porcellio scaber* at the reference site (open columns) and at the smelter site (filled columns). All females sampled monthly during a two-year period (24 times in total), were classified according to their head width. The x-axis indicates the mean weight of each head width size class and the bars represent the standard error.

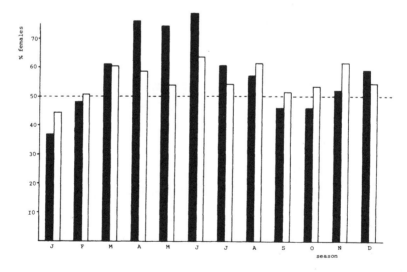

FIGURE 4. Sex ratio in two *Porcellio scaber* populations, expressed as the percentage of females in the total population. Observations from two successive years were taken together for each month. Open columns: reference population; filled columns: smelter population.

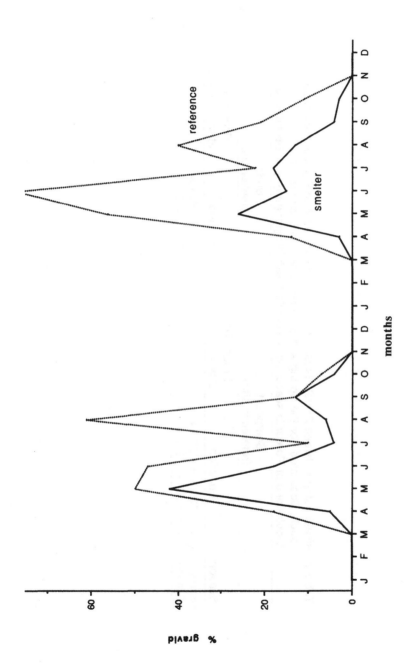

FIGURE 5. Seasonal occurrence of gravidity in *Porcellio scaber*, expressed as the percentage of gravid females in the total female population on 24 sampling occasions through a two-year period for a reference (····) and a smelter population (——).

Table 2. Relationship Between Female Fresh Weight and the Number of Eggs in the Brood Pouch, Separated For the Spring and Summer Breeding Seasons, For Two Populations of *Porcellio scaber* (w = Female Fresh Weight (mg), y = Brood Size)

| Season | Santpoort (reference site) | | Budel (smelter site) | |
	Equation	Brood size at w = 30 mg	Equation	Brood size at w = 30 mg
Spring	y = 13.23 + 0.720 w	35	y = 16.91 + 0.895 w	44
Summer	y = 6.81 + 0.319 w	16	y = 15.27 + 0.468 w	29

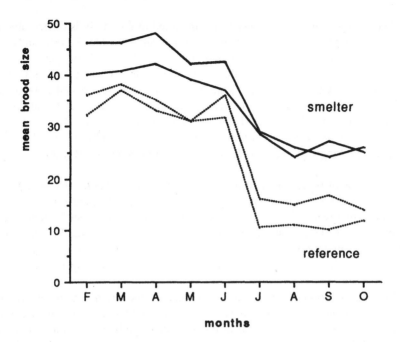

FIGURE 6. Mean brood size (number of eggs in brood pouch) for *Porcellio scaber* in relation to the season (months) during two years, for reference (····) and smelter population (——).

cover may covary with metal contamination.[20-23] In the case of *P. scaber* there is additional information available, indicating that at the smelter site the large individuals are affected by metals. Firstly, the concentration of zinc found in large isopods, especially the males, closely approaches the lethal concentration as found by Hopkin.[11,24] Secondly, the zinc concentration and to a lesser extent the cadmium concentration, are negatively correlated with survival time under starvation.[24] A third indication of metals affecting *P. scaber* at the smelter site, is suggested from the finding that increased metal concentration is accompanied by reduced energy reserves and by changes in the tissue distribution of metals.[24] Therefore, we suggest that a causal relationship does exist between the high level of metals in the litter layer, and the life-history dynamics of *P. scaber* at the smelter site.

Although the increased reproductive effort of smelter isopods can be seen as an adaptive response, the population at the smelter site is not adapted in the sense that it is less susceptible to metal intoxication. Earlier research has shown that isopods from a smelter strain have shifted their growth optimum to a higher concentration of cadmium in the food, but suffer from cadmium toxicity to the same extent as a reference population,[25] however, smelter isopods do seem to be able to reduce zinc accumulation by an increased zinc excretion efficiency and, by doing this, they retard the moment of zinc intoxication.[26]

The demography of *P. scaber* as documented in this study is comparable to other demographic studies on isopods at other sites. In the present study, *P. scaber* populations consisted mainly of two year-classes. Third year animals formed only a small fraction of the population. These findings are similar to those obtained by Brereton[27] for *P. scaber* in England, who estimated a maximum age for isopods in the field of between three and four years, but found only a few four-year-old individuals. Heeley[28] obtained similar results for the isopod *Armadillidium vulgare* in field populations in England, and Paris and Pitelka[29] estimated the longevity for *A. vulgare* from the San Francisco Bay area as three years.

Sex ratios of isopod populations seem to be rather variable. Brereton[27] suggested that male *P. scaber* do not survive as long as females; Paris and Pitelka[29] concluded the reverse with respect to *A. vulgare*, while McQueen[30] found equal survivorship for the two sexes in *Porcellio spinicornis*. It is possible that some of these interpretations result from behavioral differences between the sexes which may influence sampling efficiency, rather than diverging mortality patterns.[5]

In several isopod species mortality and reproduction are inversely related.[29,31,32] In the present study the sex ratio was found to approach unity during most of the sampling period, but in the smelter isopods the sex ratio deviated in favor of females during early spring. The reason for the disappearance of the males at that time is not known. Gere[33] showed that isopods hardly feed from January to April, followed by intensive feeding activity in spring, which will result, as a logical consequence, in a fast increase in metal concentration. As a result, perhaps, the body zinc concentration rapidly exceeds the critical level in the males, resulting in death due to zinc poisoning.[11]

Gravid females of *P. scaber* were found from March until October, which represents a longer season than that reported by Heeley[28] for *P. scaber* in England. The observations of Brereton[27] and Heeley[28] imply that *P. scaber* produces two broods per year which agrees with the present study, especially in the control population. In the smelter population few females succeeded in producing a second brood. This is most probably due to their reduced growth rate.

With respect to fertility the number of eggs was found to increase with female weight, as has been reported for other isopod studies.[29,31,32] Some authors have observed brood size to decrease under unfavorable conditions.[34] In the present study, however, brood size was larger in the smelter population, when females of equal weight were compared (Table 2). This can be seen as a trade-off between survival and reproduction in terms of life-history theory.[3] Increased mortality in older individuals of the smelter population favors investment in reproductive activity; this increase may also be due to brood size being determined by age rather than by weight. Since the smelter isopods grew more slowly, females of equal weight were of different ages, which suggests that the same relationship between brood size and female weight does hold for both smelter and reference isopods.

The work presented here shows that *P. scaber* is able to maintain a population at a highly metal polluted site, but that this population differs from a reference population in life-history characteristics. The smelter isopods respond to metal pollution not only with a smaller body size but female isopods also produced a larger number of offspring. Life-history patterns, determined in contaminated areas, are the product of selection for tolerance, as well as toxic effects on growth and reproduction. Nevertheless, analysis of demographic parameters provides a more sensitive tool for the investigation of sublethal effects of pollution, compared to studies noting only the abundance of species.

V. ABSTRACT

The presence of an organism at a metal contaminated site does not preclude that it is sublethally affected by the contamination. This study aimed to investigate sublethal effects in the terrestrial isopod *Porcellio scaber* by demographic analysis of field populations. Isopods were sampled monthly from a reference and a metal contaminated smelter site for a period of two years, and the populations were characterized on the basis of size structure, sex ratio, percentage of gravid females, and brood size. At the metal-contaminated site, isopods had a smaller body size and the percentage of gravid females was lower. Significant mortality among males was noted in spring, which caused a deviation in the sex ratio. In relation to body size, isopods at the smelter site carried larger broods. It is concluded from life-history patterns that *P. scaber* is sublethally affected at the smelter site, although the isopods are able to maintain their presence.

ACKNOWLEDGMENTS

This research was supported by the Foundation for Technical Research (STW), under project nr. VB 122.0294. The authors are indebted to Prof. Dr. E.N.G. Joosse who initiated the research and provided stimulation throughout the study. Désirée Hoonhout kindly typed the manuscript.

REFERENCES

1. **Bengtsson, G. and Tranvik, L.,** Critical metal concentrations for forest soil invertebrates. A review of the limitations, *Water, Air, Soil Pollut.,* 47, 381, 1989.
2. **Van Straalen, N. M.,** Soil and sediment quality criteria derived from invertebrate toxicity data, in *Ecotoxicology of Metals in Invertebrates,* Dallinger, R. and Rainbow, P. S., Eds., Lewis Publishers, Boca Raton, FL, 1993, ch. 21.
3. **Southwood, T. R. E.,** Tactics, strategies and templets, *Oikos,* 52, 3, 1988.

4. **Maltby, L.**, Pollution as a probe of life-history adaptation in *Asellus aquaticus* (Isopoda), *Oikos,* 61, 11, 1991.

5. **McQueen, D. J.**, The influence of climatic factors on the demography of the terrestrial isopod *Tracheoniscus rathkei, Can. J. Zool.,* 54, 2185, 1976.

6. **Hassall, M. and Dangerfield, J. M.**, Density-dependent processes in the population dynamics of *Armadillidium vulgare* (Isopoda: Oniscidae), *J. Anim. Ecol.,* 59, 941, 1990.

7. **Dallinger, R.**, The flow of copper through a terrestrial food chain. III. Selection of an optimum copper diet by isopods, *Oecologia (Berlin),* 30, 273, 1977.

8. **Beyer, N. W., Miller, G. W., and Cromartie, E. J.**, Contamination of the Dekalb O_2-soil horizon by zinc smelting and its effect on woodlouse *Porcellio scaber* survival, *J. Environ. Qual.,* 13, 247, 1984.

9. **Hopkin, S. P.**, *Ecophysiology of Metals in Terrestrial Invertebrates,* Elsevier Applied Science, London, 1989.

10. **Prosi, F. and Dallinger, R.**, Heavy metals in the terrestrial isopod *Porcellio scaber* Latreille. II. Histochemical and ultrastructural characterization of metal containing lysosomes, *Cell Biol. Toxicol.,* 4, 81, 1988.

11. **Hopkin, S. P.**, Species-specific differences in the net assimilation of zinc, cadmium and lead, copper and iron by the terrestrial isopods *Oniscus asellus* and *Porcellio scaber, J. Appl. Ecol.,* 27, 460, 1990.

12. **Edley, M. T. and Law, R.**, Evolution of life histories and yields in experimental populations of *Daphnia magna, Biol. J. Linn. Soc.,* 34, 309, 1988.

13. **Van Straalen, N. M. and Van Wensem, J.**, Heavy metal content of forest litter arthropods as related to body-size and trophic level, *Environ. Pollut. (Ser. A),* 42, 209, 1986.

14. **Brody, M. S.**, Reproductive Adaptations to Food Shortages in the Terrestrial Isopod *Armadillidium vulgare,* Ph.D. thesis, University of Texas, Austin, 1980.

15. **Sutton, S. L., Hassall, M., Willows, R., Davis, R. C., Grundy, A., and Sunderland, K. D.**, Life histories of terrestrial isopods; a study of intra- and interspecific variation, in *Symp. Zool. Soc. London,* Sutton, S. L. and Holdich, D. H., Eds., Oxford Science Publishers, London, 53, 269, 1984.

16. **Sibly, R. M. and Calow, P.**, A life-cycle theory of responses to stress, *Biol. J. Linn. Soc.,* 37, 101, 1989.

17. **Bengtsson, G.**, The optimal use of life strategies in transitional zones or the optimal use of transitional zones to describe life strategies, in *Proc. 3rd Eur. Cong. Entomology,* Velthuis, H. H. W., Ed., NEV, Amsterdam, 193, 1986.

18. **Posthuma, L.**, Life history patterns in metal-adapted Collembola, in preparation.

19. **Sutherland, W. J.**, Evolution and fisheries, *Nature,* 344, 814, 1990.

20. **Bengtsson, G. and Rundgren, S.**, Ground-living invertebrates in metal-polluted forest soils, *Ambio,* 13, 29, 1984.

21. **Hopkin, S. P., Watson, K., Martin, M. H., and Mould, M. L.**, The assimilation of heavy metals by *Lithobius variegatus* and *Glomeris marginata* (Chilopoda, Diplopoda), *Bijdr. Dierk.,* 55, 88, 1985.

22. **Read, H. J., Wheater, C. P., and Martin, M. H.**, Aspects of the ecology of Carabidae (Coleoptera) from woodlands polluted by heavy metals, *Environ. Pollut.,* 48, 61, 1987.

23. **Hågvar, S. and Abrahamsen, G.**, Microarthropoda and Enchytraeidae (Oligochaeta) in naturally lead-contaminated soil: a gradient study, *Environ. Entomol.,* 19, 1263, 1990.

24. **Donker, M. H.,** Energy reserves and distribution of metals in populations of the isopod *Porcellio scaber* from contaminated sites, *Functional Ecol.*, 6, 445.

25. **Donker, M. H. and Bogert, C. G.,** Adaptation to cadmium in three populations of the isopod *Porcellio scaber*, *Comp. Biochem. Physiol.*, 100C, 143, 1991.

26. **Donker, M. H., Raedecker, M. H., and van Straalen, N. M.,** Responses to zinc contaminated food in two populations of the isopod *Porcellio scaber*, submitted.

27. **Brereton, J. G.,** A Study of Some Factors Controlling the Population of Some Terrestrial Isopods, Ph.D. thesis, University of Oxford, 1956.

28. **Heeley, W.,** Observations on the life histories of some terrestrial isopods, *Proc. Zool. Soc. London,* B11, 79, 1941.

29. **Paris, O. H. and Pitelka, F. A.,** Population characteristics of the terrestrial isopod *Armadillidium vulgare* in California grassland, *Ecology,* 43, 229, 1962.

30. **McQueen, D. J.,** *Porcellio spinicornis* demography. II. A comparison between field and laboratory data, *Can. J. Zool.,* 49, 667, 1976.

31. **Sunderland, K. D., Hassall, M., and Sutton, S. L.,** The population dynamics of *P. muscorum* in a dune grassland ecosystem, *J. Anim. Ecol.,* 45, 487, 1976.

32. **Sutton, S. L.,** The population dynamics of *Trichoniscus pusillus* and *Philoscia muscorum* in limestone grassland, *J. Anim. Ecol.,* 37, 425, 1968.

33. **Gere, G.,** Nahrungsverbrauch der Diplopoden und Isopoden in Freilanduntersuchungen, *Acta Zool. Hung.,* 8, 384, 1962.

34. **Brody, M. S. and Lawlor, L. R.,** Adaptive variation in offspring size in the terrestrial isopod, *Armadillidium vulgare, Oecologia,* 61, 55, 1984.

35. **Crommentuijn, T.,** personal communication.

CHAPTER 20

Metal Bioaccumulation in a Host Insect (*Lymantria dispar* L., Lepidoptera) During Development — Ecotoxicological Implications

Johanna Ortel, Susanne Gintenreiter, and Herbert Nopp

TABLE OF CONTENTS

0-87371-734-1/93/$0.00 + $.50
© 1993 by Lewis Publishers

I. INTRODUCTION

Metal-specific dynamics of uptake, accumulation, and excretion at the whole body level and the tissue level have been demonstrated in terrestrial invertebrate species.[1-5] Few studies, however, have addressed the effects of developmental stages metal accumulation.[6-8] Such effects may be important, for instance, in metal loaded terrestrial insects of which different developmental stages can be infested by different parasitoid species. Under such circumstances, the degree of metal-transfer between the host and the parasitoid may depend on the accumulation capacity of a specific host stage. Such peculiarities may lead to changes in host-parasitoid relationships and to shifts in the equilibrium between herbivores and their regulators.[9,10] However, studies on metal transfer between hosts and parasites are rare.[8]

A detailed life cycle study was carried out on the forest pest insect *Lymantria dispar* L. (Lymantriidae, Lepidoptera) exposed to metals under laboratory conditions. The study was aimed at determining possible effects of the metal-contaminated host *(L. dispar)* on one of its main antagonists, the gregarious endoparasitoid *Glyptapanteles liparidis* Bouché (Braconidae, Hymenoptera). Since in most terrestrial environments invertebrates are exposed to sublethal rather than

Table 1. Metal Concentrations (μg/g) in Food
(dw) in the Artificial Diet of
Lymantria dispar Larvae

Metal	Concentration level			
	1	2	3	4
Cadmium	2	10	50	250
Lead	4	20	100	500
Copper	10	50	250	1250
Zinc	100	500	2500	12500

to acutely toxic metal concentrations, the present study was also focused on sublethal effects of metals on *L. dispar* by establishing No Observed Effect Concentrations (NOECs). Four metals (cadmium, lead, zinc, and copper) were applied in increasing concentrations, and their effects on the developmental rate of both species as well as on growth and reproduction of the gypsy moth were investigated. Special emphasis was placed on the life cycle of *L. dispar* by determining the body concentrations of metals of each stage.

II. MATERIALS AND METHODS

A. Animals and Rearing Conditions

L. dispar larvae from laboratory cultures were fed an artificial diet (main components: 80% water, 10% agar and wheat germ) modified according to Bell et al.[11] The diet was contaminated separately with four metals (applied as nitrates), each at four different concentrations as summarized in Table 1. For each concentration level two parallel control groups were reared, one feeding on the uncontaminated artificial diet, the other on oak leaves (natural diet). Each experimental group was started with 40 newly hatched larvae. From 3 to 20 individuals (depending on the larval stage) were kept in petri dishes at constant conditions of 20°C and L:D = 12:12. Every second day food was replenished and newly molted individuals transferred to other petri dishes.

Cocoons of *G. liparidis* were collected in an oak forest (Burgenland, Austria); the eclosed imagoes were kept in polystyrol boxes (up to 30 specimens/box) under a light and temperature regime of L:D = 16:8 and 15:7°C. Adult wasps were fed an agar-honey mixture and received fresh water daily. For controls and for each metal, at the two lowest concentrations (Table 1), 40 to 50 host larvae in the first premolting period were offered to female *G. liparidis* for parasitization. Infected larvae were kept under the same conditons as the unparasitized individuals.

The developmental rate of hosts and parasitoids and the number of eclosed individuals of *G. liparidis* per *L. dispar* larvae were recorded, and metal concentrations of the parasitoid imagoes and the remains of the host larvae after parasitoid eclosion were measured.

Mortality, developmental period, weight, head capsule width of all larval stages, and the reproductive success of *L. dispar* were recorded for each metal concentration used (Table 1). Additionally, metal analyses were performed on all developmental stages of *L. dispar* and on the corresponding feces, exuviae (larval skins and exuviae), and head capsules at the two lowest applied concentrations of each metal.

B. Metal Analyses and Statistics

Single individuals and pooled samples (feces, exuviae, head capsules, first instar larvae, and parasitoids) were dried to constant weight at 60°C, digested in 0.2 to 1.5 ml nitric acid (depending on sample weight), and diluted with distilled water (1:1). Metal concentrations were measured on a Varian Spectr AA30 spectrophotometer by flame (Zn and Cu) and furnace (GTA 96: Cd, Pb, and Cu).

Statistical analyses of the results were carried out using the *t*-test, Kruskal-Wallis analysis of variance, Mann-Whitney-U-test, Spearman rank-correlation-test, and the Kendall-τ-test.

III. RESULTS

A. Stage- and Metal-Specific Bioaccumulation

1. Cadmium

Cadmium accumulation in successive developmental stages of *L. dispar* displayed similar patterns in both concentrations applied (2 μg/g and 10 μg/g Cd; Figure 1 shows only the control and the Cd10 group). Body concentrations of the two contaminated groups differed by a factor of about five, reflecting the differences of Cd levels in the food. In the early larval stages (L1 to L3), Cd concentrations increased at first, but then decreased again toward pupation (range of concentration factors (Cf: 2 to 4). Due to the weight loss during metamorphosis, cadmium concentrations in the imagoes were higher than in the pupae. In larvae and pupae of controls a negative correlation of weight and cadmium concentration was observed. The head capsules and exuviae (except those of the pupae) were only slightly contaminated. Metal concentrations in the feces (data available from L3 to L6 only) increased with consecutive larval stage in inverse proportion to the metal concentrations in the larvae. Controls showed a similar relationship, but at much lower levels (mostly <0.6 μg/g).

2. Lead

In contrast to cadmium, assimilation of lead was negligible in *L. dispar* (concentration factor <0.4). Values of the controls and of most individuals of the Pb4 group were below the detection limit. First instar larvae contaminated with 20 μg/g showed the highest lead concentrations of all stages (see Figure 2). Lead concentrations of the subsequent stages decreased progressively to pupation. There was a significant negative correlation between weight and lead concentrations in the larvae. Again, body concentrations in the imagoes exceeded those in the pupae (concentration effect due to weight loss). Lead concentrations in exuviae and head capsules differed from one stage to another, but without a specific pattern. Lead concentrations in the feces of the Pb20 group (data available from L3 to L6 only) exceeded those of the larvae by a factor of 8 to 14, while no lead could be detected in control feces.

3. Copper

In contrast to the nonessential metals cadmium and lead, the accumulation of copper followed different patterns at the different concentrations applied (Figure 3). Body concentrations of the control and the Cu10 group increased from L1 to L5 stage, whereas in the Cu50 group the concentrations decreased from the L3 to the L4 stage, then rose again to pupation (range of concentration factors: 2 to 5). This might be explained either by metal elimination or a dilution effect due to gain in weight. In all groups copper was largely eliminated by defecation with the meconium, high copper concentrations in pupal exuviae being due to meconium residues. Thus, only with regard to copper did the pupal body concentrations exceed those of the imagoes.

In the Cu10 and control group the metal levels in feces decreased in consecutive larval stages, copper concentrations of head capsules and exuviae always being higher than those of feces. The reverse seemed to be true for the Cu50 group: in general, copper levels in head capsules and exuviae were even lower than those of both other groups.

4. Zinc

The accumulation of Zn through consecutive larval stages also followed different patterns at the different concentrations of the metal in the food (range of concentration factors: 0.7 to 3.5). At the two concentrations applied, the metal concentrations were almost equal in the first instars, but the Zn500 individuals showed an increase and the Zn100 larvae a decrease in Zn concentration during further larval development (Figure 4). The latter was also true in the controls. Negative correlations of weight and body concentrations were found for control and Zn100 larvae, but in Zn500 larvae the correlation was positive.

In the exuviae of consecutive larval stages zinc showed a decreasing trend in all groups, whereas Zn concentrations in head capsules showed an increasing trend from one developmental stage to another. Zn concentrations in whole bodies and feces followed inverse patterns (compare with copper).

A

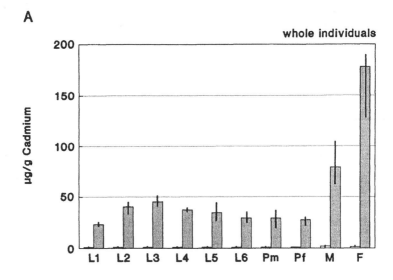

B

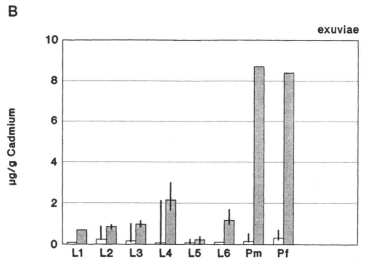

FIGURE 1. Cadmium concentrations (μg/g dw) of the consecutive developmental stages of *Lymantria dispar:* Dotted = contaminated with 10 μg/g Cd (dw in food) from hatching onwards, open = control (uncontaminated diet); A: whole individuals, B: exuviae (larval skins and exuviae), C: feces, and D: head capsules (note different scaling!); data of feces were only available from L3 on; block height: median (n = 3 to 10), lower and upper quartiles are shown; a block without quartiles indicates the median of a sample number <5. L1 to L6 = 1st to 6th larval stage, Pm = male pupae, Pf = female pupae, M = male moth, F = female moth.

C

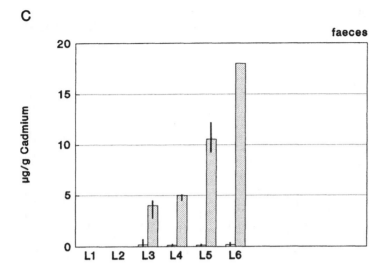

D

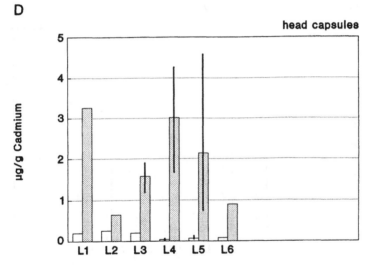

B.   Effects of Metals on Vitality of *L. dispar*

1.   *Developmental Rate*

In all groups reared on the artificial diet the developmental period was retarded as compared to that of the oak leaf group (Figure 5). This seems to be a consequence of the different food quality. Metal concentrations observed which had no effect (NOECs) on the developmental period are shown in Table 2, although the real no-effect levels might be slightly higher (compare Figure 5). For determination of the developmental rate, only the period from hatching to

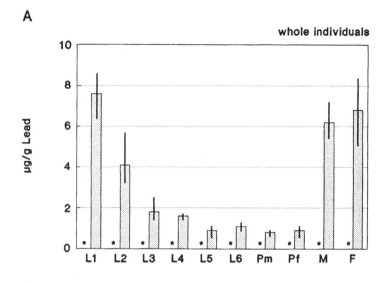

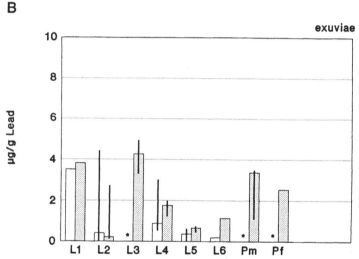

FIGURE 2. Lead concentrations (μg/g dw) of the consecutive developmental stages of *Lymantria dispar:* Dotted = contaminated with 20 μg/g Pb (dw in food) from hatching onwards, open = controls (uncontaminated diet); A: whole individuals, B: exuviae (larval skins and exuviae), C: feces, and D: head capsules (note different scaling!); data of feces were only available from L3 on; * = concentrations below detection limit; for further explanations see Figure 1 caption.

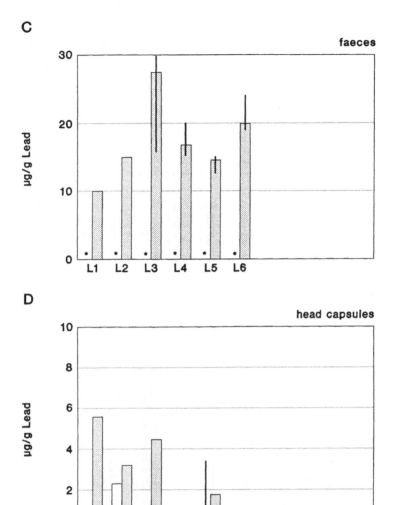

the fifth instar molting was considered. The reason for this is that after the fifth instar molting, the larvae usually exhibit sex specific differences in the number of stages until pupation (males: 1; females: 2). However, it should be mentioned that, in comparison to controls, metal contamination caused higher numbers of larval stages in nearly all experimental groups. Depending on the metal applied and its concentration, up to ten larval stages were observed. L9 and L10 larvae, however, were not able to pupate. Only the individuals of the Cu10 and Zn100 group had the same number of larval stages (5 to 6, see above) as the controls.

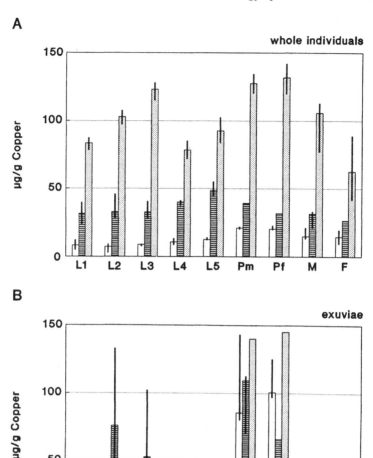

FIGURE 3. Copper concentrations (μg/g dw) of the consecutive developmental stages of *Lymantria dispar:* Striped = contaminated with 10 μg/g Cu (dw in food), dotted = 50 μg/g Cu from hatching onwards, open = controls (uncontaminated diet); A: whole individuals, B: exuviae (larval skins and exuviae), C: feces, and D: head capsules (note different scaling!); block height: median (n = 3 to 10), lower and upper quartiles are shown; a block without quartiles indicates the median of a sample number <5. Missing blocks = no data available; L1 to L5 = 1st to 5th larval stage, Pm = male pupae, Pf = female pupae, M = male moth, F = female moth.

C

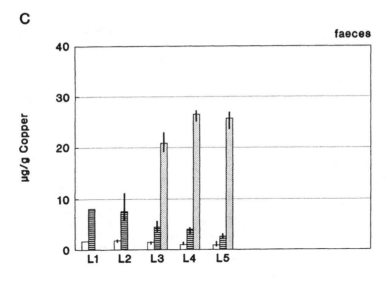

D

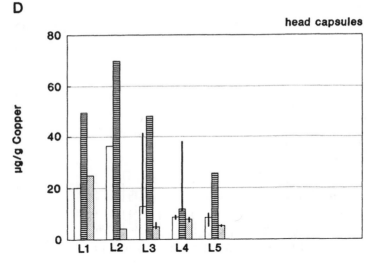

2. Mortality

For all groups exposed to the highest metal concentrations mortality was higher than that of the controls ($p < 0.01$). This was also true for the Zn2500, Zn500, and Cu250 groups. Due to the high toxicity, few or no imagoes resulted from these groups. In the Zn12500 group the mortality of first instar larvae was nearly 100%. For all metals, mortality NOEC levels are shown in Table 1.

3. Growth

In general, head capsule width decreased with increasing metal concentrations, even if not always significantly. The head capsule width of individuals

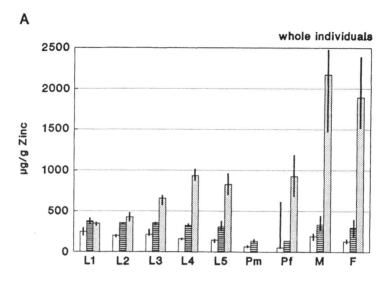

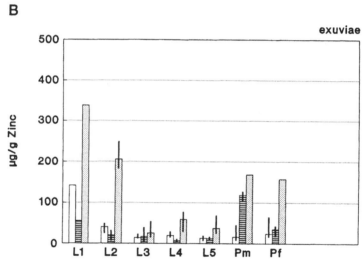

FIGURE 4. Zinc concentrations (μg/g dw) of the consecutive developmental stages of *Lymantria dispar:* Striped = contaminated with 100 μg/g Zn (dw in food), dotted = 500 μg/g Zn from hatching onwards, open = controls (uncontaminated diet); A: whole individuals, B: exuviae (larval skins and exuviae), C: feces, and D: head capsules (note different scaling!); missing blocks = no data available; for further explanation see Figure 3 caption.

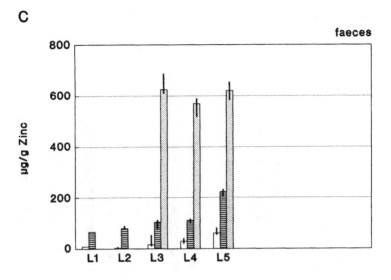

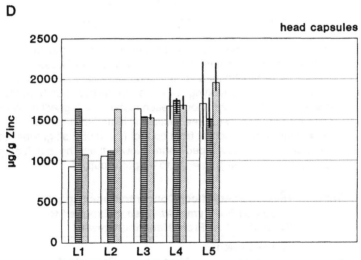

fed oak leaves invariably exceeded those reared on an artificial diet. The growth parameter for *L. dispar* proved to be more sensitive when estimated from head capsule width than from larval weight. Head capsule width was still affected significantly at the lowest concentration of each metal in at least some stages, whereas larval weight was not. Pupal weights decreased with increasing metal concentrations in the larval diet (except for Pb20 females and males). Significant differences in weight were even found between the control groups (artificial diet) of the four batches.

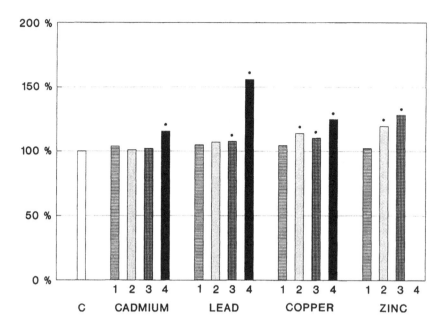

FIGURE 5. Average development rate (estimated from hatching to fifth instar molting) of metal contaminated *Lymantria dispar* larvae in % of the control; block height: median; missing blocks indicate 100% mortality before fourth instar molting; C = control (artificial diet) 100%, 1 = lowest applied concentration of each metal, 4 = highest applied concentration of each metal (see section on Materials and Methods, Table 1); ● significantly different from control ($p < 0.05$); all groups feeding on artificial diet developed more slowly than the oak-leaf group ($p < 0.01$; not shown); n = 22 to 45 per group.

Table 2. No-Observed-Effect Concentrations (μg/g in Food dw) For *Lymantria dispar* Exposed to Several Metals From Hatching Onward

Metal	Development rate	Mortality	Reproduction
Cd	50	50	10
Pb	20	100	—
Cu	10	50	10
Zn	100	100	100

Note: Compare with Figure 5.

4. Reproduction

Since females deposit only one egg cluster after copulation, the reproductive success of *L. dispar* could be estimated from the number of hatched larvae per egg cluster. The highest concentration of each metal caused high larval and pupal mortality and thus totally inhibited reproduction. This also applied to the Zn2500 and Cu250 group. Although imagoes did appear in the Zn500 group, they were not able to copulate successfully, the reason for this probably being their abnormalities in habit and/or behavior.

Imagoes from all other groups mated successfully. Metal concentrations which did not significantly reduce the number of hatched larvae per egg cluster are given in Table 2. No NOEC could be established for lead, since reproduction was affected by 20 μg/g lead but not by 100 μg/g.

C. Effects of Metals on the Parasitoid

1. Development Rate (Parasitization-Eclosion of Imagoes) and Developmental Success (Sum of Eclosed Imagoes)

Parasitoids of the Zn500 and Cu50 larvae showed a clear retardation of development as compared to parasitoids of the control larvae ($p < 0.0001$; Figure 6A) and those of the Zn100 and Cu10 larvae ($p < 0.002$; not shown). In all groups, the developmental rate of the parasitoid was positively correlated with that of the host ($p < 0.005$). Additionally, a negative relationship was found between the number of eclosed *G. liparidis* per host larva and the developmental rate of the gypsy moth ($p < 0.02$). In all less contaminated groups (except lead) more parasitoids per host larva eclosed than in the groups exposed to the higher metal concentrations ($0.05 > p < 0.001$; Figure 6B).

2. Metal Concentrations of the Host and the Parasitoid

All parasitoids eclosed from contaminated hosts (except the Zn500 group) had higher metal concentrations than the controls ($0.05 > p < 0.0001$), although in most cases concentration factors were well below 1. Table 3 summarizes the median metal concentrations of hosts and parasitoids as well as concentration factors.

IV. DISCUSSION

A. Dynamics of Metals

The life-cycle study of *Lymantria dispar* revealed stage- and metal-specific accumulation (Figures 1A to 4A). The concentration-dependent accumulation patterns of the essential metals indicate the existence of regulatory mechanisms in the strict sense, rather than the concentration-independent course observed

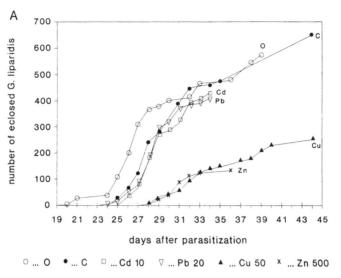

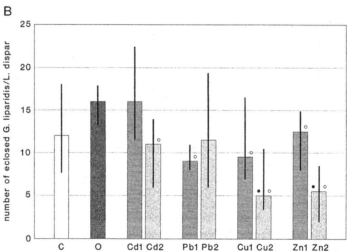

FIGURE 6. A: Eclosion frequency of *Glyptapanteles liparidis* eclosed from its metal-contaminated host *Lymantria dispar*. C = control (artificial diet), O = control (natural diet, oak leaves); other groups, e.g., Cd10 = 10 µg/g in food (dw) of the host larvae; n = 32 to 40 parasitized *L. dispar* larvae per group. B: Average number of eclosed *Glyptapanteles liparidis* per *Lymantria dispar* larva. Block height: median, lower and upper quartiles are shown; n = 32 to 40 parasitized *L. dispar* larvae per group; ● above blocks significantly different from C (*p* <0.05); ○ above blocks significantly different from O (*p* <0.01).

Table 3. Metal Concentrations (μg/g dw) of Male *Glyptapanteles liparidis* (G.I.) and of 4th Instar Larvae of *Lymantria dispar* (L.d.; Remains After Eclosion of the Parasitoids)

	Cadmium			Lead			Copper			Zinc		
	L.d.	G.I.	CF	L.d.	G.I.	CF	L.d.	G.I.	CF	L.d.	G.I.	CF
1	19.7	0.07	0.003	0.61	0.41	0.67	41.0	46.3	1.13	236	70.4	0.30
	(18.6–22.7)	(0.06–0.08)		(0.50–0.74)	(0.31–0.72)		(35.8–43.0)	(43.7–48.9)		(156–273)	(67.8–74.9)	
2	48.0	0.15	0.003	1.62	0.34	0.21	129.0	18.8	0.15	1132	96.6	0.09
	(35.7–65.2)	(0.15–0.22)		(1.37–1.91)	(0.16–0.45)		(124–174)	(18.0–19.9)		(941–1236)	(91.4–106)	

Note: The 4th larval stage was chosen, because the vast majority of the parasitoids eclosed from this stage. Below the medians, lower and upper quartiles are shown in parentheses.
CF: concentration factors (G.I./L.d.).
1.: Low contamination (artificial diet of *L. dispar* contained 2 μg/g Cd, 4 μg/g Pb, 10 μg/g Cu, 100 μg/g Zn dw).
2: High contamination (10 μg/g Cd, 20 μg/g Pb, 50 μg/g Cu, 500 μg/g Zn dw).

for the nonessential metals (at least within the range investigated). Basically, regulation in the strict sense implies the existence of negative feedback loops,[12] although different physiological mechanisms originating from alterations in the dynamics of uptake, distribution and storage, and excretion may also lead to phenomena resembling regulation.

1. Uptake

The uptake of trace metals by terrestrial invertebrates usually takes place from food via the gut. Cutaneous uptake, which in aquatic invertebrates often is of equal or greater importance than uptake via the gut,[12,13] may occur only occasionally in some groups of terrestrial invertebrates such as soil-dwelling organisms and endoparasitoids.[14-16] Metal uptake is not only determined by the degree of contamination in the food. Some organisms are able to discriminate between contaminated and uncontaminated food, adjusting their consumption rate accordingly.[13,17,18] Moreover, the assimilation efficiency may be influenced by the period of exposure, the metal itself (its concentration and its binding state), the temperature, the gut transit time, and the binding mechanisms within the animal.[4,5,14,19-23] In the case of *L. dispar*, lead seems to provide an example of weak assimilation, which is in agreement with the well-known limited mobility of lead in organisms.[20,24] This is confirmed by the similarity of lead concentrations in food and feces, and the low concentrations of this metal in the body (compare results for lead).

2. Distribution and Storage

There is a paucity of detailed information concerning trace metal resorption from gut content in invertebrates (for vertebrates see Tacnet et al.[25]). It can be assumed that the same physiological mechanisms are involved as for the passage of other ions (e.g., diffusion, facilitated diffusion, codiffusion, and active transport). It could be speculated that active transport mechanisms have evolved mainly for essential metals; in contrast, nonessential metals may cross membranes by diffusion or by active transport due to competition with essential metals (e.g., Pb-Ca, Zn-Cd).[24,26] Little is also known about transport forms within body fluids and tissues. In crustaceans, metals are sometimes bound to respiratory pigments.[27]

The distribution between tissues and cellular compartments can vary by orders of magnitude,[1-5,20] but even highest levels of contamination need not necessarily imply damage, if the metal concerned is physiologically inactivated by complexing with metal-binding proteins (MBP)[27-30] or by inclusion into vesicles or lysosomes.[31-36] On the other hand, small amounts of metals may become toxic in the absence of inactivating mechanisms. Generally, tissues with high metabolic rates like the excretory organ, alimentary tract, gonads, etc.[1,4,5,31] are heavily contaminated. Intracellular compartmentalization of metals is a possible means of detoxification (see above), but probably could also represent, for essential metals, a reserve for times of low supplies or enhanced requirement. Metal incorporation into extracellular structures may be regarded as a kind of inactivation, but may also serve functional purposes. For example, the large amounts of zinc found in mandibles[37] and head capsules of insects (present study) may contribute towards hardening these structures. Basically, trace metals are deposited either in mineral concretions for the whole life span (storage-excretion in arachnids and the housefly)[38,39] or temporarily, to be remobilized for metabolic use and/or excretion.

The latter variant is supported by studies in which the concentration of certain metals was found to change in various tissues after termination of exposure. These experiments also proved the existence of metal- and tissue-specific dynamics for uptake, storage, and elimination within single organism.[3,4] Individual differences in these dynamics due to variable physiological states (e.g., development and nutrition), as well as experimental conditions (e.g., temperature), may also cause considerable variation of metal loads within one species.

3. Excretion, Elimination

Chronic exposure to metals normally leads to a steady state, shown to be concentration-dependent in some species.[14,40] In such cases continuous uptake has to be compensated for by continuous elimination. However, Knutti et al.[41] stated that a decrease in growth efficiency may lead to an increase in body concentrations of metal even if the concentration in the food and the ratio between metal absorption and assimilation efficiency remain constant. But Janssen et al.[39] reported that in different arthropods the steady state concentration of Cd depends much more on the mode of excretion than on assimilation efficiency.

So far, it is not clear whether a metal which has been enclosed in vesicles or bound to specific proteins can be remobilized for excretion or for turnover (and thus, probably becoming toxic).

Apart from storage-excretion, two types of elimination can be distinguished: a continuous loss on the one hand and episodic elimination on the other.[8,29,42,43] Continuous excretion of metals is based on the activity of excretory organs and specialized gut epithelial cells. It should be stressed that metal in the feces can be of totally different origins. In addition to the metal not assimilated from food, feces may contain certain amounts from secretion, exocytosis of vesicles, and extrusion of degenerated cells.[17,32,39,44] Therefore, if metal concentrations in the feces exceed those in food this may be due either to a concentration effect in the lumen or to secondary elimination of once-incorporated metals. In the case of arthropods, molts have also to be considered as a regular mode of excretion.

The constant lead load in feces of *L. dispar* plus the low body concentrations of larvae indicate that lead is hardly assimilated at all, whereas cadmium concentrations of feces and larvae display an inverse relationship, which rather implies secondary metal elimination. Surprisingly, copper concentrations of head capsules and exuviae were lower in the Cu50 group than in the control and Cu10 groups (see Figure 3). The larvae of the two latter seem to deposit copper into the exoskeleton and to lose it by molting rather than via the feces, whereas individuals of the Cu50 group eliminate considerable amounts of copper with their feces. Obviously, copper elimination via feces is enhanced when the metal load in the animal exceeds a certain level. Additionally, an episodic elimination of copper takes place via the deposition of the meconium. Preliminary data (not shown) suggest that zinc and cadmium are also eliminated via the meconium. Similar results have been reported for other species.[8,29,42,43] Under circumstances where elimination and/or detoxification are inadequate, toxic effects may become apparent.

B. No-Observed-Effect Concentrations (NOECs)

Ultimately, all of the damages diagnosed as arising from metal intoxication can be explained on the basis of interference with metabolic processes and are not necessarily lethal. The assessment of toxicity at sublethal doses is based on parameters concerning the vitality (e.g., survival, consumption, assimilation, respiration, growth, and developmental rate) and/or reproductive success (fertility, fecundity, and mortality of F1-generation).[13,14,40,43,45,46] In this context, the establishment of NOECs is a useful approach.

In general, the assessment of NOEC levels for a particular species on the basis of experiments involving long-term metal exposure presents difficulties: parameters sensitive to stress, such as fecundity or developmental rate, are very complex quantities, not uniformly applied, and their components (e.g., for fecundity: number of eggs, fertilization or hatching success) are variously affected. Therefore, NOECs based on laboratory studies will vary according to the parameter investigated. Moreover, some experiments revealed considerable differences in NOEC levels for a particular parameter, depending upon whether assessment was made at the individual or population level. Hence, the parameter with the lowest NOEC is not necessarily the most decisive one for the development of the individual and/or population development.[47]

In the present study, the establishment of NOECs was rendered very difficult by the great variability between the different batches. All four control groups, for instance, exhibited significantly different developmental rates. This may be due to different starting conditions for the newly hatched larvae, depending on the quality and quantity of the yolk provided by the parental generation. Diapause conditions (duration, temperature regime) may also be of importance for the developmental rate of larvae.[48,49]

In the present study the effects of metals on *L. dispar* were determined by recording changes in developmental rate, mortality, and reproduction. For cadmium and copper the most sensitive of these parameters proved to be reproduction, whereas for zinc, all parameters investigated proved to have the same NOEC (100 μg/g Zn). Using zinc, the same NOECs for growth and reproduction have been reported for other lepidoptera,[18,50,51] whereas *L. dispar* seems to be more susceptible to copper and lead than other species.[52,53]

The reproductive success of insects depends, among other factors, on the synchronized eclosion of males and females. Thus, the limited number of individuals in the present study and the great variability of individual rates of development as intensified by metal stress may have adversely affected reproductive success. NOECs for the growth of *L. dispar* could not be assessed, since the sensitivity of the different larval stages proved to be inconsistent.[47] But, it can be assumed that the sublethal impairment of a particular stage can be compensated for during further development.

As shown in Table 2 the lowest NOEC levels for each parameter measured in *L. dispar* are 10 μg/g for cadmium, 10 μg/g for copper, 20 μg/g for lead, and 100 μg/g for zinc. These NOECs are probably the most decisive concentrations to be considered for risk assessment on an individual level. However, even strong impairment at the individual level can be compensated for to some extent at the population level. This means that the risk for field populations can only be judged if additional information about their ecology and life-history is available.

C. Ecotoxicological Implications

Metal uptake and accumulation in an animal depends primarily on the species-specific physiology and biology.[19,39,41,55] Thus, if one species is affected by metal contamination, the implications for successive trophic links in the food chain is practically impossible to estimate. Usually, metal transfer in a food chain is described by means of concentration factors based on concentration ratios of two consecutive trophic links. Therefore, the information content of such factors depends on the degree of additional knowledge about life history, ecology, and physiology of the organisms concerned. Thus, concentration factors provide only limited information in the case of extraintestinal digestion, or feeding on hemolymph or certain tissues.

In the present study *G. liparidis* — eclosed from metal-contaminated hosts — showed elevated metal concentrations in comparison to controls, but concentration factors were mostly far below 1. Since *G. liparidis* feeds exclusively on hemolymph, concentration ratios of hemolymph/parasitoid would have more relevance and will be investigated in a further experiment. The demonstrated stage- and metal-specific bioaccumulation of *L. dispar* implies that the potential hazard for parasitoids differs depending on the host stage parasitized and on the feeding habits of the parasitoids. *G. liparidis* does not seem to be affected directly by the metals, since the body concentrations of the parasitoid were rather low. It is more likely that metal stress impairs the hemolymph quality of the host, and hence adversely affects the trophic situation for the parasitoid. In contrast, tissue-feeding parasitoids may rather be negatively affected by direct intoxication with metals, depending on the tissues they prefer. However, indirect disadvantages to parasitoids on account of nutritional conditions within the host hemolymph are of more general interest — considering that many parasites belonging to different systematic groups feed on host hemolymph. In the case of *G. liparidis*, *L. dispar* serves only as the principle summer host; for winter quiescence, the parasitoid depends at least on one additional host,[48] in which the metal pathways may be different. Moreover, developmental rates of *G. liparidis* and *L. dispar* proved to be positively correlated. A retardation in the development of *L. dispar* due to metal contamination may interfere with the synchronization of eclosion of *G. liparidis* and the presence of its winter host. A consequence of this could be the collapse of a population of a divoltine species such as *G. liparidis*.

Generally, an impairment of vitality observed in the laboratory may be enhanced in the field by natural stress factors such as drought, temperature, etc. and the presence of antagonists and competitors. Additionally, field organisms have, as a rule, to face mixtures of contaminants which interfere synergistically or antagonistically.[6,13,31,56] In contrast to most laboratory studies in which combinations of contaminants are only seldomly applied.

REFERENCES

1. **Hopkin, S. P. and Martin, M. H.,** The distribution of zinc, cadmium, lead and copper within the woodlouse *Oniscus asellus* (Crustacea, Isopoda), *Oecologia (Berlin)*, 54, 227, 1982.
2. **Coughtrey, P. J. and Martin, M. H.,** The distribution of Pb, Zn, Cd and Cu within the pulmonate mollusc *Helix aspersa* Müller, *Oecologia (Berlin)*, 223, 315, 1976.
3. **Dallinger, R. and Wieser, W.,** Patterns of accumulation, distribution and liberation of zinc, copper, cadmium and lead in different organs of the land snail, *Helix pomatia, Comp. Biochem. Physiol.,* 79C, 117, 1984.
4. **Hörth, E.,** Untersuchungen zur Verteilung der Metalle Cadmium, Blei und Zink in den Organen der Schabe *Periplaneta americana,* Diplomarbeit, Universtiät Wien, 1989.
5. **Berger, B. and Dallinger, R.,** Accumulation of cadmium and copper by the terrestrial snail *Arianta arbustorum* L.: kinetics and budgets, *Oecologia (Berlin)*, 79, 60, 1989.
6. **Vogel, W. R.,** Zur Aufnahme und Auswirkung der Schwermetalle Zink und Cadmium beim Mehlkäfer *Tenebrio molitor* L. (Col., Tenebrionidae) unter Berücksichtigung möglicher Wechselwirkungen, *Zool. Anz.,* 220, 25, 1988.
7. **Scharnagl, A. and Vogel, W.,** Schwermetalleinflüsse auf ein Beute-Räuber-System (*Drosophila melanogaster — Chrysopa carnea*). *Forschungsinitiative gegen das Waldsterben,* Symp. 1988, Führer, E. und Neuhuber, F., Hrsg., BMfWuF, Wien, 1988, 295.
8. **Ortel, J.,** Einflüsse von Cadmium und Blei auf eine Wirt — Parasitoidbeziehung (*Galleria mellonella* L., Lepidopt. — *Pimpla turionellae* L., Hym.), Dissertation, Universität of Wien, 1989.
9. **Führer, E.,** Air pollution and the indices of forest insect problems, *Z. Angew. Entomol.,* 99, 371, 1985.
10. **Heliövaara, K.,** Occurrence of *Petrova resinella* (Lepidoptera, Torticidae) in a gradient of industrial air pollutants, *Silva Fenn.,* 20, 83, 1986.

11. **Bell, R. A., Owens, C. D., Shapiro, M., and Tardif, J. R.**, Mass rearing and virus production, in The Gypsy Moth: Research Toward Integrated Pest Management, Doane, C. C. and McManus, M. L., Eds., Tech. Bull. 1584, U.S. Department of Agriculture, Washington, D.C., 1981, 599.

12. **Depledge, M. H. and Rainbow, P. S.**, Models of regulation and accumulation of trace metals in marine invertebrates, *Comp. Biochem. Physiol.*, 97C, 1, 1990.

13. **Van Capelleveen, E.**, *Ecotoxicity of Heavy Metals for Terrestrial Isopods*, Free University Press, Amsterdam, 1987.

14. **Streit, B.**, Effects of high copper concentrations on soil invertebrates (earthworms and oribatid mites): Experimental results and a model, *Oecologia (Berlin)*, 64, 381, 1984.

15. **Vinson, S. B. and Barras, D. J.**, Effects of the parasitoid, *Cardiochiles nigriceps*, on the growth, development, and tissues of *Heliothis virescens*, *J. Insect Physiol.*, 16, 1329, 1970.

16. **Edson, K. M. and Vinson, S. B.**, Nutrient absorption by the anal vesicle of the braconid wasp, *Microplitis croceipes*, *J. Insect Physiol.*, 23, 5, 1977.

17. **Joosse, E. N. G. and Verhoef, S. C.**, Lead tolerance in Collembola, *Pedobiologia*, 25, 11, 1983.

18. **Gahukar, R. T.**, Effects of dietary zinc sulphate on the growth and feeding behavior of *Ostrinia nubilalis* Hbn. (Lep, Pyraustidae), *Z. Ang. Entomol.*, 79, 352, 1975.

19. **Van Hook, R. I. and Yates, A. J.**, Transient behaviour of cadmium in a grassland arthropod food chain, *Environ. Res.*, 9, 76, 1975.

20. **Hopkin, S. P., Watson, K., Martin, M. H., and Mould, M. L.**, The assimilation of heavy metals by *Lithobius variegatus* and *Glomeris marginata* (Chilopoda, Diplopoda), *Bijd. Dierk.*, 55, 88, 1985.

21. **Ireland, M. P. and Wooton, R. J.**, Variations in the lead, zinc and calcium content of *Dendrobaena rubida* (Oligochaeta) in a base metal mining area, *Environ. Pollut.*, 10, 201, 1976.

22. **Dallinger, R. and Wieser, W.**, The flow of copper through a terrestrial food chain. I. Copper and nutrition in isopods, *Oecologia (Berlin)*, 30, 253, 1977.

23. **Wieser, W., Dallinger, R., and Busch, G.**, The flow of copper through a terrestrial food chain. II. Factors influencing the copper content of isopods, *Oecologia (Berlin)*, 30, 265, 1977.

24. **Fangmaier, A. and Steubing, L.**, Cadmium and lead in the food web of a forest ecosystem, *Atmospheric Pollutants in Forest Areas*, 1986, 223.

25. **Tacnet, F., Watkins, D. W., and Ripoche, P.**, Zinc binding in intestinal brush-border membrane isolated from pig, *Biochim. Biophys. Acta*, 1063, 51, 1991.

26. **Webb, M.**, Interactions of cadmium with cellular components, in *The Chemistry, Biochemistry and Biology of Cadmium*, Webb, M., Ed., Elsevier/North-Holland, Amsterdam, 1979.

27. **Donker, M. H., Koevoets, P., Verkleij, J. A. C., and Van Straalen, N. M.**, Metal binding compounds in hepatopancreas and haemolymph of *Porcellio scaber* (Isopoda) from contaminated and reference areas, *Comp. Biochem. Physiol.*, 97C, 119, 1990.

28. **Maroni, G. and Watson, D.**, Uptake and binding of cadmium, copper and zinc by *Drosophila melanogaster* larvae, *Insect Biochem.*, 15, 55, 1985.

29. **Aoki, Y. and Suzuki, K. T.**, Excretion of Cd and changes in the relative ratio of iso-cadmium-binding proteins during metamorphosis of fleshfly *(Sarcophaga peregrina)*, *Comp. Biochem. Physiol.*, 78, 315, 1984.

30. **Dallinger, R., Berger, B., and Bauer-Hilty, A.**, Purification of cadmium-binding proteins from related species of terrestrial Helicidae (Gastropoda, Mollusca): A comparative study, *Mol. Cell. Biochem.*, 85, 135, 1989.

31. **Jeantet, A. Y., Ballan-Dufrancais, C., and Martoja, R.**, Insect resistance to mineral pollution. Importance of spherocrystals in ionic regulation, *Rev. Ecol. Biol. Sol.*, 14, 563, 1977.

32. **Humbert, W.**, The mineral concretions in the midgut of *Tomocerus minor* (Collembola): microprobe analysis and physiological significance, *Rev. Ecol. Biol. Sol.*, 14, 71, 1977.

33. **Brown, B.**, The form and function of metal-containing "granules" in invertebrate tissues, *Biol. Rev.*, 57, 621, 1982.

34. **Prosi, F., Storch, V., and Janssen, H. H.**, Small cells in the midgut glands of terrestrial Isopoda: sites of heavy metal accumulation, *Zoomorphologie*, 102, 53, 1983.

35. **Dallinger, R. and Prosi, F.**, Heavy metals in the terrestrial isopod *Porcellio scaber* Latreille. II. Subcellular fractionation of metal-accumulating lysosomes from hepatopancreas, *Cell Biol. Toxicol.*, 4, 97, 1988.

36. **Hopkin, S. P. and Martin, M. H.**, The distribution of zinc, cadmium, lead and copper within the hepatopancreas of a woodlouse, *Tissue & Cell*, 14, 703, 1982.

37. **Hillerton, J. E., Robertson, B., and Vincent, J. F. V.**, The presence of zinc and manganese as the predominant metal in the mandibles of adult, stored-product beetles, *J. Stored Prod. Res.*, 20, 133, 1984.

38. **Sohal, R. S. and Lamb, R. E.**, Storage-excretion of metallic cations in the adult housefly, *Musca domestica*, *J. Insect Physiol.*, 25, 119, 1979.

39. **Janssen, M. P. M., Bruins, A., De Vries, T. H., and Van Straalen, N. M.**, Comparison of cadmium kinetics in four soil arthropod species, *Arch. Environ. Contam. Toxicol.*, 20, 305, 1991.

40. **Bengtsson, G., Gunnarson, T., and Rundgren, S.**, Influence of metals on reproduction, mortality and population growth in *Onychiurus armatus* (Collembola), *J. Appl. Ecol.*, 22, 967, 1985.

41. **Knutti, R., Bucher, P., Stengl, M., Stolz, M., Tremp, J., Ulrich, M., and Schlatter, C.**, Cadmium in the invertebrate fauna of an unpolluted forest in Switzerland, in *Cadmium*, Environ. Toxin Ser. 2, Stoeppler, M. and Piscator, M., Eds., Springer-Verlag, New York, 1988, 171.

42. **Vogel, W. R.**, Zur Schwermetallbelastung der Borkenkäfer, *Entomol. Exp. Appl.*, 42, 259, 1986.

43. **Ernst, W. H. O. and Joosse, E. N. G.**, *Umweltbelastung durch Mineralstoffe*, Gustav Fischer, Jena, 1983.

44. **Humbert, W.**, The midgut of *Tomocerus minor* Lubbock (Insecta, Collembola): ultrastructure, cytochemistry, aging and renewal during a moulting cycle, *Cell Tissue Res.*, 196, 39, 1979.

45. **Bengtsson, G., Gunnarsson, T., and Rundgren, S.**, Effects of metal pollution on the earthworm *Dendrobaena rubida* (Sav.) in acidified soils, *Water, Air, Soil Pollut.*, 28, 361, 1986.

46. **Ortel, J. and Vogel, W.,** Effects of lead and cadmium on oxygen consumption and life expectancy of the pupal parasitoid, *Pimpla turionellae, Entomol. Exp. Appl.,* 52, 83, 1989.

47. **Van Straalen, N. M., Schobben, J. H. M., and De Guede, R. G. M.,** Population consequences of cadmium toxicity in soil microarthropods, *Ecotoxicol. Environ. Saf.,* 17, 190, 1989.

48. **Schopf, A.,** The effect of host age of *L. dispar* larvae L. (Lep., Lymantriidae) on the development of *Glyptapanteles liparidis* Bouché (Hym., Braconidae), *Entomophaga,* in press.

49. **Capinera, J. L., Barbosa, P., and Hagedorn, H. H.,** Yolk and yolk depletion of gypsy moth eggs: implications for population quality, *Ann. Entomol. Soc. Am.,* 70, 40, 1977.

50. **Miyoshi, T., Shimizu, O., Miyazawa, F., Machida, J., and Ito, M.,** Effect of heavy metals on the mulberry plant and silkworm. III. Cooperative effects of heavy metals on silkworm larvae *Bombyx mori* L., *J. Seric. Sci. Japan,* 47, 77, 1978.

51. **Salama, H. S.,** Zinc sulphate induces sterility in the cotton leafworm *Spodoptera littoralis* Boisduval, *Experientia,* 28, 1318, 1972.

52. **Weisman, L. and Svatarakova, L.,** The influence of lead on some vital manifestations of insects, *Biologia (Bratislava),* 36, 147, 1981.

53. **Sivapalan, P. and Gnanapragasma, N. C.,** Influence of copper on the development and adult emergence of *Homona coffearia* (Lepidoptera: Tortricidae) reared *in vitro, Entomol. Exp. Appl.,* 28, 59, 1980.

54. **Künig, M.,** Beeinflussung der Entwicklung und Fortpflanzung von *Lymantria dispar* L. (Lep.) durch Blei, Cadmium, Kupfer und Zink, Diplomarbeit, Universität Wien, 1989.

55. **Van Straalen, N. M. and Van Wensem, J.,** Heavy metal content of forest litter arthropods as related to body-size and trophic level, *Environ. Pollut.,* 42A, 209, 1986.

56. **Ortel, J.,** Effects of lead and cadmium on chemical composition and total water content of the pupal parasitoid, *Pimpla turionellae, Entomol. Exp. Appl.,* 59, 93, 1991.

CHAPTER 21

Soil and Sediment Quality Criteria Derived From Invertebrate Toxicity Data

Nico M. van Straalen

TABLE OF CONTENTS

427

I. INTRODUCTION

The study of metals in invertebrates has centered mostly around questions of physiological and biochemical implications, such as metal uptake, regulation, and subcellular distribution. Metal pathways in invertebrates have been excellently reviewed by Hopkin.[1] The picture of how metals are assimilated, distributed over certain organs, bound to cellular constituents, and excreted, is becoming more and more clear.[2] Still, the quantitative study of dose-related adverse effects has received much less attention.

For many invertebrates, especially in the terrestrial environment, even the most basal toxicological data (LC_{50}) are lacking. Only earthworms have been investigated to some extent, and standardized test procedures using the species *Eisenia foetida* have been proposed.[3] However, several toxicity studies using invertebrates have applied only a single dose level, or two metals, each at a single dose and in combination. Although these studies are useful in that they demonstrate that metals have an intrinsic toxicity, they ignore the fundamental theorem of toxicology, that is, each effect depends on the dose. Carefully designed toxicity studies, using a geometric series of exposure levels with a ratio smaller than ten are very rare. Yet, these types of studies are most useful when it comes to establishing maximum acceptable concentrations of metals in the environment.

This chapter attempts to review the few toxicological data of metals in invertebrates, and to use these in a recently developed framework for derivation of soil quality criteria.[4] A comparison will be made between criteria for soils, derived from terrestrial invertebrate data, and criteria for sediments, derived from data on benthic invertebrates.

II. MECHANISMS OF INTOXICATION

On the biochemical level, several targets for metals have been identified which may lead to physiological lesions. These include binding to proteins, binding to nucleic acids, catalysis of free radicals, and replacement of essential metals. However, it is often not clear which lesions are responsible for decreased performance, or disturbed behavior, of the intact animal.

In some cases, changes in the internal distribution of metals may be indicative of intoxication. For example, Hopkin and Martin[5] suggested that zinc becomes toxic to isopods *(Oniscus asellus)* when the storage capacity of the hepatopancreas is saturated, allowing the metal to pass to organs where it normally does not belong. Van Straalen et al.[6] demonstrated that cadmium uptake in an oribatid mite *(Platynothrus peltifer)* was correlated with an equimolar loss of zinc. Janssen and Dallinger[7] demonstrated that when slugs *(Arion lusitanicus)* were exposed to a high cadmium dose, cadmium in the midgut gland was not only bound to a MT-like protein, but was also present in high molecular weight fractions, possibly due to a "spillover effect".

Studies like these may contribute to finding physiological explanations for metal intoxication in invertebrates. However, the total metal burden of an animal does not determine its suceptibility to toxic effects. Among the animals which concentrate metals to a high degree, some are very resistant in terms of lethality (e.g., isopods). Comparing the lethal body burdens of cadmium in springtails and mites, Van Straalen et al.[6] concluded that springtails die sooner than mites, but at a much lower internal concentration. Clearly, there is no relation between the bioconcentration potential of a species and its disposition to metal intoxication.

Metals usually have a low acute toxicity. Unlike insecticides, lesions caused by metals will not directly kill an animal, but will first cause physiological disturbances leading to decreased growth and reproduction. For metals one may expect the LC_{50} to be considerably greater than EC_{50} values for sublethal criteria. This paper will therefore not be concerned with LC_{50}s, but with sublethal effects only, expressed as NOECs (No Observed Effect Concentrations).

III. TOXICITY DATA FOR SOIL AND SEDIMENT INVERTEBRATES

Data on toxicity of metals to soil and sediment invertebrates were collected from the literature. Only those studies were selected from which a no observed effect concentration (NOEC) could be derived, using reproduction (egg-laying and cocoon production) as a criterion. In some cases other criteria (growth, consumption, and development) were considered as well, where it was clear that this would affect reproduction. The NOEC was taken as the highest concentration in soil or food causing no adverse effect, out of a series of concentrations applied in the experiment. The NOEC, expressed in μg/g dry mass, was either taken as stated by the author, or it was read from tables or figures. Some studies using unusual substrates (e.g., shred paper or sewage sludge), or inadequate replication, were rejected. In a few cases details of experimental conditions were confirmed with the authors.

Table 1 provides a list of the raw data for soil invertebrates, as well as some calculated figures to be explained later. Cadmium, copper, and lead are the most thoroughly investigated metals, fewer data are available for mercury, zinc, nickel, and chromium. Among the animals, most data concern earthworms, followed by slugs and springtails. Hardly any data are known for oribatid mites, although this is the most species-rich group of soil invertebrates.

Table 1. Survey of No-Effect Concentrations of Seven Metals, to Various Soil Invertebrates

Metal	Species	NOEC	L	H	NÔEC	Ref.
Cadmium	*Dendrobaena rubida*	100	0	5.7	154	8
	Lumbricus rubellus	10	17	3.4	13.6	9
	Eisenia foetida	25	0	50	13.8	10
	Helix aspersa	10	0	85.5	3.64	11
	Porcellio scaber	10	0	95	3.34	12
	Orchesella cincta	56	0	95	18.7	6
	Folsomia candida	73	10	8	91.5	13
	Platynothrus peltifer	2.9	0	95	0.97	6
Copper	*Dendrobaena rubida*	122	0	5.7	238	8
	Lumbricus rubellus	30	17	3.4	39.6	9
	Eisenia foetida	60	10	8	83.7	3
	Allolobophora caliginosa	50	0	1	115	14
	Arion ater	25	0	95	12.5	15
	Onychiurus armatus	2608	0	95	1304	16
	Platynothrus peltifer	168	0	95	84	17
Lead	*Dendrobaena rubida*	560	0	5.7	855	8
	Lumbricus rubellus	200	17	3.4	241	9
	Eisenia foetida	1000	0	50	850	10
	Allolobophora caliginosa	1000	0	1	1667	14
	Arion ater	1000	0	95	586	15
	Onychiurus armatus	1096	0	95	642	18
	Aiolopus thalassinus	100	0	0	170	19
	Platynothrus peltifer	431	0	95	253	17
Zinc	*Eisenia foetida*	1000	0	50	1120	10
	Arion ater	100	0	95	72.7	15
	Porcellio scaber	398	0	95	290	12
Mercurcy	*Eisenia foetida*	3.25	3	59	3.14	20
	Octochaetes pattoni	0.25	25	10	0.25	21
	Arion ater	10	0	95	8.30	15
	Aiolopus thalassinus	0.12	0	0	0.18	19
Nickel	*Eisenia foetida*	100	0	50	350	10
	Lumbricus rubellus	50	17	3.4	64.8	9
Chromium	*Octochaetes pattoni*	1.0	25	10	1.0	21

Note: NOEC = no observed effect concentration (highest concentration in the experiment causing no adverse effect); L = lutum content (particles <2 μm) of the soil, in w/w percentages; H = organic matter content of the soil, in w/w percentages; NÔEC = NOEC adjusted to standard conditions (L = 25, H = 10), according to Equation (1) and Table 2.

Variation between studies is not only due to different species being used, but also to the animals being tested under different conditions. Properties of the substrate, such as organic matter content and pH, will greatly influence the bioavailability of the metals and hence their toxicity. In an attempt to reduce some of the variation due to different substrates being used in the experiments,

Table 2. System of Linear Equations Relating Background Concentrations of Metals in Soil to Clay Content and Humus Content[22]

Metal	Equation	Reference value for "standard soil" (L = 25, H = 10) ($\mu g\ g^{-1}$)
Cd	R = 0.4 + 0.007 (L + 3 H)	0.8
Cu	R = 15 + 0.6 (L + H)	36
Pb	R = 50 + L + H	85
Zn	R = 50 + 1.5 (2L + H)	140
Hg	R = 0.2 + 0.0017 (2L + H)	0.3
Ni	R = 10 + L	35
Cr	R = 50 + 2 L	100

Note: R = background concentration in $\mu g\ g^{-1}$ (reference value); L = lutum content of soil (w/w %); H = organic matter content of soil (w/w %).

we adopted the system of reference values as it is used in the Netherlands to adjust for soil factors.

In this system, background levels of metals in uncontaminated soils were related to the clay content of the soil and its humus content. Sandy soils with little humus tend to have lower metal concentrations than clayey, peaty soils; criteria for clean soils were qualified accordingly. The qualification takes the form of a metal-specific linear equation between the reference value, the lutum content (L), and the organic matter content (H) of the soil. Table 2 lists the equations.[22] It can be seen, for example, that for cadmium the influence of L and H is not so great (coefficient = 0.007), while for copper the influence is much stronger (coefficient = 0.6). This more or less reflects the stronger adsorption of copper to soil, compared to cadmium.

The data in Table 1 were adjusted for differences in substrates between studies by means of the equations from Table 2, in the following way:

$$\text{NÔEC} = \text{NOEC (L, H)}\frac{R(25,\ 10)}{R\ (L,\ H)}$$

where: NÔEC = no observed effect concentration adjusted to "standard conditions", i.e., 25% clay and 10% humus;

NOEC (L, H) = no observed effect concentration as taken from the experiment using a substrate with L% clay and H% humus;

R (25, 10) = reference value for "standard soil", derived from Table 2 using L = 25 and H = 10; and

R (L, H)   = reference value associated with the experimental
                 conditions (L, H).

For most of the studies cited in Table 1, clay contents and organic matter contents of the soils used could be derived from the descriptions given in the papers; in a few cases an educated guess had to be made. Several experiments, however, did not use contaminated soil, but used contaminated food with an inert substrate (e.g., plaster). To allow a comparison with soil, it was assumed that food could be equated to a soil with 95% organic matter and no clay. In this way the concentrations for soil are lower than the corresponding ones for food; this reflects the fact that saprotrophs feeding on soil organic matter are exposed to a higher concentration than the average concentration for the total soil, since metals are concentrated in soil organic matter.

This system of normalization does not pretend to be the final answer. It is one of the possible ways to correct for the influence of soil factors on metal toxicity. Another important soil factor, pH, is not taken into account here. For sediments, metal concentrations may be normalized on the basis of the acid-volatile sulfide content.[23] Further research is needed to better quantify the relations between toxicity and soil characteristics.

For benthic invertebrates a similar exercise was done. Only those studies were selected where metals were applied to a sediment and concentrations could be expressed as $\mu g \ g^{-1}$ dry mass of sediment. This is not to mean that animals directly ingest sediment, or are exposed via the sediment only: many benthic invertebrates take up metals from the pore water.[24]

For nonionic organic chemicals, sediment quality criteria can be derived from aquatic toxicity experiments by assuming that the partitioning of the chemical between sediment organic carbon and pore water is at equilibrium.[25] For metals, however, relations between pore water and sediment are extremely complex and depend on a range of sediment characteristics.[26] For the purpose of the present paper, it was decided to include only those studies reporting NOEC values expressed per unit of sediment.

When the clay and humus contents of the experimental sediments were not given in the literature, it was assumed that H = 4% and L = 3.5%, which are the figures for the sediment used by Hooftman and Adema.[27] The raw data, as well as the adjusted NOECs are given in Table 3. Among benthic invertebrates, chironomid larvae are most widely investigated. Many other animals are used in aquatic toxicity experiments, but these data are not considered here, as explained above.

For cadmium, lead, and copper the data allow some statistical treatment; for the other metals the data are too limited to draw concrete conclusions. Since response curves in toxicology are usually expressed on a logarithmic concentration scale, it is relevant to calculate geometric means, instead of arithmetic means. This is presented in Table 4.

Table 3. Survey of No-Effect Concentrations of Three Metals in Sediments, to Benthic Invertebrates

Metal	Species	NOEC ($\mu g\ g^{-1}$)	L	H	NÔEC ($\mu g\ g^{-1}$)	Ref.
Cadmium	*Rhepoxynius abronius*	3.5	3.5	4	5.51	28
	Chironomus tentans	32	3.5	4	50.3	27
	Chironomus riparius	32	3.5	4	50.3	27
	Chironomus plumosus	100	3.5	4	157	27
Copper	*Protothaca staminea*	4.1	0	0.09	9.8	29
	Gammarus lacustris	488	3.5	3	930	30
	Hyaella azteca	875	3.5	3	1670	30
	Chironomus tentans	100	3.5	4	185	27
	Chironomus riparius	1000	3.5	4	1846	27
	Chironomus plumosus	100	3.5	4	185	27
	Chironomus decorus	750	0	95	375	31
Mercury	*Chironomus tentans*	100	3.5	4	137	27
	Chironomus riparius	3200	3.5	4	4390	27
	Chironomus plumosus	3200	3.5	4	4390	27

Note: NOEC = no observed effect concentration (highest concentration in the experiment causing no adverse effect); L = lutum content (particles <2 μm) of the sediment, in w/w percentages; H = organic matter content of the sediment, in w/w percentages; NÔEC = NOEC adjusted to standard conditions (L = 25, H = 10), according to Equation (1) and Table 2.

Table 4. Summary Statistics for Toxicity Data of Metals to Soil and Sediment Invertebrates

	x_m	s_m	$\overline{NOEC}$ ($\mu g\ g^{-1}$)	$\overline{NOEC}$ ($\mu mol\ g^{-1}$)
Soil				
Cd	2.523	1.701	12.5	0.111
Cu	4.637	1.449	103	1.63
Pb	6.238	0.780	512	2.47
Sediment				
Cd	3.650	1.402	38.5	0.342
Cu	5.775	1.813	322	5.07

Note: x_m = mean of ln (NÔEC) in Tables 1 and 3; s_m = standard deviation of ln (NÔEC); $\overline{NOEC}$ = mean no effect concentration after back-transformation to the original scale: $\overline{NOEC}$ = exp (x_m).

When the atomic masses of the metals are taken into account, cadmium, on the average, is 15 times more toxic than copper and 22 times more toxic than lead, in the case of soil invertebrates. For sediments, cadmium likewise appears to be 15 times more toxic than copper (Table 4).

The variability among species seems to be smallest for the least toxic metal, lead. Earthworms and snails are very sensitive to copper, but springtails are very insensitive. Oribatid mites are extremely sensitive to cadmium, but earthworms are rather insensitive to this metal. For lead such differences are less clear, all species reacting more or less within the same order or magnitude. This could indicate that the physiological lesions caused by lead are common to all invertebrates, while the receptors for cadmium and copper are different for each species.

Comparing soil and sediment invertebrates, the average NOECs are higher for the latter group. This may relate to a lower bioavailability of metals in sediments, compared to soils, e.g., due to the precipitation of sulfides.[23] Especially in the case of mercury, toxicity in sediments seems to be considerably lower than toxicity in soils (compare Tables 1 and 3); this correlates with the extremely low solubility product of mercury sulfide. For cadmium and copper, the difference between soils and sediment is not so large, and actually appears to be statistically nonsignificant when a *t*-test for comparison of means is applied (Table 4).

IV. DERIVATION OF MAXIMUM ACCEPTABLE CONCENTRATIONS

The data in Tables 1 and 3 suggest that, for each metal, NOEC values are dispersed according to a frequency distribution, with both very sensitive and very insensitive species being relatively rare. Although the data do not allow a precise determination of the frequency function, one may assume a symmetric distribution on a logarithmic scale, such as the lognormal, or the loglogistic. This assumption is confirmed when tested on more extensive data sets.[32,33] The loglogistic distribution, as elaborated in Kooijman[32] hardly differs from the lognormal distribution.[34] From a mathematical point of view, the log logistic distribution is simpler since it can be integrated analytically, while the log normal distribution cannot.

Assuming a loglogistic frequency distribution of sensitivities, Van Straalen and Denneman[4] proposed to consider a quantity called HCp, that is: hazardous concentration for p% of the species. HCp is to be interpreted as a concentration in the environment such that, at most p% of the species in that environment has an NOEC smaller than HCp. Following Kooijman,[32] the estimate for HCp includes an uncertainty margin due to the fact that only a limited sample of species is used to estimate the frequency distribution of sensitivities.

A detailed account of the theory is given in Kooijman,[32] Van Straalen and Denneman,[4] and Van Straalen.[35] Recently, the way to estimate the uncertainty

Table 5. Values for the Factor d_m in Equation (2), For Various m (= nr. of Species Tested)

m	d_m	m	d_m
2	3.72	11	2.56
3	3.40	12	2.53
4	3.22	13	2.51
5	3.06	14	2.50
6	2.93	15	2.49
7	2.82	20	2.44
8	2.72	30	2.30
9	2.65	∞	1.814
10	2.59	—	—

From Kooijman, S. A. L. M., *Water Res.*, 21, 269, 1987. With permission.

margin has been refined by Aldenberg and Slob[36] and Wagner and Løkke.[34] According to their arguments, the confidence interval for HCp should be larger than is assumed in Van Straalen and Denneman,[4] based on Kooijman.[32] These statistical arguments will have to be settled by further development of the theory. For the purpose of the present paper; the original approach in Van Straalen and Denneman[4] will suffice.

According to the log logistic distribution, HCp is given by:

$$HCp = \exp(x_m - s_m d_m k_p) \tag{2}$$

where: x_m = mean of m NÔEC-data, each transformed to natural logarithms,

s_m = standard deviation of the m ln (NÔEC),

d_m = a factor depending on the sample size m, given in Table 5,

k_p = a factor depending on the percentage of unprotected species p, given in Table 6.

The estimated HCp decreases with decreasing mean NÔEC, with increasing standard deviation of NÔEC, with decreasing number of tested species (see Table 5), and with decreasing p (percentage of species with NÔEC < HCp). The accepted value for p in the Dutch environmental policy is 5%; accordingly, HCp is called HC5.

Application of the approach requires a reasonable set of data, preferably more than five species. If m <5, then the influence of the uncertainty margin becomes increasingly large. Another condition for applying Equation 2 is that the available toxicity data (Tables 1 and 3) can be considered as a representative sample from the community of invertebrates. This is of course very doubtful, since the taxonomic diversity is rather limited. Criteria for "representativeness"

Table 6. Values for the Factor k_p in Equation (2), For Various Values of p (= Percentage of Species With NOEC Smaller Than HCp)

p (%)	k_p
10	0.668
5	0.895
1	1.397
0.1	2.099

Table 7. Estimated Hazardous Concentrations for 5% of Soil and Sediment Invertebrates, Calculated According to Equation (2)

Element	HC5 (μg g^{-1})		EC guide value (μg g^{-1})
	Soil	Sediment	
Cadmium	0.20	0.68	1.0
Copper	2.66	3.32	50
Lead	76.6	—	50

have to be developed further. Yet, as this situation will not improve soon, it is illustrative to apply the model using the best of present data.

Table 7 provides some HC5 values, calculated according to Equation 2, using Tables 4, 5, and 6. HC5 is compared with the EC guide values. It is interesting to note that the HC5 estimates for soil and sediment do not differ too much, in the case of cadmium and copper. There is no reason to believe that quality criteria derived for soil would not be valid for sediments as well. This is an important conclusion in the context of environmental policy, which, at least in the Netherlands, tends to keep to a single list of criteria for both soils and sediments. This facilitates the evaluation of dredged material, temporarily inundated soils, marshes, etc.

The HC5 values for cadmium are in the background range[37] and do not differ much from the EC value. For copper, HC5 is rather low, and would be too low from the viewpoint of plant nutrition. This indicates the fact that the "window of essentiality", in the sense of Hopkin,[1] will not be the same for every species: there may be concentrations of copper in soil that are deficient to some species, but are mildly toxic to other, very sensitive species. The HC5 for copper is especially influenced by the sensitivity of both earthworms and molluscs for this element.

The HC5 for lead is somewhat above the EC value. Apparently, soil invertebrates are not particularly susceptible to this element. Concentrations higher than the HC5 are very common in urban areas (see Dallinger et al.[37]).

V. DISCUSSION

The role of invertebrate ecotoxicology for providing a framework for setting environmental standards so far has been rather limited. As this chapter has shown, this is mainly due to a lack of adequate toxicological data. Many studies cited in Tables 1 and 3 were not designed for estimating NOECs; variability in the data is not only due to different species being investigated, but also due to variation in experimental conditions, bioavailability of the metal, etc. Instead of correcting for these differences, as was attempted in this paper, it may be better to standardize the toxicity tests themselves, especially with respect to the substrates used. A promising development is the availability of reproduction toxicity tests using earthworms and springtails exposed in "artificial soil", a substrate with a reproducible composition.[3,13,38]

The influence of scientific progress in test results is illustrated by comparing the presently established HC5 values with those given in Van Straalen.[35] When the recently reported data for *Folsomia candida*[13] and *Aiolopus thalassinus*[19] were added, HC5 for cadmium increased from 0.16 to 0.20 $\mu g\ g^{-1}$, while HC5 for lead decreased from 112 to 77 $\mu g\ g^{-1}$. This is a result of both the new data (*F. candida* is relatively insensitive to cadmium, while *A. thalassinus* is rather sensitive to lead), and the decrease of the uncertainty margin.

Another way to improve the role of ecotoxicology in contributing to environmental regulation is to better develop ecotoxicology itself. There are three issues at stake here, that require some consideration.

In the first place, the relation between metal kinetics and metal dynamics is still not very clear. To predict toxicities and modes of action it is not only sufficient to know the kinetics of the metal and the internal distribution, but also the physiological processes that are susceptible to disturbance by metals.[39] Uptake and excretion efficiencies vary considerably between various invertebrate groups;[40] a high body burden may be the result of either a low excretion efficiency or a high assimilation efficiency, with no apparent relation between the two. To understand the dynamic action of metals, the behavior of the metal should be approached as a rate process, not as a static distribution.

Secondly, metals in the environment elicit a much more complex response than can be studied in laboratory toxicity tests. One point in particular is the effect of interactions between species that may be critical in the field. For example, Tranvik and Eijsackers[41] demonstrated that metal pollution changed the feeding preference of Collembola grazing on fungi. Van Wensem et al.[42] showed that an organotin pesticide affected detritivore-microbial activity more than microbial activity alone. To better understand the role of ecological interactions in toxicological responses, a more fundamental approach is needed, and soil microecosystems may be useful instruments.[43]

In the third place, selection for the most tolerant genotypes will be a continuous process in the field, and will invalidate predictions from simple tests. The evidence for the presence of genetic variation for metal tolerance in natural populations is restricted to a few examples.[44-46] These studies demonstrate that, under intense selection at mining sites, or at sites under heavy industrial influence, both sediment and soil invertebrates will develop tolerant populations. The consequences of resistance, and the underlying physiological mechanisms deserve further study.

Metal ecotoxicology using invertebrates has developed into an extensive field of research. Although dose-response relations are still limited, a framework for applying results in environmental legislation is already available. To further strengthen the position of the field, fundamental issues require more attention.

VI. ABSTRACT

Quantitative dose-response relations provide the basis for establishing environmental quality criteria. For metal toxicity to soil invertebrates, only few data are available. This chapter reviews these data and applies a recently developed model to arrive at maximum acceptable concentrations of metals in soils. These concentrations, expressed as "hazardous concentrations for 5% of the species" (HC5), are estimated as 0.20 μg g^{-1} for Cd, 2.7 μg g^{-1} for Cu, and 77 μg g^{-1} for Pb. For Zn, Hg, Ni, and Cr the data set is too limited to derive criteria. Preliminary HC5 values are also established for sediments, based on benthic invertebrate toxicity data. There seems to be no systematic difference between criteria derived for soils and sediments, in the case of Cd and Cu. The HC5 values for Cd and Cu are below the EC guide values, while HC5 for Pb is somewhat higher. The role of invertebrate ecotoxicology for providing a framework for setting environmental standards is discussed, emphasizing the need for standardization of soil toxicity tests, for more comparative toxicity data, and for a better understanding of ecological responses to metal pollution.

ACKNOWLEDGMENTS

Theo Traas, Carl Denneman, and John Schobben performed various literature studies providing the basis for this paper. I have appreciated discussing this material with Joop Vegter, Joke van Wensem, Martien Janssen, Leo Posthuma, and Marianne Donker. Désirée Hoonhout is acknowledged for typing the text. The editors of this book provided valuable comments to an early version of the manuscript.

REFERENCES

1. **Hopkin, S. P.**, *Ecophysiology of Metals in Terrestrial Invertebrates*, Elsevier Applied Science, London, 1989.
2. **Dallinger, R.**, this volume, chap. 14.
3. **Van Gestel, C. A. M., Van Dis, W. A., Van Breemen, E. M., and Sparenburg, P. M.**, Development of a standardized reproduction toxicity test with the earthworm species *Eisenia foetida andrei* using copper, pentachlorophenol, and 2,4-dichloroaniline, *Ecotoxicol. Environ. Saf.*, 18, 305, 1989.
4. **Van Straalen, N. M. and Denneman, C. A. J.**, Ecotoxicological evaluation of soil quality criteria, *Ecotoxicol. Environ. Saf.*, 18, 241, 1989.
5. **Hopkin, S. P. and Martin, M. H.**, Heavy metals in woodlice, *Symp. Zool. Soc. London*, 53, 143, 1984.
6. **Van Straalen, N. M., Schobben, J. H. M., and De Goede, R. G. M.**, Population consequences of cadmium toxicity in soil microarthropods, *Ecotoxicol. Environ. Saf.*, 17, 190, 1989.
7. **Janssen, H. H. and Dallinger, R.**, Diversification of cadmium-binding proteins due to different levels of contamination in *Arion lusitanicus*, *Arch. Environ. Contam. Toxicol.*, 20, 132, 1991.
8. **Bengtsson, G., Gunnarsson, T., and Rundgren, S.**, Effects of metal pollution on the earthworm *Dendrobaena rubida*, (Sav.) in acidified soils, *Water, Air, Soil Pollut.*, 28, 361, 1986.
9. **Ma, W.-C.**, *Regenwormen als bioindicators van bodemverontreiniging*, Soil Protection Series, Vol. 15, SDU,'s-Gravenhage, 1982.
10. **Malecki, M. R., Neuhauser, E. F., and Loehr, R. C.**, The effect of metals on the growth and reproduction of *Eisenia foetida* (Oligochaeta, Lumbricidae), *Pedobiologia*, 24, 129, 1982.
11. **Russell, L. K., Dehaven, J. I., and Botts, R. P.**, Toxic effects of cadmium on the garden snail *(Helix aspersa), Bull. Environ. Contam. Toxicol.*, 26, 634, 1981.
12. **Van Capelleveen, H. E.**, Ecotoxicity of Heavy Metals for Terrestrial Isopods, Ph.D. thesis, Vrije Universiteit, Amsterdam, 1987.
13. **Crommentuijn, T., Brils, J., and Van Straalen, N. M.**, Population growth of *Folsomia candida* (Willem) in contaminated soils, as related to toxic effects on life-history characteristics, *Ecotoxicol. Environ. Saf.*, 1992.
14. **Martin, N. A.**, Toxicity of pesticides to *Allolobophora caliginosa* (Oligochaeta, Lumbricidae), *N.Z. J. Agric. Res.*, 29, 699, 1986.
15. **Marigomez, J. A., Angulo, E., and Saez, V.**, Feeding and growth responses to copper, zinc, mercury and lead in the terrestrial gastropod *Arion ater* (Linné), *J. Mol. Stud.*, 52, 68, 1986.
16. **Bengtsson, G., Gunnarsson, T., and Rundgren, S.**, Growth changes caused by metal uptake in a population of *Onychiurus armatus* feeding on metal polluted fungi, *Oikos*, 40, 216, 1983.
17. **Denneman, C. A. J. and Van Straalen, N. M.**, The toxicity of lead and copper in reproduction toxicity tests using the oribatid mite *Platynothrus peltifer*, *Pedobiologia*, 35, 305, 1991.
18. **Bengtsson, G., Gunnarsson, T., and Rundgren, S.**, Influence of metals on reproduction, mortality and population growth in *Onychiurus armatus* (Collembola), *J. Appl. Ecol.*, 22, 967, 1985.

19. **Schmidt, G. H., Ibrahim, H. M. M., and Abdallah, M. D.**, Toxicological studies on the long-term effects of heavy metals (Hg, Cd, Pb) in soil on the development of *Aiolopus thalassinus* (Fabr.) (Saltatoria: Acrididae), *Sci. Total Environ.*, 107, 109, 1991.

20. **Beyer, W. N., Cromartie, E., and Moment, G. B.**, Accumulation of methyl-mercury in the earthworm, *Eisenia foetida*, and its effect on regeneration, *Bull. Environ. Contam. Toxicol.*, 35, 157, 1985.

21. **Abassi, S. A. and Soni, R.**, Stress-induced enhancement of reproduction in earthworm *Octochaetes pattoni* exposed to chromium (VI) and mercury (II) — implications in environmental management, *Int. J. Environ. Stud.*, 22, 43, 1983.

22. **MPV**, Milieuprogramma 1988-1991. Voortgangsrapportage, Tweede Kamer, vergaderjaar 1987-1988, 20202, nrs. 1 and 2, 's-Gravenhage, 1987.

23. **Di Toro, D. M., Mahony, J. D., Hansen, D. J., Scott, K., Hinks, M. B., Mayr, S. M., and Redmond, M. S.**, Toxicity of cadmium in sediments: the role of acid volatile sulfide, *Environ. Toxicol. Chem.*, 8, 1487, 1989.

24. **Van Hattum, B., Korthals, G., Van Straalen, N. M., Govers, H. A. J., and Joosse, E. N. G.**, Accumulation patterns of trace metals in freshwater isopods in sediment bioassays. Influence of substrate characteristics, temperature and pH, submitted.

25. **Di Toro, D. M., Zarba, C. S., Hansen, D. J., Swartz, R. C., Cowan, C. E., Pavlou, S. P., Allen, H. E., Thomas, N. A., and Paquin, P. R.**, Technical basis for establishing sediment quality criteria for nonionic organic chemicals using equilibrium partitioning, *Environ. Toxicol. Chem.*, 10, 1541, 1991.

26. **Salomons, W. and Förstner, U.**, *Metals in the Hydrocycle*, Springer-Verlag, Berlin,

27. **Hooftman, R. N. and Adema, D. M. M.**, Ontwikkeling van een Set Toets-methoden voor de Toxicologische Analyse van de Verontreinigingsgraad van de Onderwaterbodems, TNO report 88-16105, TNO, Delft, 1988.

28. **Swartz, R. C., Ditsworth, G. R., Schultz, D. W., and Lamberson, J. O.**, Sediment toxicity to a marine infaunal amphipod: cadmium and its interaction with sewage sludge, *Mar. Environ. Res.*, 18, 133, 1986.

29. **Phelps, H. L., Pearson, W. H., and Hardy, J. T.**, Clam burrowing behaviour and mortality related to sediment copper, *Mar. Pollut. Bull.*, 16, 309, 1985.

30. **Carins, M. A., Nebeker, A. V., Gakstatter, J. H., and Griffis, W. L.**, Toxicity of copper spiked sediments to freshwater invertebrates, *Environ. Toxicol. Chem.*, 3, 435, 1984.

31. **Kosalwat, P. and Knight, A. W.**, Chronic toxicity to a partial life cycle of the midge, *Chironomus decorus, Arch. Environ. Contam. Toxicol.*, 16, 275, 1987.

32. **Kooijman, S. A. L. M.**, A safety factor for LC50 values allowing for differences in sensitivity among species, *Water Res.*, 21, 269, 1987.

33. **Volmer, J., Kördel, W., and Klein, W.**, A proposed method for calculating taxonomic-group-specific variances for use in ecological risk assessment, *Chemosphere*, 17, 1493, 1988.

34. **Wagner, C. and Løkke, H.**, Estimation of ecotoxicological protection levels from NOEC toxicity data, *Water Res.*, 10, 1237, 1991.

35. **Van Straalen, N. M.**, New methodologies for estimating the ecological risk of chemicals in the environment, in *Proc. 6th Int. Cong. Int. Assoc. Engineering Geology*, Price, D. G., Ed., A. A. Balkema, Rotterdam, 1990, 165.

36. **Aldenberg, T. and Slob, W.**, Confidence Limits for Hazardous Concentrations Based on Logistically Distributed NOEC Toxicity Data, RIVM Report 719102002, RIVM, Bilthoven, 1991.
37. **Dallinger, R., Berger, B., and Birkel, S.**, Terrestrial isopods: useful biological indicators of urban metal pollution, *Oecologia (Berlin)*, 89, 32, 1992.
38. **Jancke, G.**, Modelversuche zur subakuten und subletalen Wirkung von Herbiziden auf Collembolen im Hinblick auf ein Testsystem für Umweltchemikalien, *Zool. Beitr. N.F.*, 32, 261, 1989.
39. **Depledge, M. H.**, New approaches in ecotoxicology: can inter-individual physiological variability be used as a tool to investigate pollution effects?, *Ambio*, 19, 251, 1990.
40. **Janssen, M. P. M., Bruins, A., De Vries, T. H., and Van Straalen, N. M.**, Comparison of cadmium kinetics in four soil arthropod species, *Arch. Environ. Contam. Toxicol.*, 20, 305, 1991.
41. **Tranvik, L. and Eijsackers, H.**, On the advantage of *Folsomia fimetarioides* over *Isotomiella minor* (Collembola) in a metal polluted soil, *Oecologia (Berlin)*, 80, 195, 1989.
42. **Van Wensem, J., Jagers op Akkerhuis, G. A. J. M., and Van Straalen, N. M.**, Effects of the fungicide triphenyltin hydroxide on soil fauna mediated litter decomposition, *Pestic. Sci.*, 32, 307, 1991.
43. **Van Wensem, J.**, A terrestrial micro-ecosystem for measuring effects of pollutants on isopod-mediated litter decomposition, *Hydrobiologia*, 188/189, 507, 1989.
44. **Klerks, P. L. and Levinton, J. S.**, Rapid evolution of metal resistance in a benthic oligochaete inhabiting a metal-polluted site, *Biol. Bull.*, 176, 135, 1989.
45. **Posthuma, L.**, Genetic differentiation between populations of *Orchesella cincta* (Collembola) from heavy metal contaminated sites, *J. Appl. Ecol.*, 27, 609, 1990.
46. **Donker, M. H. and Bogert, C. G.**, Adaptation to cadmium in three populations of the isopod *Porcellio scaber*, *Comp. Biochem. Physiol.*, 100C, 143, 1991.

INDEX

INDEX

T - #0069 - 101024 - C0 - 234/156/26 [28] - CB - 9780873717342 - Gloss Lamination